Geotechnical Engineering Design

Geotechnical Engineering Design

Ming Xiao
Associate Professor of Civil Engineering
College of Engineering
The Pennsylvania State University, University Park
USA

with contributions from

Daniel Barreto
Lecturer
School of Engineering & the Built Environment
Edinburgh Napier University
UK

WILEY Blackwell

Library of Congress Cataloging-in-Publication Data applied for

ISBN: 9780470632239

A catalogue record for this book is available from the British Library.

Wiley also publishes its books in a variety of electronic formats. Some content that appears in print may not be available in electronic books.

Set in 10/12.5pt, AvenirLTStd by Laserwords Private Limited, Chennai, India

1   2015

Contents

Contents

Preface

This book presents the fundamental design principles and approaches in geotechnical engineering, including an introduction to engineering geology, subsurface explorations, shallow and deep foundations, slope stability analyses and remediation, filters and drains, Earth retaining structures, geosynthetics, and basic seismic evaluations of slope stability, lateral earth pressures, and liquefaction. It is intended for use as a textbook in the geotechnical design courses for senior undergraduate and M.S. graduate students. Therefore, the topics covered in this book are presented to meet this level. This book applies the principles of soil mechanics and focuses on the design methodologies in geotechnical engineering. The readers of this book are expected to have undertaken a soil mechanics course and already understood the principles of engineering properties of soils, stresses in soils, seepage in soils, soils shear strength, and consolidation.

The book was completed after I have taught geotechnical engineering for 9 years as faculty of civil engineering. Although excellent textbooks on the principles of geotechnical engineering and textbooks on foundation engineering are available to students, instructors and students have few options in selecting textbooks that cover geotechnical design aspects other than foundation engineering, particularly in senior undergraduate and M.S. graduate courses. This prompted me to embark on writing this textbook. While writing this book, I remained mindful of how a student can best and most easily grasp the content. Each chapter opens with an introduction on why the topic is important in the engineering practice, and graphical illustrations are appropriately included to offer visual images of the engineering applications. Ample graphical illustrations on field applications and design approaches are provided throughout the book. In Chapters 3–9 where designs are presented, a sample problem and its solution are included at the end of each topic. The homework problems at the end of each chapter are designed to test the student's basic understanding of the concepts and design approaches and to challenge the student to solve real-world design issues.

A unique aspect of this book is the inclusion of Eurocode 7: Geotechnical design, the European Standard for the design of geotechnical structures. The design approaches of many topics in this book use both allowable stress design (in the United States) and limit state design (in Europe), and two sets of solutions for many sample problems are provided to explain both the design methodologies. Both the America Society for Testing and Materials (ASTM) standards and the British standards are referred to in Chapter 1 (Introduction to Engineering Geology) and Chapter 2 (Geotechnical Subsurface Exploration). The inclusion of Eurocode allows the international audience to preliminarily understand the commonalities and differences in geotechnical engineering designs on a global scale, particularly in Europe and North America.

Considering the targeted level of readers and the typical duration of a course in which this textbook would be used, some topics are not presented in great depth. For example, Chapter 3 (Shallow Foundation Design) and Chapter 4 (Introduction to Deep Foundation Design) present only the fundamentals of foundation design; the topic of drilled shafts is not presented. Chapter 8

Preface

(Introduction to Geosynthetics Design) presents only the basics of geosynthetics and three common field applications using geosynthetics: mechanically stabilized earth walls, reinforced soil slopes, and filtration and drainage. Chapter 9 (Introduction to Geotechnical Earthquake Design) presents the basic seismology and earthquake characteristics and three basic seismic evaluations: slope stability, lateral earth pressures, and liquefaction. Special topic courses on these individual topics may require other available textbooks.

I am indebted to many people who helped and supported the long process of writing this book. Jennifer Welter, Madeleine Metcalfe, and Harriet Konishi of John Wiley and Sons had been patient, supportive, and instrumental in the development of this book. Benjamin T. Adams, my undergraduate, master's, and doctoral student and friend, provided valuable thoughts and help. Many professors, practitioners, and agencies generously provided photos and graphs for this book; the acknowledgements of them are included in the figure captions. I particularly appreciate my wife, Shasha, for her continuous support and sacrifice in the pursuit of this book and in life.

Ming Xiao
The Pennsylvania State University
University Park
USA

About the Authors

Ming Xiao

Dr. Ming Xiao is an Associate Professor in the Department of Civil and Environmental Engineering at the Pennsylvania State University, USA. He received a B.S. degree in Civil Engineering from Shandong University (China), an M.S. degree in Geotechnical Engineering from Zhejiang University (China), and both M.S. degree in Computer Science and Ph.D. degree in Geotechnical Engineering from Kansas State University. Before joining the Pennsylvania State University, Dr. Xiao was an Assistant and then Associate Professor in the Department of Civil and Geomatics Engineering at the California State University, Fresno, from 2005 to 2013. Dr. Xiao's research involves seepage and erosion, performances of earthen structures (such as dams, levees, and geosynthetically reinforced bridge supports) under in-service conditions and extreme events, and innovative alternative materials, biogeochemically treated soils, recycled materials, and their engineering applications. Dr. Xiao has received a number of research grants as principal or co-principal investigator from the National Science Foundation (NSF), the National Aeronautics and Space Administration (NASA), the Federal Highway Administration (FHWA), the California Department of Resources Recycling and Recovery, the California Department of Transportation, the Pennsylvania Department of Transportation, and other companies. He has written two books and published over 50 peer-reviewed journal and conference papers. He is a registered Professional Engineer of Civil Engineering in the states of California and Ohio. He is a member of the American Society of Civil Engineers (ASCE), the International Society of Soil Mechanics and Geotechnical Engineering (ISSMGE), the Consortium of Universities for Research in Earthquake Engineering (CUREE), Tau Beta Pi National Engineering Honor Society (USA), and Chi Epsilon National Civil Engineering Honor Society (USA). He is the Chair of the Geotechnics of Soil Erosion technical committee and a member of the Embankments, Dams, and Slopes technical committee of the ASCE Geo-Institute. He is an Editorial Board Member of the ASTM Journal of Testing and Evaluation, and he has served as an international advisory board member and session chair for a number of international conferences.

Daniel Barreto

Dr. Daniel Barreto is a Lecturer in Geotechnical Engineering in the School of Engineering and the Built Environment at Edinburgh Napier University, UK. He received a B.Eng. degree in Civil Engineering from Universidad de los Andes (Colombia), an M.Sc. degree in Soil Mechanics and Engineering Seismology from Imperial College London (UK), and completed his Ph.D. in Soil Mechanics also at Imperial College London. Before the start of his postgraduate studies, Dr. Barreto worked as Graduate Civil (Geotechnical) Engineer in Colombia and contributed to multiple projects related to the design of various geotechnical structures. Dr. Barreto's research involves advanced laboratory testing on soil and use of the discrete element method (DEM) for geotechnical applications. His particular research interests include soil anisotropy, mechanical behavior of soft rocks and dissolving soils, among others. Dr. Barreto has received a number of research grants as principal or co-principal investigator from institutions such as the Royal Academy of Engineering and the British Council. He has published over 25 peer-reviewed journal and conference papers. He is a member of the British Geotechnical Association (BGA), the International Society of Soil Mechanics and Geotechnical Engineering (ISSMGE), and the International Society of Rock Mechanics (ISRM). Dr Barreto is also a Teaching Fellow at Edinburgh Napier University, a title that highlights excellence in teaching practice, and is a Fellow of the Higher Education Academy (FHEA).

About the Companion Website

This book's companion website www.wiley.com/go/Xiao provides you with a solutions manual, resources and downloads to further your understanding of geotechnical engineering design:

- Solutions to the end-of-chapter exercises, including the full workings
- A suite of editable spreadsheets which map onto the worked examples in the book, showing how they are solved.
- Colour versions of the book's many photographs and figures
- PowerPoint slides for tutors

Chapter 1

Introduction to Engineering Geology

1.1 Introduction

Engineering geology involves description of the structure and attributes of rocks that are associated with engineering works, mapping, and characterization of all geologic features and materials (rocks, soils, and water bodies) that are proximate to a project and the identification and evaluation of potential natural hazards such as landslides and earthquakes that may affect the success of an engineering project. It is different from geology, which concerns the present and past morphologies and structure of the Earth, its environments, and the fossil records of its inhabitants (Goodman 1993).

1.2 Structure of the Earth and geologic time

The Earth is divided into three main layers: crust, mantle, and core (Figure 1.1). The crust is the outer solid layer of the Earth and comprises the continents and ocean basins. The crust varies in thickness from 35 to 70 km in the continents and from 5 to 10 km in the ocean basins. It is composed mainly of aluminosilicates. The mantle, a highly viscous layer about 2900 km thick, is located beneath the outer crust. It includes the upper mantle (about 35–60 km thick) and the lower mantle (about 35–2890 km thick) (Jordan 1979). The mantle is composed mainly of ferro-magnesium silicates. Large convective cells in the mantle circulate heat and may drive the plate tectonic processes. Beneath the mantle and at the center of the Earth are the liquid outer core and the solid inner core. The outer core is an extremely low viscosity liquid layer, about 2300 km thick, and composed of iron and nickel, with an approximate temperature of 4400 °C. The inner core is solid, about 1200 km in radius, and is entirely composed of iron, with an approximate temperature of 5505 °C (Engdahl et al. 1974). The Earth's magnetic field is believed to be controlled by the liquid outer core.

Geologic time is a chronological measurement of the rock layers in the history of the Earth. Evidence from radiometric dating indicates that the Earth is about 4.57 billion years old. The geologic time scale is shown in Figure 1.2. The rocks are grouped by age into *eons*, *eras*, *periods*, and *epochs*. Among the various periods, the Quaternary period (from 1.6 million years ago to the present) deserves special attention as the top few tens of meters of the Earth's surface, which geotechnical engineers often work with, developed during this period (Mitchell and Soga 2005).

Geotechnical Engineering Design, First Edition. Ming Xiao.
© 2015 John Wiley & Sons, Ltd. Published 2015 by John Wiley & Sons, Ltd.
Companion Website: www.wiley.com/go/Xiao

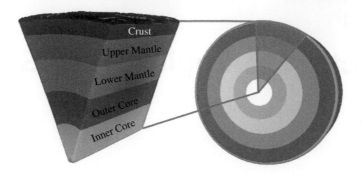

Fig. 1.1   Structure of the Earth.

Eon	Era	Period		Epoch	Time (million of years)	Some important events
Phanerozoic	Cenozoic	Quaternary		Holocene	0.01	Last glacial period
				Pleistocene	1.6	
		Tertiary	Neogene	Pliocene	5.3	
				Miocene	23.7	
			Paleogene	Oligocene	36.6	
				Eocene	57.8	
				Paleocene	68.4	Mountains form in Europe
	Mesozoic	Cretaceous			144	Mountains form in western North America
		Jurassic			208	
		Triassic			245	Dinosaurs appear
	Paleozoic	Permian			286	
		Pennsylvanian			320	
		Mississippian			360	First land vertebrates
		Devonian			408	
		Silurian			438	
		Ordovician			505	Ice age at end of period. Major diversification of life
		Cambrian			570	
Proterozoic					2500	
Archean					3800	Simple single-cell life
Hadean					4570	Formation of Earth

Fig. 1.2   Geologic time scale.

1.3   Formation and classification of rocks

The main rock-forming minerals are silicates, and the reminders are carbonates, oxides, hydroxides, and sulfates. There are three major categories of rocks: igneous rocks, sedimentary rocks, and metamorphic rocks.

Fig. 1.3   Igneous rock in the Yosemite National Park, California, USA.

1.3.1   Igneous rocks

Igneous rocks are formed due to igneous activities, i.e., the generation and movement of silicate magma. There are two kinds of igneous rocks: *extrusive or volcanic rocks* that are formed by cooling of lava from volcanic eruption, and *intrusive or plutonic rocks* that are formed by the slow cooling of magma beneath the surface. Cooling below the Earth's surface is slow and results in large crystals, whereas cooling on the surface is rapid and results in small crystal size. Therefore, the extrusive rocks are usually fine-grained, and the intrusive rocks are coarse-grained. The intrusive rocks are found in many great mountain ranges that were brought to the surface due to erosion of the overlying material and relative tectonic plate movements. The main intrusive rock type is the light-colored granite. The main extrusive rock type is the dark-colored basalt. Coarse-grained intrusive rocks generally have lower strength and abrasion resistance as compared to fine-grained extrusive rocks (West 1995). Figure 1.3 shows Half Dome in the Yosemite National Park, California; Half Dome is an igneous rock.

1.3.2   Sedimentary rocks

Sedimentary rocks are formed by the accumulated and hardened deposits of soil particles and weathered rocks transported by wind, streams, or glaciers. The accumulated deposits are hardened due to overburden pressure and cemented by minerals such as iron oxide (FeO_2) and calcium carbonate ($CaCO_3$). Among sedimentary rocks, the most widespread are shale, sandstone, limestone, siltstone, mudstone, claystone, and conglomerates. They all display the characteristic stratification resulting from the gradual accumulation of layers of compacted and cemented deposits. The three main rocks that comprise 99% of sedimentary rocks are shale (46%), sandstone (32%), and limestone (22%) (West 1995). Sedimentary rocks are extremely diverse in their texture and mineral composition due to their diverse origin, transportation, and formation environment. Shale, claystone, and mudstone usually have low strength and low abrasion resistance

Fig. 1.4   Sedimentary rocks in the Grand Canyon National Park, Arizona, USA.

and can be problematic in engineering works. Figure 1.4 shows Grand Canyon National Park, Arizona. The Grand Canyon is mainly composed of sedimentary rocks, which were eroded by the Colorado River for millions of years leading to the current formation of the Canyon.

1.3.3   Metamorphic rocks

Metamorphic rocks are formed when igneous or sedimentary rocks are subjected to the combined effects of heat and pressure, resulting in compaction, cementation, and crystallization of the rock minerals; the extreme pressure and heat transform the mineral structure. The metamorphic rocks commonly encountered in nature include marble, slate, schist, gneiss, and quartzite. These rocks include foliated and nonfoliated rocks. The layering within metamorphic rocks is called "foliation." The foliated metamorphic rocks include slates, phyllites, schists, and gneisses. The nonfoliated metamorphic rocks include marble (which is metamorphosed limestone) and quartzite (which is metamorphosed sandstone). Foliated metamorphic rocks exhibit directional properties. Strength, permeability, and seismic velocity of metamorphic rocks are strongly affected by the direction of foliation (West 1995). Figure 1.5 shows metamorphic rocks in Marble Canyon in Death Valley National Park, California.

1.4   Engineering properties and behaviors of rocks

1.4.1   Geotechnical properties of rocks

The geotechnical properties of rocks include basic properties, index properties, hydraulic properties, and mechanical properties (Hunt 2005).

Fig. 1.5 Metamorphic rocks in Marble Canyon, Death Valley National Park, California, USA. (Photo courtesy of Benjamin T. Adams.)

Rocks can be categorized into *intact rock specimens* and *in situ rock mass*, and their *basic properties* and *index properties* are tested accordingly. Intact rock specimens are fresh to slightly weathered rock samples that are free of defects. The *in situ* rock mass may consist of rock blocks, ranging from fresh to decomposed, separated by discontinuities.

- The basic properties of intact rocks include specific gravity, density, porosity, hardness (for excavation resistance), and durability and reactivity (for aggregate quality). The basic property of rock mass is its density.
- The index property tests for intact rocks include the uniaxial compression test, point load index test, and sonic velocities that provide a measure of the rock quality.
  The index properties of rock mass are the sonic-wave velocities and the rock quality designation (RQD).
- The hydraulic property of rock is its permeability.
- The mechanical properties of rocks are the rupture strength and deformation characteristics. The rupture strength includes the uniaxial compressive strength, uniaxial tensile strength, flexural strength, triaxial shear strength, direct shear strength, and borehole shear test.

1.4.2 Comparison of the three types of rocks

Typically, igneous and metamorphic rock formations are hard and durable. Sedimentary rock formations can also be sound and durable; however, when compared to igneous and metamorphic rocks, more inconsistencies are expected in sedimentary rocks because of the presence and inclusion of foreign materials at the time of formation or because of weak bonding and cementing. Shale and mudstone generally soften when soaked in water, settle significantly under load, and yield under relatively low stresses. Such weak rocks can provoke unpredictable difficulties in excavations and foundations.

1.5 Formation and classification of soils

1.5.1 Soils formation

Soils are formed by the weathering of rocks. Weathering is a process that breaks down rocks into smaller pieces due to mechanical, chemical, and biological mechanisms. All types of rocks are subjected to weathering.

Mechanical weathering

The following physical mechanisms contribute to mechanical weathering:

- *Temperature*: It includes extremely high or low temperatures that cause the shrinking and expansion of rocks.
- *Pressure*: It includes high overburden stresses in the subsoil or in the ocean floor.
- *Unloading*: When the overburden stress that exerts on rocks is removed, cracks and joints in rocks may form.
- *Water in rocks*: Water may seep into the tiny voids in rocks; volume expansion of ice in the voids can disintegrate rocks.
- *Flow of water*: Rivers and ocean waves can agitate rocks, causing surface erosion.
- *Glaciers*: Movement of glaciers over rocks can cause significant abrasion, which breaks down rocks.
- *Wind*: Wind can blow against the surface of rocks, causing abrasive erosion.

Chemical weathering

The following chemical reactions contribute to chemical weathering:

- *Hydrolysis*: Molecules of water (H_2O) are split into hydrogen ions (H^+) and hydroxide (OH^-) such that an acid and a base are formed. The free hydrogen ions and hydroxide ions can react with the mineral compositions of rocks, causing the weakening and breaking down of rocks.
- *Reduction/oxidation* (redox): Oxidation is the loss of electrons by a molecule, atom, or ion. For example:

$$4\,Fe^{2+} + O_2 \rightarrow 4\,Fe^{3+} + 2\,O^{2-} \text{ (oxidation of ferrous iron into ferric iron)}$$

$$C + O_2 = CO_2 \text{ (oxidation of carbon)}$$

Reduction is the gain of electrons by a molecule, atom, or ion. For example:

$$C + 2H_2 = CH_4 \text{ (reduction of carbon to yield methane)}$$

$$3Fe_2O_3 + H_2 \rightarrow 2Fe_3O_4 + H_2O \text{ (reduction of } Fe_2O_3 \text{ with hydrogen)}$$

The redox processes can change the composition of rocks and, therefore, weaken and disintegrate rocks.
- *Solution of rock minerals due to carbonation*: Carbonation is the process of dissolving carbon dioxide in water:

$$CO_2 + H_2O = H_2CO_3 \text{ (carbonic acid)}$$

Carbonic acid can react with and dissolve rock minerals.

Biological weathering

Biological processes can also break down rocks and include the physical penetration and growth of roots in rocks, digging activities of animals, and metabolisms of microorganisms in rocks. Biological weathering actually involves both mechanical weathering and chemical weathering agents.

1.5.2 Soil types

Soils can be largely divided into mineral soils and organic soils.

Mineral soils

Mineral soils originate from rocks and possess similar physical compositions of the parent materials. On the basis of the mineral types and grain sizes, the mineral soils can be classified into four types: gravel, sand, silt, and clay. Gravel, sand, and silt share common mineral compositions; they only differ in grain sizes. Clays are categorized not only by their small sizes (usually smaller than 0.005 mm) but also by their unique clay minerals. The common clay minerals include the following:

- Kaolinite minerals: Kaolinite has low shrink–swell capacity. It is soft, earthy, and usually white in color.
- Smectite minerals: Montmorillonite is the most common mineral of the smectite group. Because montmorillonite has very high shrink–swell capacity, its volume can significantly increase when absorbing water. Bentonite, a highly plastic and swelling clay and used for a variety of purposes in engineering practices, consists mostly of montmorillonite. Bentonite is usually used as drilling slurry during subsoil exploration, in slurry walls, or as protective liners for landfills.
- Illite mineral: Illite is non-expanding.

Organic soils

Organic soils contain high percentage of organic matter as a result of decomposition of plant and animal residues. Peat is a major type of naturally occurring organic soil. Peats are the partly decomposed and fragmented remains of plants that have accumulated under water and fossilized. In the United States, peat deposits are found in 42 states, and their properties are significantly different from those of inorganic soils. Some of the distinctive characteristics of fibrous peats include (Mesri and Ajlouni 2007) the following:

- High initial permeability;
- High compressibility;
- Permeability decreases dramatically as peats compress under embankment load;
- High secondary consolidation;
- Exceptionally high frictional resistance: the effective friction angle can be as high as 60°;
- Exceptionally high value of undrained shear strength to consolidation pressure ratio;
- Exceptionally low values of undrained Young's modulus to undrained shear strength ratio.

1.5.3   Residual and transported soils

Residual soils

Residual soils develop *in situ* and have not been moved. Therefore, their characteristics depend on the bedrocks underneath and the extent of their weathering.

Transported soils

The residual soils or weathered bedrocks are transported by various physical mechanisms and then deposit elsewhere; they are called transported soils and include the following:

- *Aeolian soils*: Aeolian soils are transported by wind and deposit elsewhere. They include sand dunes and loess. Loess is windblown silt and usually forms the characteristic vertical cut. Both sand dunes and loess are very unstable material and are usually avoided as a foundation soil.
- *Alluvial (fluvial) soils*: Alluvial soils, or alluvium, include all sediments that are transported and deposited by streams. The deposits are usually stratified into layers of clay, silt, sand, and gravel. In mountainous areas, alluvium consists largely of boulders; in areas dominated by sluggish streams, alluvium typically is clayey silt.
- *Colluvial soils*: When soils and weathered bedrocks slowly move downslope due to gravity, they form colluvial soils or colluvium. Typical colluvium consists of unstratified, seemingly randomly oriented angular blocks of bedrock in a clayey mix. In geologic time, colluvium is actively moving. Colluvium is widespread and can reach 100 m thickness in humid areas (Rahn 1996).
- *Glacial soils*: Glacial soils, or glacial drifts, include all deposits formed by glaciers. There are two types of glacial drifts: glacial till (nonstratified drift) and stratified drift. Glacial till is directly deposited by glacial ice, is typically nonstratified and unsorted, and contains angular to subrounded rock particles of all sizes. It generally has favorable engineering characteristics. Sediments that are deposited by streams of water from melting glaciers are called "stratified drift." It is, in fact, alluvium and, therefore, has the properties of alluvium.
- *Marine soils*: Marine soils are the coastal deposits that are transported by rivers and deposit in the ocean. They include clay, silt, sand, and gravel.
- *Lacustrine soils*: Lacustrine soils are the fine-grained sediments that are deposited in lakes. Lacustrine deposits are usually underconsolidated and can be very troublesome for foundation.

For example, many of the foundation problems in the Chicago area originated from thin lacustrine sediments deposited in Lake Michigan when the Great Lakes were more extensive than today (Rahn 1996).

● *Engineered fills*: Engineered fills are transported by human to construction sites; they typically have satisfactory strength and deformation.

Formations of residual soils typically have strength-related characteristics distinctly different from deposits of transported soils. Residual soil formations change gradationally with depth, from soil to weathered rock to sound rock. Therefore, the residual soil would display stratification and possess many of the textural and strength properties of the parent rock. Transported soils tend to be more uniform in texture and type in each soil stratum.

1.6   Maps used in engineering geology

An engineering geologist should investigate all possible sources of published information before extensive fieldwork. This includes topographic maps, geologic maps, soils maps, aerial photographs, and remote-sensing data. In this section, topographic maps and geologic maps are introduced.

1.6.1   Topographic maps

A topographic map is used to scale and indicate the locations of natural and man-made features on the Earth's surface. The man-made features include buildings, roads, freeways, train tracks, and other civil engineering works. The natural features include mountains, valleys, plains, lakes, rivers, sea cliffs, beaches, and vegetation. The main purpose of a topographic map is to indicate ground-surface elevations with respect to the sea level. Therefore, a unique feature of topographic maps is the use of *contour lines* to portray the shapes and elevations of the land. Topographic maps render the three-dimensional ups and downs of a terrain on a two-dimensional surface. This information can be used to determine the major topographic features at the site and to plan subsurface exploration. Using different colors and shadings, a topographic map can also indicate older versus newer development at the site.

1.6.2   Geologic map

A geologic map is a special-purpose map that shows the distribution, relationship, and composition of Earth materials such as rocks and surficial deposits (landslides, sediments) and shows structural features of the Earth (faults, folded strata). It provides fundamental and objective information to identify, protect, and use water, land, resources, and to avoid risks of natural hazards. A geologic map shows geologic features that are identified by colors, letter symbols, and lines of various types. See Figure 1.6 as an example.

Colors

The most important feature of a geologic map is color. Each standard color represents a geologic unit. A geologic unit is a volume of rock with identifiable origin and age range. Some geologic units have not yet been named; therefore, they are identified by the type of rock in the unit,

Fig. 1.6   Geologic map. *Portion of the "Geologic Map of Monterey 30′ × 60′ Quadrangle and Adjacent Areas, California." California Geological Survey CD 2002–04.* (Photo courtesy of California Department of Conservation, USA.)

such as "sandstone and shale," or "unnamed sandstone." However, all the units, named and unnamed, are indicated by a specific color on the geologic map. The standard color code that represents different geologic times is identified by the United State Geological Survey (USGS). Color conventions for geologic maps around the world are generally similar. For example, the British Geological Survey (BGS) uses pastel colors for Quaternary surface deposits, darker colors are generally used for bedrock units, and red is used for igneous intrusions or lava flows.

Letter symbols

In addition to color, each geologic unit is sometimes assigned a set of letters to represent the geologic time of the rock formation (refer to USGS map in Figure 1.6). The most common division of time used in letter symbols on geologic maps is the *period*. Rocks of the four most recent periods are J (Jurassic – about 208 to 142 million years ago), K (Cretaceous – about 144 to 68 million years ago), T (Tertiary – about 68 to 1.6 million years ago), and Q (Quaternary – about 1.6 million years ago until present day). The most recent period, Quaternary, includes the Holocene and Pleistocene Epochs; the Tertiary period includes the Pliocene, Miocene, Oligocene, Eocene, and Paleocene Epochs (Figure 1.2). Usually a letter symbol is the combination of an initial capital letter followed by one or more lower-case letters. The capital letter represents the geologic age of the rock formation. Occasionally, the age of a rock unit may span more than one period, i.e., the time span that was required to create a body of rock falls on both sides of a time boundary. In that case, both capital letters, representing the two periods, are used, with the first letter representing the later period. For example, QT indicates that the rock unit began to form in the Tertiary period and was completed in the Quaternary period. A formation is usually named after a geographic feature (mountain, canyon, or town) near the area where the unit was first identified. The geologic units that were formed at an unknown time do not have capital letters in the symbols. The small letters indicate either the name of the rock formation, if known, or the type of the rock formation, if it has no name. For example, on the map in Figure 1.6, the area labeled "Kp" shows the Panoche formation of the Cretaceous period (Panoche formation is named after the Panoche Pass, a landform feature within a mountainous area of San Benito County, California); the area labeled "Mvqa" represents an unnamed formation composed of a volcanic rock ("vq") called andesite ("a") of the Miocene Epoch in the Tertiary Period (between 23.7 to 5.3 million years ago); the areas labeled "Qls" show landslide ("ls") deposits of the Quaternary period and the arrows in these deposits indicate the directions of the landslides.

Lines on the map

- *Contact lines*
  A contact is where two geologic units meet. The main types of contacts on most geologic maps are *depositional contacts*, *faults*, and *folds axis*.

  Depositional contact: All geologic units are formed over, under, or beside other geologic units. For example, lava from a volcano may flow over a rock; when the lava hardens into rock, the interface between the lava-turned-rock and the rocks underneath is a depositional contact. A depositional contact is shown on a geologic map as a *thin line*.

  Faults: In geologically active areas such as Southern and Northern California, geologic units tend to be broken up and move along faults. When different geologic units have been moved next to each other subsequent to their formation, the contact is a fault contact. A fault is shown on a geologic map as a *thick line*. Faults can cut through a single geologic unit. These faults are shown with the same thick lines on the map, but have the same geologic unit on both sides. Not all faults are still active and are likely to cause an earthquake. Rocks can preserve records of faults that have been inactive for millions of years, and knowing the locations of faults is the first step toward finding the ones that can move.

  Fold axis: Another kind of line shown on most geologic maps is a fold axis. In addition to being moved by faults, geologic units can also be bent and warped by the Earth's surface movement into wavelike shapes, which are called folds. A line that follows the crest or trough of the fold is called the "fold axis." This is marked on a geologic map with a *line of thickness between depositional contact and fault*.

- *Solid, dashed, and dotted lines*
  All lines of different thicknesses can also be modified using solid, dashed, or dotted line styles. If a line is precisely located, it is shown as solid; if its location is uncertain, it is dashed. The shorter the dash is, the more uncertain the location is. A dotted line is the most uncertain of all. Some contact lines can be covered by another geologic unit and are difficult to locate. The lines on the map may also be modified by other symbols on the line, such as triangles, small tic marks, arrows, and so on; these give more information about the line. For example, faults with triangles on them show that the side with the triangles has been thrust up and over the side without the triangles; this type of fault is called a reverse fault or a thrust fault. On the map in Figure 1.6, the solid thick line in the Mvqa formation represents a precisely located fault; the dashed thick line in the Mvqa formation shows a fault with less certainty; the dotted medium-thick lines in the lower left of the map show folds axes with relative uncertainty; the thick, solid line in the upper left between Kp and Mlt is a depositional contact for two different rock formations.

Chapter 1

Homework Problems

1. How many major layers are in the Earth's internal structure? What are their thicknesses?
2. A geologic map in Monterey, CA, is shown below (Figure 1.7).
    (1) From the map, identify the following rocks (types and names): Mvqa, Qls, Mlt, and Kp.
    (2) Identify what geologic features the following four lines represent.

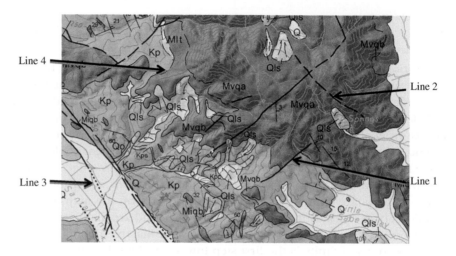

Fig. 1.7   Geologic map for problem 2.

3. Name the three major classes of rocks, and briefly describe how they were formed.
4. Provide generalized comments on the hardness, soundness, and durability associated with the three basic types of rocks.
5. What types of bedrock formations would offer good foundation support for structures? What types of rock formations are likely to be poor foundation materials?
6. Describe the three major types of rock weathering.
7. What are the four major types of mineral soils? List them in decreasing order of sizes.
8. Briefly describe the types of transported soils.
9. What soil types would be expected in a river or stream delta?
10. For each of the following multiple-choice questions, select all of the correct answers.
    (1) The smallest geologic time unit is:
        A. Epoch
        B. Age

C. Period
D. Era
E. Eon
(2) What are the important features on a geologic map?
   A. Color
   B. Shade
   C. Letter symbols
   D. Contact lines
   E. Contour lines
(3) The following mechanisms play roles in the weathering of rocks:
   A. Pressure or stress on the rock
   B. Cementation and crystallization
   C. Movement of glaciers
   D. Oxidation
   E. Igneous activities
(4) Granite is:
   A. Intrusive rock
   B. Extrusive rock
   C. Igneous rock
   D. Sedimentary rock
   E. Metamorphic rock
(5) The sedimentary rocks are formed due to:
   A. Compaction
   B. Crystallization
   C. Cementation
   D. Cooling of lava
   E. Extreme heat
(6) The mechanical properties of rocks include:
   A. Rupture strength
   B. Deformation characteristics
   C. Reactivity
   D. Density
   E. Texture
(7) Shale is:
   A. Igneous rock
   B. Sedimentary rock
   C. Metamorphic rock
   D. Stratified in its structure
   E. A sound foundation material
(8) Loess is:
   A. A transported soil
   B. An Aeolian soil
   C. An Alluvial soil
   D. A mineral soil
   E. An organic soil

Chapter 1

References

Engdahl, E.R., Flynn, E.A., and Massé, R.P. (1974). "Differential PkiKP travel times and the radius of the core." *Geophysical Journal*. Royal Astronomical Society. 40 (3): 457–463.

Goodman, R. (1993). *Engineering Geology: Rock in Engineering Construction*. John Wiley & Sons, Inc., Hoboken, NJ.

Hunt, R.E. (2005). *Geotechnical Engineering Investigation Handbook*, 2nd edition. CRC Press, Taylor & Francis Group, Boca Raton, FL.

Jordan, T.H. (1979). "Structural geology of the Earth's interior." *Proceedings of National Academy of Science,* 76 (9): 4192–4200.

Mesri, G., and Ajlouni, M. (2007). "Engineering properties of fibrous peats." *ASCE Journal Geotechnical and Geoenvironmental Engineering*. 133 (7): 850–866.

Mitchell, J.K., and Soga, K. (2005). *Fundamentals of Soil Behavior*, third edition. John Wiley & Sons, Inc. Hoboken, NJ.

Rahn, P.H. (1996). *Engineering Geology: An Environmental Approach*, 2nd edition. Prentice Hall, Inc., Englewood Cliffs, New Jersey.

West, T.R. (1995). *Geology Applied to Engineering*. Prentice Hall, Inc., Englewood Cliffs, NJ.

Chapter 2
Geotechnical Subsurface Exploration

2.1 Framework of subsoil exploration

In order to perform a sound geotechnical project design, the subsurface profile information must be obtained. Subsurface exploration, also known as "geotechnical investigation," usually entails drilling holes (also known as "boreholes") in the ground, retrieving soil or rock samples through the boreholes at predetermined depths, and conducting field testing. The extent of subsurface exploration depends on the type and the size of the project (roads, bridges, buildings, etc.). The cost of geotechnical field exploration usually ranges between 0.5% and 1% of the total cost of the construction project (Kumar 2005).

The scope of a subsurface exploration may include the following aspects:

- Determination of the subsurface conditions, such as soil strata (depth, thickness, and types of soils).
- Recovery of soil and rock samples for further laboratory testing.
- Determination of relevant engineering properties (shear strength, compressibility, plasticity, permeability, expansion and collapse potential, and frost susceptibility).
- Field testing to obtain the *in situ* properties of the soil or rock.
- Determination of the depth of the groundwater table.
- Identification of the existence of any problematic soils, such as soft or expansive soils.

2.2 Field drilling and sampling

2.2.1 Information required before drilling and sampling

The initial phase of a field investigation should consist of a detailed review of geological conditions at the site. This should include reviews of preexisting data, including remote sensing imagery, aerial photography, topographic and geologic maps, and general field reconnaissance. Elevation of the groundwater table, particularly confined aquifers, should be known. Drilling through a confined aquifer may cause water to spring to the ground surface and cause local flooding. The information obtained should be used as a guide in planning the subsurface exploration. Before subsurface drilling can be conducted, the underground utilities should be clearly identified to avoid damage and injury. For example, in California, the Underground Service Alert

Geotechnical Engineering Design, First Edition. Ming Xiao.
© 2015 John Wiley & Sons, Ltd. Published 2015 by John Wiley & Sons, Ltd.
Companion Website: www.wiley.com/go/Xiao

(USA), an organization that links the excavation community and the owners of underground utility lines, should be contacted. Overhead utility lines should also be checked before setting up a drill rig.

Subsurface drilling and sampling should be strategically planned in order to efficiently obtain the most subsurface information within a specific budget. The Soil Mechanics Design Manual by the Naval Facilities Engineering Command (NAVFAC 1986) provides the guidelines for the boring layouts (the number and locations of boreholes) and the boring depths based on the type and size of the projects, as summarized in Table 2.1 and Table 2.2. Similar recommendations are also provided in most design codes (e.g., Annex B of Eurocode 7, BS EN 1997–2:2007). The boring depth is also controlled to a great degree by the characteristics and sequence of the subsurface materials encountered; in the cases where unfavorable conditions exist (i.e., weak strata below the strata of higher bearing capacity), greater depth of investigation is always required.

Table 2.1   Requirements for boring layout (NAVFAC, soil mechanics design manual 7.01. 1986).

Project type	Boring layout
New site of wide extent	Space preliminary borings 60 to 150 m apart such that the area between any four borings includes approximately 10% of the total area. In detailed exploration, add borings to establish geological sections at the most useful orientations.
Development of site on soft compressible strata	Space borings 30 to 60 m apart at possible building locations. Add intermediate borings when building sites are determined.
Large structure with separate closely spaced footings	Space borings approximately 15 m apart in both directions, including borings at possible exterior foundation walls and at machinery or elevator pits. Add borings to establish geologic sections at the most useful orientations.
Low-load warehouse building of large defined area	Minimum of four borings at corners plus intermediate borings at interior foundations.
Isolated rigid foundation 232 to 930 m^2 in area	Minimum of three borings around perimeter. Add interior borings depending on initial drilling and sampling results.
Isolated rigid foundation, less than 930 m^2 in area	Minimum of two borings at opposite corners, add more for erratic conditions.
Major waterfront structures, such as dry docks	If definite site is established, space borings generally not farther than 15 m. Add intermediate borings at critical locations, such as deep pump-well gate seat, tunnel, or culverts.
Long bulkhead or wharf wall	Preliminary borings on line of wall at 60 m spacing. Add intermediate borings to decrease spacing to 15 m. Place certain intermediate borings inboard and outboard of wall line to determine materials in scour zone at toe and in active wedge behind wall.
Slope stability, deep cuts, high embankments	Provide three to five borings in the critical direction to provide geological section for analysis. Number of geological sections depends on extent of stability problem. For an active slide, place at least one boring up slope of sliding area.
Dams and water retention structures	Space preliminary borings approximately 60 m apart over foundation area. Decrease spacing on centerline to 30 m with intermediate borings. Include borings at locations of cutoff, critical spots in abutment, spillway and outlet works.

Table 2.2   Requirements for boring depth (NAVFAC, soil mechanics design manual 7.01 1986).

Project type	Boring depth
Large structure with separate, closely spaced footings	Extend to depth where increase in vertical stress for combined foundations is less than 10% of *effective overburden stress*. Generally, all borings should extend to no less than 10 m below the lowest part of foundation unless rock is encountered at shallower depth.
Isolated rigid foundations	Extend to depth where vertical stress decreases to 10% of *bearing pressure*. Generally, all borings should extend no less than 10 m below the lowest part of foundation unless rock is encountered at shallower depth.
Long bulkhead or wharf wall	Extend to depth below dredge line between 0.75 and 1.5 times unbalanced height of wall. Where stratification indicates possible deep stability problem, selected borings should reach the top of hard stratum.
Slope stability	Extend to an elevation below active or potential failure surface and into hard stratum, or to a depth where failure is unlikely because of geometry of cross section.
Deep cuts	Extend to depth between 0.75 and 1 time base width of narrow cuts. Where cut is above groundwater in stable materials, depth of 1.2 to 2.4 m below base may suffice. Where base is below groundwater, determine the extent of pervious strata below base.
High embankments	Extend to depth between 0.5 to 1.25 times horizontal length of side slope in relatively homogeneous foundation. Where soft strata are encountered, borings should reach hard materials.
Dams and water retention structures	Extend to depth of half of base width of earth dams or 1 to 1.5 times height of small concrete dams in relatively homogeneous foundations. Borings may terminate after penetration of 3 to 6 m in hard and impervious stratum if continuity of this stratum is known from reconnaissance.

2.2.2   Drill rigs

Drill rigs are machines that can create holes (also known as "boreholes") in the ground that can be subsequently used for soil or rock sampling. As shown in Figure 2.1, the drill rigs that are commonly used in the geotechnical subsurface exploration include: (1) truck-mounted drill rigs, (2) track-mounted off-road (all-terrain vehicle) drill rigs, (3) over-water drill rigs that are mounted on customized barges, and (4) portable drill rigs that are operated by machine or manually (also known as "hand auger/sampler"), which can usually drill up to 7.5 m. The type of drill rigs that are used on a project depends on the project requirements and the availability of drill rigs.

2.2.3   Drilling methods and augers

In order to sample soils or conduct field soil testing at various depths, a hole should be drilled first. The drilling methods that are commonly used in geotechnical and geoenvironmental subsurface exploration include:

(a) Auger drilling
(b) Mud rotary drilling (also known as "rotary wash drilling")

Chapter 2

(a)   (b)

(c)   (d)

Fig. 2.1 Common types of drill rigs used in subsurface investigation. (a) Truck-mounted drill rig, (b) Track-mounted drill rig, (c) Barge-mounted drill rig, (d) Portable drill rig. (Photos (b) and (c), courtesy of Martin McIlroy of Taber Consultants, Sacramento, CA, USA.)

(c) Wash boring
(d) Rock coring

   The drilling methods are normally specified in local standards such as "ASTM D1452-09 Standard Practice for Soil Exploration and Sampling by Auger Borings," "ASTM D6286–98 (2006) Standard Guide for Selection of Drilling Methods for Environmental Site Characterization," and "BS EN ISO 22475–1:2006 Geotechnical Investigation and testing–Sampling methods and groundwater measurements–Part 1: Technical principles for execution."
   In general terms, the choice of sampling method will also determine the quality of soil specimens and the type of properties that can be determined from those samples. For example, BS

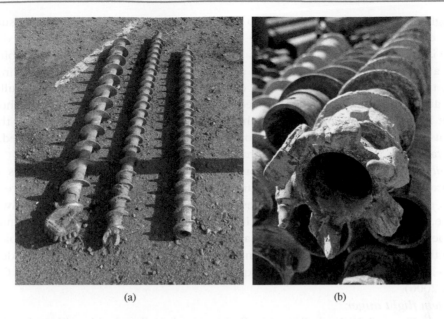

(a)   (b)

Fig. 2.2 Solid-stem and hollow-stem flight augers. (a) Solid-stem flight augers. (b) Hollow-stem flight augers (Photo courtesy of Martin McIlroy of Taber Consultants, Sacramento, CA, USA.)

Chapter 2

EN1997-2:2007 divides the sampling methods into three categories (A, B, and C). Category A sampling methods provide the best sampling quality, while category C sampling methods provide the lowest quality samples. In practical terms, this means that sampling methods of category C can only be used for the determination of water contents and the extension/depth of soil/rock strata. Category A methods, on the other hand, provide the best quality samples that can also be used for the determination of permeability, compressibility, and shear strength properties of soils and rocks from laboratory tests.

Auger drilling

Auger drilling uses various types of augers to drill a borehole through the subsurface. The augers that are commonly used in the drilling include the following:

- Solid-stem flight auger (category C)
- Hollow-stem flight auger (category B)
- Mud rotary auger (category B)
- Coring auger (Category A or B, depending on the diameter and wall thickness)
- Hand auger (Category C)

Each type of auger is suited for a particular soil or rock condition; thus, more than one method may be used on a given project based on the subsurface soil strata and rock formations.

- *Solid-stem flight auger*

  A solid-stem flight auger, as shown in Figure 2.2(a), includes a drill bit and multiple sections of solid auger flights. The drill bit cuts and loosens the soil, and the spiral projections of the auger flight help advance the auger into the ground and bring the soil to the surface. The diameter of such augers ranges from 10 to 36 cm (4 to 14 inches). During the drilling, the borehole is advanced by lowering the rotating drill head. The auger and the drill bit penetrate the soil, and flights of augers are added as required. When the specified sampling depth is reached, the borehole is cleared of soil cuttings, the auger and drill bit are pulled out by a cable, and a sampling tool is connected to a drill rod and lowered into the hole to take a sample below the bottom of the borehole. After the sampling is completed, the drill bit and flights of auger are lowered back down the hole and the drilling is resumed. Frequent withdrawal of the auger is required in order to take soil samples.

Applicability: This drilling method is ordinarily used for shallow explorations above the groundwater table in partly saturated sand and silt and in soft to stiff cohesive soils. The borehole can collapse in soft cohesionless soils and in soils below the groundwater table. Therefore, this method cannot be used in collapsible soils, and a casing is required when drilling below the groundwater table. In addition, due to the high sample disturbance resulting from this type of drilling, some design codes only allow its use for soil/rock identification purposes and the determination of the sequence of soil/rock layers (i.e., category C in EN ISO 22475–1:2006).

- *Hollow-stem flight auger*

  As shown in Figure 2.2(b), a hollow-stem flight auger is hollow inside with an inside diameter between 5 and 30 cm (2 and 12 inches); the most common inside diameters are between 7.6 and 15.2 cm (3 and 6 inches). Each auger section is 1.5 m (5 ft) in length. Hollow-stem continuous flight auger drilling is similar to the solid-stem method, in that the hole is advanced and the soil is removed by the flight on the augers. The primary difference is that the auger flight is hollow; this allows the sampler to be lowered down through the auger, thus eliminating the need to remove the auger from the borehole and thereby reducing sample disturbance. As such EN ISO 22475–1:2006 classifies this method as category B. The auger contains an outer bit (teeth) at the end of the hollow auger and an inner bit (or center bit). The inner bit is connected to a drill rod. During drilling, the inner bit and the outer bit rotate together, cutting and loosening the soil. To sample, the drill rod with the center bit is withdrawn, the center bit is disconnected from the drill rod, and a sampler is connected to the drill rod and lowered inside the hollow stem into the borehole to take a sample at the bottom of the borehole. After the sampling, the center bit is reconnected to the drill rod and lowered back down the hole, and the drilling is resumed. The hollow-stem flight auger is specified by "ASTM D6151-08 Standard Practice for Using Hollow-Stem Augers for Geotechnical Exploration and Soil Sampling."

  Applicability: Hollow-stem flight augers keep a borehole open and make the sampling of sand below the groundwater table possible. The method can be used to retrieve disturbed soil or undisturbed clay or rock coring through the hollow stem. However, this method is not suitable for undisturbed sampling in sand and silt (NAVFAC 1986). EN ISO 22475–1:2006 allows the use of this sampling method for layer sequence determination, moisture content measurement, and determination of particle sizes, Atterberg limits, particle density, and organic content.

- *Mud rotary augers* are used in mud rotary drilling, and *coring augers* are used in rock drilling and sampling, i.e., rock coring. They are described in the subsequent sections.

- *Hand auger*

  Hand augers (Figure 2.3) are operated manually. They are portable and are used to drill and sample surficial soils, usually up to a depth of 7.5 m.

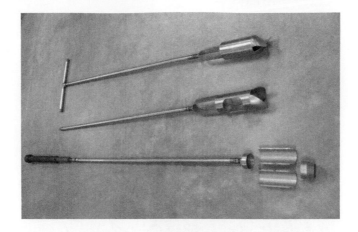

Fig. 2.3   Hand auger and sampler.

Applicability: Hand auger drilling is usually used for shallow explorations above the ground-water table. The borehole can collapse in soft soils and in soils below the groundwater table; therefore, hand auger drilling is not suitable in such soils. Hand auger drilling and sampling result in highly disturbed samples (i.e., category A).

Mud rotary drilling (rotary wash drilling)

Mud rotary or rotary washing drilling discharges a mixture of water and bentonite at the drill bit within the borehole to facilitate soil cutting. Figure 2.4 shows a mud rotary drilling in progress.

The water–bentonite mixture is also known as "mud or slurry." The use of slurry has two purposes: (1) the slurry suspends the soil cuttings at the bottom of the borehole and carries them to the top, where they are collected in the mud tank and (2) the slurry can keep the hole from collapsing, particularly below the groundwater table. The consistency of the slurry is important: if it is too thick, it can clog the pump; if it is too thin, then it cannot provide sufficient buoyancy to the soil cuttings and consequently leaves them at the bottom of the hole. The density of slurry should be 10%~20% higher than the density of water.

Mud rotary drilling uses several types of rotary bits: drag bits (Figure 2.5(a)), roller bits (with three rotating teeth) (Figure 2.5(b)), and plug bits. Drag bits are commonly used in clay and loose sands. Roller bits are used to penetrate dense coarse-grained granular soils, cemented soils, and soft or weathered rock. Plug bits are used to drill soft rocks. Each type of bit has a hole to discharge the mud at a high speed, either from the face of the bit (face discharge) or from the side of the bit (side discharge). The face discharge provides higher impact on the soil and may significantly disturb the soil. Mud rotary drilling can be performed with or without a casing. When mud rotary drilling is used in cohesionless or erodible soil, an outer casing is generally used. A hollow-stem flight auger can be used as an outer casing along with mud rotary drilling.

When the specified sampling depth is reached, circulation of drilling mud is first stopped, and the borehole is cleared of soil cuttings. Then the drill rod with a rotary bit is withdrawn, and a sampler is quickly attached to the drill rod and lowered in the hole to take a sample below the bottom of the borehole. It is noted that once the overburden stress is removed and the soil at the bottom of the borehole is exposed to the atmosphere, the soil can expand and

Fig. 2.4   Mud rotary drilling. (Photo courtesy of Martin McIlroy of Taber Consultants, Sacramento, CA, USA.)

(a)   (b)

Fig. 2.5   Rotary bits. (a) Drag bits, (b) Roller bit. (Photo courtesy of Martin McIlroy of Taber Consultants, Sacramento, CA, USA.)

the soil characteristics can change. Therefore, attaching a sampler to the drill rod and sampling are usually done quickly. After sampling, the rotary bit is reconnected to the drill rod to resume drilling. Specific requirements for mud rotary drilling are normally described in design codes such as "ASTM D5783-95(2006) Standard Guide for Use of Direct Rotary Drilling with Water-Based Drilling Fluid for Geoenvironmental Exploration and the Installation of Subsurface Water-Quality Monitoring Devices." and "BS EN ISO 22475–1:2006 Geotechnical Investigation and testing – Sampling methods and groundwater measurements – Part 1: Technical principles for execution."

Applicability: Mud rotary drilling is applicable to all types of soils except for coarse sand and cobbles, in which the slurry can drain freely and, therefore, cannot be circulated. This drilling method is particularly useful when drilling in collapsible soils (e.g., loose sand) below the groundwater table, as the slurry can keep the borehole open under such condition.

Wash boring

The wash boring method uses chopping, twisting, and jetting actions of a light bit while circulating drilling fluid removes cuttings from the hole. If required, a casing can be used to prevent caving.

Applicability: Wash boring can be used in sands, sand and gravel without boulders, and soft to hard cohesive soils. Owing to the chopping, twisting, and jetting actions of the drill bit, a large volume of soil below the borehole is disturbed. This method is no longer commonly used.

Rock coring

Coring rocks or hard soils is similar to the mud rotary method, where slurry is used to facilitate the drilling. However, in rock coring, sampling and drilling occur simultaneously. A core barrel, as shown in Figure 2.6(a), is connected to a drill bit that is usually embedded with diamond in order to cut into rocks, as shown in Figure 2.6(b). The core barrel consists of an inner barrel and an outer barrel. The barrel is connected to a drill rod of the same diameter, and the drill rod acts as a casing for the borehole. The entire assembly rotates at high speed while weight from the drill rig is added to force the diamond bit into the rock formation. The type of bit, rotation speed, and weight on the drill bit are determined by the type of rock. During the drilling, water is pumped into the drill rods to cool the diamond bit and to remove the cuttings from the borehole. When the barrel is full, the inner barrel is removed from the outer barrel and lifted to the ground surface. Figure 2.6(c) shows a rock core sample retrieved by the coring. Once the rock sample is removed, the inner barrel can be lowered back into the core barrel and the drilling and sampling process continues. Rock coring is specified in standards such as "ASTM D2113-08 Standard Practice for Rock Core Drilling and Sampling of Rock for Site Investigation" and "BS EN ISO 22475–1:2006 Geotechnical Investigation and testing - Sampling methods and groundwater measurements - Part 1: Technical principles for execution."

Applicability: Rock coring can be used to drill and sample weathered rocks, bedrocks, and boulders.

2.2.4 Soil sampling methods

Soil sampling can obtain disturbed or relatively undisturbed soil samples. Disturbed samples are primarily used for classification tests (grain size distribution, water content, specific gravity, and

(a)　　　　　　　　　　　　　　(b)

(c)

Fig. 2.6 Core barrel used in the rock coring. (a) Rock core barrels (The devices with the "fingers" are the inner core barrels that are inserted into the outer core barrels. Inside the inner core barrel are two split sleeves that hold the rock samples.), (b) Diamond bits used for rock coring, (c) Rock sample in the split core barrel, the tube by the sample conducts drilling fluid. (Photo courtesy of Martin McIlroy of Taber Consultants, Sacramento, CA, USA.)

Atterberg limits) and must contain all of the constituents of the soil, although the structure is disturbed. Undisturbed samples are taken primarily for laboratory strength and compressibility tests. The degree of soil sample disturbance depends on the samplers used, and hence, the same principles of sample categories described for augers are used by various design codes (e.g., BS EN ISO 22475–1:2006 and BS EN 1997–2:2007)

The most common soil samplers include the following:

(a) Split spoon samplers (category B)
(b) Shelby tube samplers (thin-walled samplers) (category A or B)
(c) Piston samplers (category A or B)
(d) Pitcher barrel samplers (category A or B)

The selection of sampling devices can be referred to in "ASTM D6169-98(2005) Standard Guide for Selection of Soil and Rock Sampling Devices Used With Drill Rigs for Environmental

Investigations" and "BS EN ISO 22475–1:2006 Geotechnical Investigation and testing - Sampling methods and groundwater measurements - Part 1: Technical principles for execution" as good examples of standardization.

Split spoon sampler

A split spoon sampler, as shown in Figure 2.7, includes a driving shoe (to cut the soil), a split barrel (to contain the soil sample), an optional liner (to protect the soil), a top sleeve (to connect to the drill rod), and a plastic catcher (not shown in the figure) (to prevent the soil from falling out of the sampler). The plastic catcher is secured between the split barrel and the driving shoe. The split spoon sampler is driven into the soil by hammers of different types. Therefore, the soil collected in split spoon samplers is disturbed due to the hammering of the sampler. The split spoon sampler usually has an inside diameter between 3.5 and 11.4 cm (1-3/8 and 4.5 inches) and a length between 45.7 and 76.2 cm (18 and 30 inches).

The most commonly used split spoon samplers are the *SPT sampler* and the *modified California sampler* (Figure 2.8). The standard penetration test (SPT) sampler is used in the SPT to retrieve soil samples and to obtain preliminary *in situ* soil characteristics. According to the

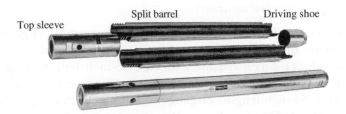

Fig. 2.7 Illustration of a split spoon sampler.

Fig. 2.8 Common split spoon samplers. (From left to right: modified California sampler, California sampler, and SPT sampler.)

ASTM D1586, a SPT sampler typically has an outer diameter of 2 inches, an inside diameter of 1.5 inches, and a length between 18 and 30 inches. If using a liner, the inside diameter is 1-3/8 inch. The modified California sampler was originally invented to sample hard and desiccated soils and soft sedimentary rocks that are common in southern California; now it is widely used in other locations and soil conditions. Similar in concept to the SPT sampler, the modified California sampler barrel has a larger diameter and is usually lined with metal tubes to contain samples. It usually has a 3-inch outside diameter and a 2.5-inch inside diameter. Both types of split spoon samplers are driven into the subsoil using various types of hammers that are described in Section 2.4 "*In situ* field testing." The split spoon samplers are specified by "ASTM D3550-01(2007) Standard Practice for Thick Wall, Ring-Lined, Split Barrel, Drive Sampling of Soils." The SPT sampler is specified by "ASTM D1586-11 Standard Test Method for Standard Penetration Test (SPT) and Split-Barrel Sampling of Soils." Another example where these samplers are specified is "BS EN ISO 22475–1:2006 Geotechnical Investigation and testing - Sampling methods and groundwater measurements - Part 1: Technical principles for execution."

After the sampling, the split spoon sampler is split open, and the samples are carefully wrapped in plastics and stored in sealed plastic or metal tubes (or jars). Project number, boring number, sample number, sampling depth, and SPT blow count (if any) should be recorded on the jar.

Shelby tube sampler (thin-walled sampler)

The Shelby tube sampler consists of a thin-walled tube with a cutting edge at the toe, as shown in Figure 2.9. The Shelby tube samplers have three standard dimensions that are specified by ASTM D1587, including: (1) 2-inch outside diameter, 36-inch length, 18-gauge (0.0478 inch) thickness; (2) 3-inch outside diameter, 36-inch length, 16-gauge (0.0598 inch) thickness; (3) 5-inch outside diameter, 54-inch length, 11-gauge (0.1196 inch) thickness. "Gauge" is a unit of steel sheet thickness. The ASTM also allows other diameters, as long as they are proportional to the standardized tube designs, and the tube length is suited for field conditions. The sampler contains a check valve and pressure vents to facilitate the entering of soil samples into the sampler. To sample, the sampler is attached to the drill rod and is lowered in the borehole and rests at the bottom. Then, it is steadily pushed into the soil at a constant speed by hydraulics on the drill rig for a penetration of generally 6 inches less than the length of the tube. Further, the

Fig. 2.9   Shelby tube sampler.

drill rod and the sampler are rotated by one revolution to shear the soil. The sampler is then pulled out of the ground. The vacuum created by the check valve and the cohesion of the sample in the tube keep the soil sample retained in the tube while the tube is withdrawn. *The Shelby tube sampler is generally used to obtain undisturbed cohesive soil samples.* After the sampling, melted wax or o-ring packers are used to seal the soil inside the Shelby tube sampler. It is then transported to the laboratory for testing. The Shelby tube sampler is specified by "ASTM D1587-08 Standard Practice for Thin-Walled Tube Sampling of Soils for Geotechnical Purposes." An equivalent standard used in Britain and similarly in other European countries is "BS EN ISO 22475–1:2006 Geotechnical Investigation and testing - Sampling methods and groundwater measurements - Part 1: Technical principles for execution."

Piston sampler

When sampling soft soils, the soil sometimes may enter the sampler before the sampler is pushed into the soil. In addition, when a sampler containing soft soils is lifted out of the borehole, the soil may fall out of the sampler. In order to avoid these two problems, a piston sampler was invented (Figure 2.10). A piston sampler is a thin-walled metal tube (such as a Shelby tube) that contains a piston at the tip. When the sampler is first introduced into the borehole, the piston prevents the soil from entering the sampler before the sampling starts. As the sampler is steadily

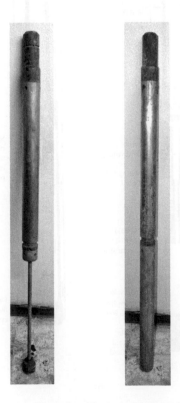

Fig. 2.10 Piston sampler (the sampler on the left without outer barrel; the sampler on the right with outer barrel). (Photo courtesy of Martin McIlroy of Taber Consultants, Sacramento, CA, USA.)

pushed into the soil at a constant speed by hydraulics on the drill rig, the soil sample enters the sampler and pushes the piston upward. When the sampler is pulled out of the borehole, the tendency of the soil to fall creates a vacuum, which keeps the soil in the sampler. The piston sampler is mainly used to retrieve undisturbed soft soil samples. The piston sampler is specified by "ASTM D6519-08 Standard Practice for Sampling of Soil Using the Hydraulically Operated Stationary Piston Sampler," and as for the previous examples, also in "BS EN ISO 22475–1:2006 Geotechnical Investigation and testing - Sampling methods and groundwater measurements - Part 1: Technical principles for execution."

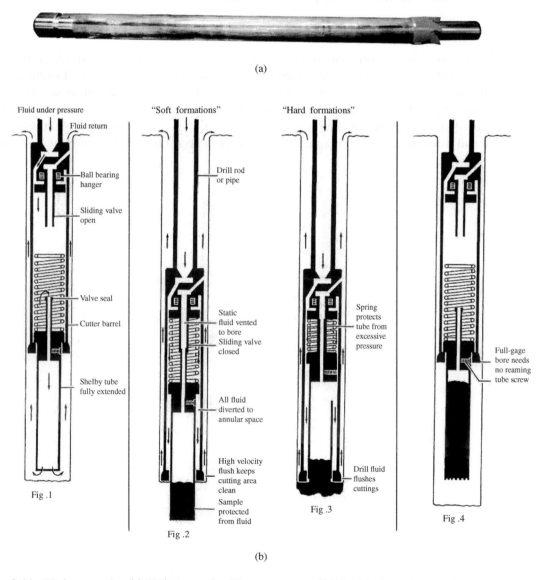

Fig. 2.11   Pitcher sampler. (a) Pitcher sampler (Photo courtesy of Michael D. DiCindio of Acker Drill Company), (b) Functioning of Pitcher sampler. (Diagram courtesy of Michael D. DiCindio of Acker Drill Company.)

Pitcher barrel sampler

Figure 2.11(a) shows a Pitcher sampler. It is a double-tube sampler (Fig. 1 in Figure 2.11 (b)). The core sizes are usually 3, 4, and 6 inches in diameter with lengths of 3 or 5 feet. The Pitcher sampler contains a high-tension spring that is located between the inner and the outer barrel above the inner head. The inner barrel (or tube) can extend or retract relative to the outer barrel, depending on the soil stiffness. For example, in softer formations (Fig. 2 in Figure 2.11 (b)), the spring extends such that the inner barrel shoe extends out of the outer barrel bit; this prevents damage to the sample by the drilling fluid and drilling action. For stiffer soils (Fig. 3 in Figure 2.11 (b)), the sampling tube is pushed back into the outer barrel by the stiff soil. Once the inner tube rests at the bottom of the borehole, the barrel retracts and the drilling fluid is diverted to the annulus between the inner and the outer barrel. This arrangement facilitates the washing of material from the inside of the sampler before sampling and circulation of drilling fluid to remove cuttings during sampling. The inner barrel with soil sample can be easily removed from the sampler and stored, and a new empty inner barrel can be installed into the sampler (Fig. 4 in Figure 2.11 (b)). The Pitcher sampler is suitable for sampling alternating hard and soft layers or soils of variable hardness.

2.3   Geotechnical boring log

A geotechnical boring log is a record of subsurface information obtained from subsurface drilling, sampling, and analysis. The information includes topographic survey data, including boring location and surface elevation, and bench mark location and datum if available; soil description and classification including density, consistency, color, moisture, geologic origin, depths of soil or rock samples; drilling and sampling methods, field test methods and results such as SPT blow counts; and groundwater table depth (Mayne et al. 2002). Also included on the boring log are project title, location, bore hole number, drilling personnel information, and so on. The information is presented in a graphical and a tabular format. Different companies or agencies have different formats of the boring log. Figure 2.12 shows an example of a boring log. Geotechnical boring logs are specified for US practice "ASTM D5434-09 Standard Guide for Field Logging of Subsurface Explorations of Soil and Rock." "BS EN ISO 22475-1 : 2006 Geotechnical Investigation and testing – Sampling methods and groundwater measurements – Part 1: Technical principles for execution." specifies the requirements of borehole logs to be included in the site investigation report.

2.4   *In situ* field testing

2.4.1   Standard penetration test (SPT)

The SPT records the number of hammer drops (or blows) required to drive an SPT sampler into the soil by 18 inches; the number is then used to derive certain characteristics of the soil. The hammer weighs 140 lb and drops from a 30-inch height onto an anvil at the top of the drill rod that connects to the SPT sampler. The number of hammer drops, referred to as blows, in each of the three 6-inch penetrations is counted. The numbers of blows for the second and the third 6-inch penetrations are added together to derive the dimensionless standard penetration

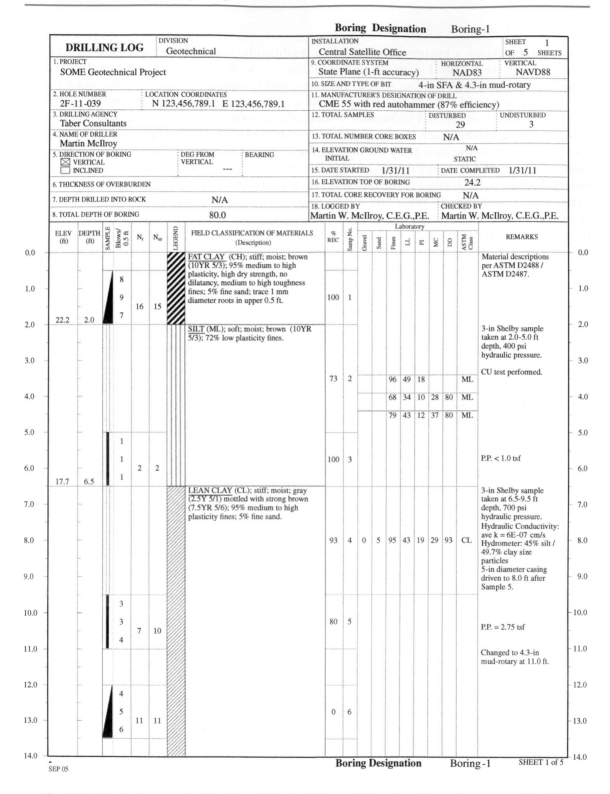

Fig. 2.12   Boring log sample. (Courtesy of Martin McIlroy of Taber Consultants, Sacramento, CA, USA.)

Chapter 2

Fig. 2.13   Standard penetration test in progress. (The SPT sampler in the photo is marked to show three 6-inch sections.)

resistance (also known as the *N* value or SPT blow count). The number of blows for the first 6-inch penetration is not included in the *N* value, because the drilling disturbs the soil at the bottom of the borehole and the first 6-inch soil sample does not represent the *in situ* soil condition. Figure 2.13 shows a photo of the standard penetration test in progress. The standard penetration test is specified by "ASTM D1586-11 Standard Test Method for Standard Penetration Test (SPT) and Split-Barrel Sampling of Soils" and "BS EN ISO 22476-03 : 2005 Geotechnical investigation and testing. Field testing. Standard penetration test."

The SPT generally cannot be used in soil deposits containing gravels, cobbles, or boulders that typically result in penetration refusal and damage to the equipment.

Factors affecting the SPT blow count

In a field standard penetration test, the blow count depends on many factors, including the hammer efficiency, borehole diameter, type of samplers, sampling depths (i.e., the drill rod length), and rate of hammer drops. The hammer efficiency is defined as the ratio of the actual hammer energy passed to the sampler to the total input energy from the hammer. In the field test, the hammer efficiency varies from 30% to 90%. If using a 140 lb hammer that drops from 30 inch, the average energy delivered to the sampler per hammer blow is 2520 in-lb, which is 60% of the theoretical maximum energy per hammer blow. A hammer efficiency of 60% is used in the standard practice in the United States and in other countries. The N value is also affected by the type of hammers, the borehole diameter, and the depth at which the SPT is conducted. Therefore, the *standard penetration number that is corrected based on the field conditions*, N_{60}, is defined as:

$$N_{60} = \frac{N \, \eta_H \, \eta_B \, \eta_S \, \eta_R}{60} \tag{2.1}$$

where:

N = measured SPT blow count,
η_H = hammer efficiency (%),
η_B = correction for borehole diameter,
η_S = sampler correction, and
η_R = correction for rod length.

Tables 2.3–2.6 list the recommended values for $\eta_H, \eta_B, \eta_S, \eta_R$, respectively, by Seed et al. (1985) and Skempton (1986).

Table 2.3   Variation of η_H in the United States.

Hammer type	Hammer release	η_H(%)
Safety hammer	Rope and pulley	60
Donut hammer	Rope and pulley	45
Automatic	Trip	73*

*Reference: Clayton 1990.

Table 2.4   Variation of η_B.

Borehole diameter	η_B
2.4 – 4.7 inch (60 – 120 mm)	1.0
6 inch (150 mm)	1.05
8 inch (200 mm)	1.15

Table 2.5   Variation of η_S.

Samplers	η_S
Standard sampler without liner (not recommended)	1.2
Standard sampler	1.0
With liner for dense sand and clay	0.8
With liner for loose sand	0.9

Table 2.6   Variation of η_R.

Rod length	η_R
> 30 ft(> 10 m)	1.0
20 – 30 ft (6 – 10 m)	0.95
12 – 20 ft (4 – 6 m)	0.85
0 – 12 ft(0 – 4 m)	0.75

The field-corrected standard penetration number, N_{60}, can be used to derive certain soil characteristics. The following shows three examples:

(a) Relative density of sand (Cubrinovski and Ishihara 1999):

$$D_r(\%) = \left[\frac{N_{60}\left(0.23 + \frac{0.06}{D_{50}}\right)^{1.7}}{9} \left(\frac{1}{\frac{\sigma_0'}{p_a}}\right) \right]^{0.5} \times 100\% \qquad (2.2)$$

where:

D_{50} = soil diameter corresponding to 50% finer by mass in the grain size distribution (mm),

σ_0' = effective overburden stress, and

p_a = atmospheric pressure ($\approx 101\ kN/m^2$ or $\approx 2000\ lb/ft^2$).

(b) Effective friction angle (Schmertmann 1975):

$$\phi' = \tan^{-1}\left[\frac{N_{60}}{12.2 + 20.3\left(\frac{\sigma_0'}{p_a}\right)} \right]^{0.34} \qquad (2.3)$$

(c) Modulus of elasticity of granular soils (Kulhawy and Mayne 1990):

$$E_S = \alpha p_a N_{60} \qquad (2.4)$$

where:

$\alpha = 5$ for sands with fines, $\alpha = 10$ for clean normally consolidated sand, and $\alpha = 15$ for clean overconsolidated sand.

In the aforementioned equations, σ_0' and p_a should use the same unit.

The drilling method also affects the N value. Mud drilling is preferred over the hollow-stem flight auger. The diameter of a borehole is usually 4-5 inches. As the borehole diameter decreases, the penetration resistance increases. High-pressure slurry discharge can also affect the penetration resistance. Therefore, when using mud rotary drilling, side discharge rather than bottom discharge should be used to reduce the disturbance to the soil beneath the bottom of the borehole.

Chapter 2

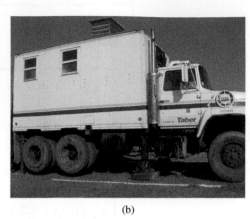

(a)   (b)

Fig. 2.14   Cone penetration test. (a) Penetrometers, (b) CPT truck. (Photo courtesy of Taber Consultants.)

2.4.2   Cone penetration test (CPT)

The cone penetration test (CPT) determines the penetration resistance (end bearing and side friction) of a pointed rod (cone) when it is steadily and slowly pushed into soil. This test method is sometimes referred to as the "Dutch cone test," as it was originally performed in the Netherlands. The pointed rod, called cone or penetrometer (Figure 2.14(a)), is of different types, including the *cone penetrometer*, which measures the end bearing resistance only; the *friction-cone penetrometer*, which measures the end bearing resistance and friction resistance; and the *piezocone penetrometer*, which measures both the end bearing and the friction resistance as well as the pore water pressure. The CPT system includes the penetrometer, a rod that connects the penetrometer and pushes it into soil, and the data acquisition system. The system is mounted in a CPT truck, as shown in Figure 2.14(b). The cone penetration test is specified by "ASTM D3441-05 Standard Test Method for Mechanical Cone Penetration Tests of Soil," "ASTM D5778-07 Standard Test Method for Electronic Friction Cone and Piezocone Penetration Testing of Soils," and "BS EN ISO 22476–1:2012 Geotechnical investigation and testing. Field testing. Electrical cone and piezocone penetration test."

The advantages and disadvantages of the SPT and CPT are compared in Table 2.7.

Many empirical relationships are available to correlate the measured tip resistance and friction resistance in the CPT to the soil characteristics. For example, Kulhawy and Mayne (1990) proposed the following relationships for normally consolidated sand:

$$D_r(\%) = 68 \left[\log \left(\frac{q_c}{\sqrt{p_a \sigma_0'}} \right) - 1 \right] \tag{2.5}$$

$$\phi' = \tan^{-1} \left[0.1 + 0.38 \ \log \left(\frac{q_c}{\sigma_0'} \right) \right] \tag{2.6}$$

Table 2.7   Comparisons of the SPT and CPT.

	SPT	CPT
Advantages	• Can obtain representative soil samples for further laboratory testing. • Can provide boreholes for other field testing, such as vane shear test. • Is able to penetrate dense granular soils. • Equipment and expertise are widely available.	• Can provide the entire subsoil profile and is considered the best technique for delineation of stratigraphy. • Fast, economical, and productive. • Can eliminate operator errors, and can provide reliable and repeatable test results. • Particularly suitable for soft soils
Disadvantages	• Does not provide continuous soil data, but rather at intervals; could miss thin and problematic soil stratum. • Samples are disturbed. • Significant variability of measured penetration resistance can occur due to considerable variation of apparatus and procedure.	• Cannot obtain soil samples for further laboratory test and verification. • Cannot penetrate dense, granular soils, such as cobbles. • High capital investment. • Requires skilled operator to run. • Electronic noise of data requires calibration.

Chapter 2

where:

$D_r(\%)$ = relative density,

φ' = effective friction angle of the soil,

q_c = tip resistance of the penetrometer,

σ_0' = effective stress at the same depth where q_c is measured, and

p_a = 1 atmosphere pressure = 101 kN/m^2.

In the aforementioned equations, q_c, σ_0', and p_a should use the same unit.

2.4.3   Vane shear test

The vane shear test measures the undrained shear strength of *in situ* soil. As shown in Figure 2.15, the vane shear test device consists of a vane comprising four blades that are 2 mm thick, 1.4–4.0 inch in diameter, and 5 inch long. The tips of the blades are either flat or tapered at 45 degrees. The height of the vane is 1.0~2.5 times the vane diameter. To conduct the test, the device is connected to the drill rod and lowered to the bottom of the borehole, and the vane is gently pushed into the soil. Then, a torque wrench with a measuring gauge is attached to the top of the drill rod to rotate the vane shear device at 6 degrees per minute, and the maximum torque (pound-inch) required to shear the soil is recorded. The vane shear test is specified by "ASTM D2573–08 Standard Test Method for Field Vane Shear Test in Cohesive Soil."

The maximum measured torque can be used to determine the *in situ* undrained shear strength (S_u) of fine-grained soils (ASTM D2573):

For a rectangular vane of $H/D = 2$ (H is height of vane, D is diameter of vane):

$$S_u = \frac{6T_{max}}{7\pi D^3}$$

(2.7)

Fig. 2.15   Vane shear test.

For a tapered vane:

$$S_u = \frac{12T_{max}}{\pi D^2 \left(\dfrac{D}{\cos i_T} + \dfrac{D}{\cos i_B} + 6H \right)}$$

(2.8)

where:

T_{max} = maximum measured torque corrected for apparatus and rod friction,
i_T = angle of taper at vane top, and
i_B = angle of taper at vane bottom.

2.4.4   Flat plate dilatometer test

The flat plate dilatometer provides information about the soil's *in situ* stratigraphy, strength, compressibility, and pore-water pressure for the design of earthworks and foundations. Figure 2.16 shows a flat plate dilatometer with data acquisition. The test is initiated by forcing the dilatometer blade (a flat steel plate with sharp cutting edge) into the soil, often using a CPT rig or a conventional drill rig. A diaphragm on the plate is inflated, and it applies a lateral force to the soil and measures the strain induced for various levels of applied stress at the desired depth intervals. The test method can be applied to sands, silts, clays, and organic soils that can be penetrated with the dilatometer blade with a static push. The flat plate dilatometer test is specified in standards such as "ASTM D6635-01 Standard Test Method for Performing the Flat Plate Dilatometer" and "BS EN ISO 22476–5:2012 Geotechnical investigation and testing. Field testing. Flexible dilatometer test."

Chapter 2

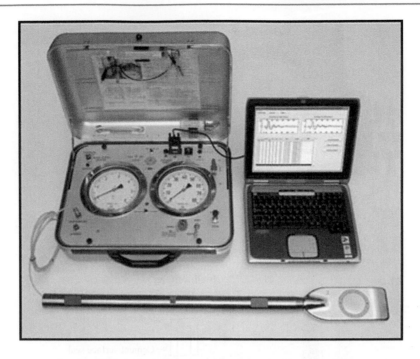

Fig. 2.16   Flat-plate dilatometer test. (Photo courtesy of In-Situ Soil Testing.)

This method is not applicable to soils that cannot be penetrated by the dilatometer blade without causing significant damage to the blade or its internal membrane.

2.4.5   Inclinometer test

An inclinometer is a device for measuring deformation normal to the axis of a pipe by passing a probe along the pipe and measuring the inclination of the probe with respect to the line of gravity. An inclinometer can determine the characteristics of landslides, slope movements, deflections in retaining walls and piles, and deformations of excavation walls, tunnels, and shafts. The measurable characteristics include (1) location of the slip plane of landslides or the depth of the movement, (2) rate of movement, (3) type of slope movement (rotational or translational), (4) magnitude of movement, and (5) direction of movement. As shown in Figure 2.17, an inclinometer system includes inclinometer casing (not shown in the figure), an inclinometer probe and control cable, and an inclinometer readout unit. The inclinometer casing is typically installed in a near-vertical borehole that passes through the zone of suspected movement and is anchored in a stable soil stratum below the suspected depth of movement. The inclinometer probe is lowered into the casing and establishes the casing's initial alignment. Subsequent ground movement will cause the casing to move away from its initial position. The inclination of the probe with respect to the direction of gravity is measured, and the measurements are converted to distances using trigonometric functions. The rate, depth, and magnitude of the movement are calculated by comparing the initial data with the data of subsequent measurements. The inclinometer test is specified by "ASTM D6230-98(2005) Standard Test Method for Monitoring Ground Movement Using Probe-Type Inclinometers" and "ASTM D7299-06 Standard Practice for Verifying Performance of a Vertical Inclinometer Probe."

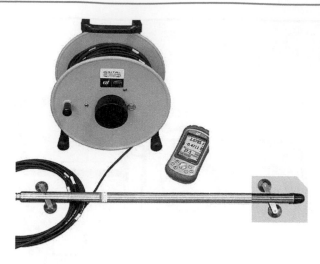

Fig. 2.17   Inclinometer.

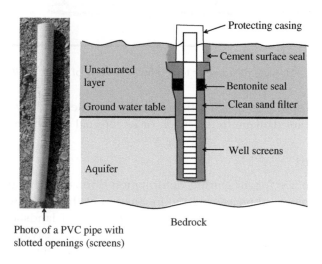

Photo of a PVC pipe with
slotted openings (screens)

Fig. 2.18   Monitoring well.

2.4.6   Groundwater monitoring well

A groundwater monitoring well, also known as "piezometer," can be installed after a borehole is drilled in order to monitor the groundwater elevation and quality. As illustrated in Figure 2.18, the piezometer is usually a 5 cm PVC pipe and generally includes a screen section that is perforated or slotted and a riser that is not perforated or slotted. The screen section is within the target monitoring zone such that groundwater can flow into the pipe and be sampled. The riser is above the monitoring zone and extends to the ground surface. After the borehole is drilled to the desired sampling/monitoring level, the piezometer is lowered and positioned in the center in the borehole; clean sand is poured into the borehole and around the pipe to serve as a filter; the top portion of the annular hole is filled with clay to prevent the seepage of surface water into the filter. The piezometer is caped at the top. A groundwater detector or sampler can be lowered into the piezometer for monitoring purpose as required. The following ASTM specifications provide the guidelines and procedures for groundwater monitoring wells.

- ASTM D5092-04(2010)e1 standard practice for design and installation of groundwater monitoring wells.
- ASTM D5521-05 standard guide for development of groundwater monitoring wells in granular aquifers.
- ASTM D5781-95(2006) standard guide for use of dual-wall reverse-circulation drilling for geoenvironmental exploration and the installation of subsurface water-quality monitoring devices.
- ASTM D5782-95(2006) standard guide for use of direct air-rotary drilling for geoenvironmental exploration and the installation of subsurface water-quality monitoring devices.
- ASTM D5783-95(2006) standard guide for use of direct rotary drilling with water-based drilling fluid for geoenvironmental exploration and the installation of subsurface water-quality monitoring devices.
- ASTM D5784-95(2006) standard guide for use of hollow-stem augers for geoenvironmental exploration and the installation of subsurface water-quality monitoring devices.
- ASTM D5787-95(2009) standard practice for monitoring well protection.
- ASTM D5872-95(2006) standard guide for use of casing advancement drilling methods for geoenvironmental exploration and installation of subsurface water-quality monitoring devices.
- ASTM D5875-95(2006) standard guide for use of cable-tool drilling and sampling methods for geoenvironmental exploration and installation of subsurface water-quality monitoring devices.
- ASTM D5876-95(2005) standard guide for use of direct rotary wireline casing advancement drilling methods for geoenvironmental exploration and installation of subsurface water-quality monitoring devices.

Examples of similar specifications used in the United Kingdom include:

- BS EN ISO 22475 – 1 : 2006 Geotechnical investigation and testing. Sampling methods and groundwater measurements. Technical principles for execution.
- BS ISO 21413 : 2005 Manual methods for the measurement of a groundwater level in a well.
- BS ISO 5667 – 22 : 2010 Water quality. Sampling. Guidance on the design and installation of groundwater monitoring points.

2.5   Subsurface investigations using geophysical techniques

Geophysical techniques avoid the destructive effects of drilling and can generate a profile of the subsurface features. They are usually fast and cost effective. The following geophysical techniques are discussed.

(a) Ground penetration radar (GPR)
(b) Electromagnetics in time domain and in frequency domain
(c) Electrical resistivity imaging
(d) Microgravity
(e) Seismic refraction and seismic reflection

Chapter 2

Chapter 2

2.5.1 Ground penetration radar (GPR)

GPR is a nondestructive method for subsurface exploration, groundwater detection, locating utilities, underground tanks, and sinkholes, and pavement and infrastructure characterization. The device consists of a radar control unit, transmitting and receiving antennas, and data storage or display devices. It sends electromagnetic waves of high frequency (10 – 2000 MHz) into the ground; the waves then bounce back from objects in the soil or from contacts between various Earth materials and are detected and stored by the receiver (Figure 2.19). In the wave penetration process, the GPR device can be pulled along the ground by a vehicle or simply by hand (Figures 2.20 and 2.21).

GPR's penetration depth depends on the electrical conductivity of subsurface materials and the frequency of the antenna used. The penetration depth decreases with increasing electrical conductivity of the subsurface medium as well as increasing frequency. For instance, GPR waves can penetrate up to 30 meters in low-electrical conductivity materials such as granite, whereas only about 1 meter in high-electrical conductivity materials, such as saline soil. For ice and air, depths of up to 300 meters can be reached. GPRs have limited capabilities in highly conductive materials such as clayey soils and salt-contaminated soils. The quality of the image generated by the GPR survey also depends on the material type and the antenna frequency. Typically, the higher the frequency, the higher is the resolution, but the lower is the depth the GPR can *see*. Figure 2.22 illustrates an example of the GPR survey results of underground pipes.

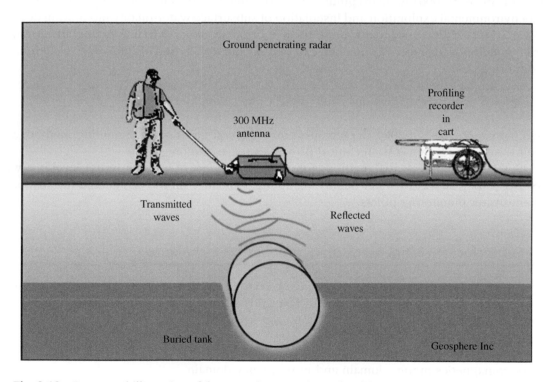

Fig. 2.19 Conceptual illustration of the ground penetration radar. (Photo courtesy of Geosphere, Inc.)

Fig. 2.20   Hand-operated GPR. (Photo courtesy of Charles Machine Works, Inc.)

Fig. 2.21   Vehicle-operated GPR truck towing 200 MHz antenna. (Photo courtesy of Paula Turner of GeoModel, Inc.)

Chapter 2

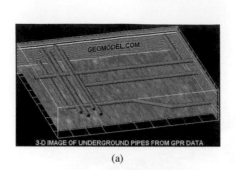
(a)

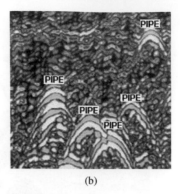
(b)

Fig. 2.22   Example of GPR survey results. (a) 3D illustration of underground pipes, (b) Subsoil image generated by GPR showing the pipes. (Photo courtesy of Paula Turner of GeoModel, Inc.)

2.5.2   Electromagnetics in frequency domain and in time domain

Electromagnetics (EM) can be used to detect bedrock discontinuities and to locate mineshafts, buried dumpsites, leachate plumes, underground streams and aquifers, and even metallic and magnetic objects. It has the advantage of covering large areas in a short amount of time. The system is composed of a transmitter and a receiver. Similarly to GPR, the penetration depth also depends on the frequency and the medium. During investigation of small objects such as underground tanks, high frequencies are usually used; moderate frequencies are optimal in detecting sinkholes; and low frequencies are effective for observing subsurface ground conditions or for locating subterranean caverns.

In *frequency-domain* electromagnetics (FDEM), the electromagnetic system is held above the ground surface, and the transmitter sends continuous low-frequency radio waves into the soil.

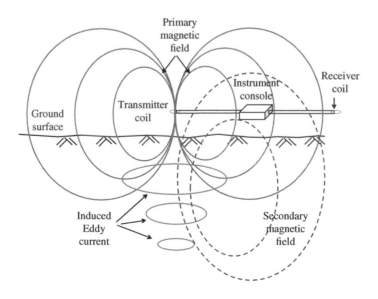

Fig. 2.23   Conceptual illustration of frequency-domain electromagnetics.

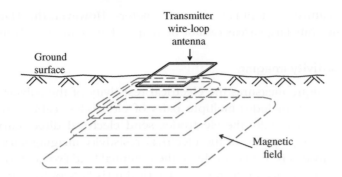

Fig. 2.24   Conceptual illustration of time-domain electromagnetics.

This primary electromagnetic field of low-frequency radio waves generates a secondary electrical current (also known as an "Eddy current") field in the ground. The eddy current creates a secondary magnetic field in the soil. The receiver detects and measures the secondary field and converts the measured data to electrical conductivity. As the electrical conductivity of soil correlates strongly with soil properties, the FDEM is a powerful tool for mapping soils and changes in soil types. The observable depths can range from 1 to 30 meters for FDEM. Figure 2.23 illustrates the concept of FDEM.

In the *time-domain* electromagnetics (TDEM), the transmitter is an insulated electrical cable in the shape of a square loop on the ground. When it is connected to alternating current (AC), the induced electrical current creates a steady-state magnetic field in the subsoil. The AC is then shut off. As the magnetic field decays with *time*, the electrical conductivity data of the subsoil are recorded by a receiver and are then used to create a vertical profile of the resistivity with the depth of the soil, which can be correlated to soil types and properties. Figure 2.24 illustrates the concept of TDEM, and Figure 2.25 shows the actual equipment. While TDEM has similar applications to FDEM, it is usually not applicable in shallow depth exploration and is better

Fig. 2.25   Time-domain electromagnetics survey. (Photo courtesy of GBG Australia.)

suited for mapping features at depths of 20–1000 meters. However, the TDEM surveys are not as rapid as FDEM and thus large areas cannot be mapped as economically as FDEM.

2.5.3 Electrical resistivity imaging

This geophysical technique measures the electrical resistivity of the subsurface materials. The electrical resistivity device includes multiple pairs of electrodes, cables, and Earth resistivity meters. As shown in Figure 2.26, the electrodes send electrical direct current (DC) into the subsoil. Figure 2.27 shows a photo of the electrical resistivity imaging survey in progress. An additional pair of electrodes is used to evaluate the potential field (voltage) that is created by the initial direct current. This device measures the electrical resistivity of the encountered materials. Then the measured resistivity is compared with known values of various materials to determine what is underground. Additionally, increasing the distance between electrodes can increase the effective penetration depth, allowing the investigators to evaluate deeper soil. This method can observe soils at a depth of approximately 250 meters below the surface and is applicable in the investigation of bedrock fracture zones, delineating tunnels, characterizing landfills, evaluating water tables, and so on. ERI is particularly effective in evaluating clayey soils and is also a preferred method in defining transitional boundaries. The measurements depend on the soil or rock type, the porosity and permeability, and the pore fluid chemistry. The ERI method may provide a better picture of the subsurface conditions than the electromagnetic surveys.

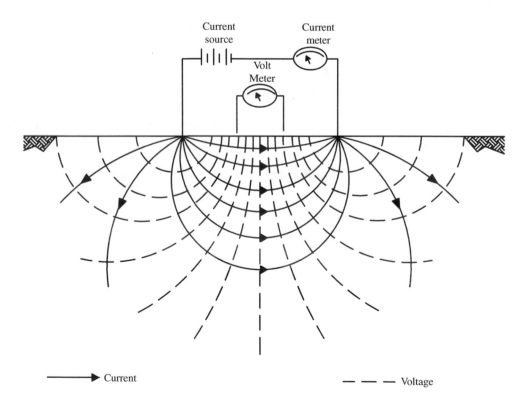

Fig. 2.26 Conceptual illustration of electrical resistivity imaging.

Fig. 2.27   Electrical resistivity imaging survey. (Photo courtesy of Dr. Michael J. Zaleha.)

2.5.4   Microgravity

The geophysical technique of microgravity involves the measurement of minute changes in the Earth's gravity field. The gravitation force varies, depending on the subsurface conditions. These changes in gravity at the Earth's surface depict the corresponding changes in the subsurface. A portable instrument is used to measure gravitational accelerations induced by the subsurface material. The instrument uses a Vernier scale to record more accurate readings, along with a sensitive spring of a constant length and a calibration spring holding a known mass. The method of microgravity can be used to locate faults, river channels, fissures, or cavities and sinkholes, and to determine the depths of bedrock or the fill thickness. This technique is inexpensive. However, several corrections must often be made to account for external conditions, such as instrument drift, topography, and tidal waves. From this method, the densities of rocks can be determined, particularly in locations where faults are present, as there are several rocks in the same area with varying densities. Microgravity is also useful in exploring poorly compacted ground, as well as discovering small voids near the Earth's surface under disturbed and varied subsurface conditions. Figure 2.28 illustrates a microgravity survey using a microgravity meter.

2.5.5   Seismic refraction and seismic reflection

Seismic refraction and reflection methods are mainly used to determine the depth of bedrock; they can also be used to detect voids, water tables, and folds. In this technique, a seismic impact is initiated at the ground surface. The dropping of a weight, the explosion of a charge, and hammering on a metal plate are common methods for creating the seismic energy, which then propagates into the ground. On reaching the bedrock, the seismic waves are refracted and eventually return to the surface where they are detected by a linear array of receivers, known as "geophones" (Figure 2.29). Figure 2.30 shows the seismic refraction and reflection equipment.

Fig. 2.28   A microgravity survey. (Photo courtesy of Scintrex Limited, Ontario, Canada.)

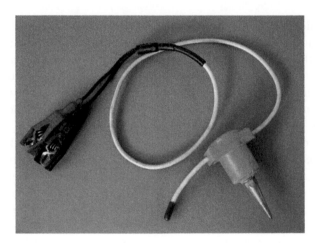

Fig. 2.29   Geophone. (Photo courtesy of Maine Geological Survey.)

In the *seismic refraction method*, a source on the ground surface provides seismic energy, and the travel time of the seismic waves refracted at the interfaces is measured. From the measured travel time, the velocities of subsurface compression waves in the bedrock are calculated. This method focuses on evaluating both the profile and the depth of bedrock and is most effective at *shallower* depths. Figure 2.31 illustrates the seismic refraction process.

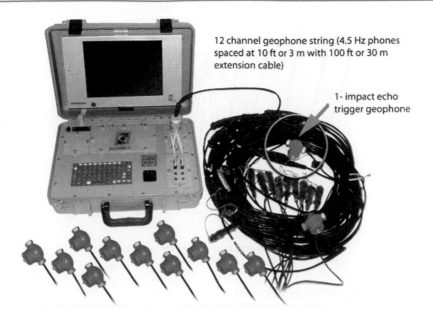

12 channel geophone string (4.5 Hz phones spaced at 10 ft or 3 m with 100 ft or 30 m extension cable)

1- impact echo trigger geophone

Fig. 2.30 Seismic refraction and seismic reflection equipment. (Photo courtesy of Olson Instruments, Inc.)

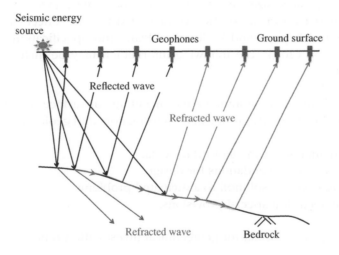

Seismic energy source

Geophones

Ground surface

Reflected wave

Refracted wave

Refracted wave

Bedrock

Fig. 2.31 Illustration of seismic refraction.

In the *seismic reflection method*, as illustrated in Figure 2.32, the seismic waves are transmitted from the Earth's surface and subsequently reflected back at the layer interfaces. High-frequency geophones detect the signals that are reflected, and the two-way travel time of seismic waves is measured. The specific target of seismic reflection must be *deep* within the Earth, allowing the reflected wave to follow the surface wave created by the impact. Moreover, there must be a difference in acoustic impedance of the different subsurface strata for the method to be effective.

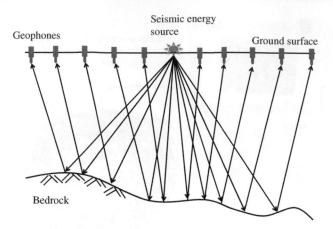

Fig. 2.32   Illustration of seismic reflection.

2.6   Geotechnical investigation report

A geotechnical investigation report is a document that thoroughly describes the site conditions and provides detailed design and construction recommendations. While the report content and format vary by project size and agency, all geotechnical investigation reports should contain certain essential information. As recommended by the "Checklist and Guidelines for Review of Geotechnical Reports and Preliminary Plans and Specifications" by the US Federal Highway Administration (2003), the following information may be included in a geotechnical investigation report.

- Summary of all subsurface exploration data, including subsurface soil profile (boring logs), laboratory or *in situ* test results, groundwater information, and geological and hydrogeological information.
- Interpretation and analysis of the subsurface data.
- Specific engineering recommendations for design.
- Discussion of conditions for solution to anticipated problems.
- Recommended geotechnical special provisions.

The following is a general format for geotechnical investigation report.

2.6.1   Site reconnaissance and description

1.1 Introduction
- Site location;
- Project description;
- Client information.
1.2 Proposed constructions, including structure type, size, and use. For example:
- Single-family dwellings and condominiums;

- Commercial and industrial sites;
- Other projects in the private sector (small private dams and power plant, privately owned roads, and parking facilities);
- Public work projects (transportation system, stadium, retention, or recharge basin);
- Essential facilities (hospital, fire, and police stations).

1.3 Site location and description, including:
- Current use of the site;
- Current topography and surface configuration;
- Presence of water course and ponds;
- Major roads accessible to the site;
- Location of buried utilities including power lines, cables, water lines, and sewers.

1.4 Geological setting and natural hazards such as landslide, earthquake, flooding, seasonal swelling and shrinkage, frost, and soil erosion.

1.5 Availability and quality of local construction materials (sand, stone, and water).

1.6 Seismic consideration (active faults, seismic zone, and liquefaction-prone sand)

1.7 Document review
- Preliminary design information (from architect and structural engineer), including building size, height, loads, and materials;
- Prior development and usage of the site, including any defects or failure of existing or former buildings due to foundation conditions;
- Geologic maps (from USGS);
- Topographic maps (from USGS);
- Aerial photo;
- Building code and other specifications;
- Groundwater conditions and detailed record of soil and rock strata.

2.6.2 Subsurface exploration (field exploration)

2.1 Borings
- Underground utilities;
- Boring layout and depths;
- Drilling rig used;
- Drilling methods and augers used.

2.2 Soil or rock sampling
- Samplers used;

2.3 Field tests, such as:
- SPT;
- CPT;
- Vane shear test;
- Flat-plate dilatometer test;
- Inclinometer installation and results;
- Groundwater monitoring well installation and results;

2.4 Boring logs.

Table 2.8   Common soil laboratory tests used in geotechnical engineering (Day 1999).

Soil characteristics	Soil properties	Specifications (ASTM)
Index tests	Classification	ASTM D 2487
	Particle size	ASTM D 422
	Atterberg limits	ASTM D 4318
	Water (or moisture) content	ASTM D 2216
	Wet density	Block samples or sampling tube
	Specific gravity	ASTM D 854
	Sand equivalent (SE)	ASTM D 2419
Settlement	Consolidation	ASTM D 2435
	Collapse	ASTM D 5333
	Organic content	ASTM D 2974
	Compaction: Standard Proctor	ASTM D 698
	Compaction: Modified Proctor	ASTM D 1557
Expansive soil	Swell	ASTM D 4546
	Expansive index test	ASTM D 4829 or UBC 18-2
Shear strength for slope movement	Unconfined compressive strength	ASTM D 2166
	Unconsolidated undrained	ASTM D 2850
	Consolidated undrained	ASTM D 4767
	Direct shear	ASTM D 3080
	Miniature vane	ASTM D 4648
Erosion	Dispersive clay	ASTM D 4647
Pavements and deterioration	Pavements: CBR	ASTM D 1883
	Pavements: R-value	ASTM D 2844
	Corrosion analysis	Chemical analysis
Permeability	Constant head	ASTM D 2434
	Falling head	ASTM D 5084

2.6.3   Laboratory testing

Relevant laboratory soil testing is conducted and reported in the geotechnical report. The laboratory tests are summarized in Table 2.8.

2.6.4   Geotechnical engineering recommendations

- Site preparations;
- Filling and compacting;
- Engineered fill;
- Foundations;
  - Types
  - Dimensions
  - Bearing capacity
  - Settlement (elastic, consolidation settlements)
- Lateral Earth pressures and retaining walls;
- Concrete slab-on-grade;

- Excavation stability;
- *R*-value testing and pavement design;
- Limitations.

2.6.5 Appendix

- Maps;
- Site plan;
- Boring logs;
- Laboratory data;
- Figures and diagrams;
- Specifications.

Homework Problems

1. Why is a geotechnical subsurface exploration conducted?
2. What are the types of *drill rigs* that are commonly used in the geotechnical subsurface exploration?
3. What are the *drilling methods* that are commonly used in the geotechnical subsurface exploration?
4. What are the types of *augers* that are commonly used in drilling?
5. What crucial information should be obtained before subsurface drilling can be conducted?
6. Under what soil conditions is mud rotary drilling used?
7. What are the types of *samplers* that are commonly used in sampling soils?
8. State the key specifications for the following parameters in the standard penetration test (SPT): hammer weight, hammer drop, and penetration depth. And, how is a SPT blow count determined?
9. What factors affect the SPT blow count (*N* value)?
10. The SPT blow count can be used to derive a wealth of subsurface information. Give three examples and show the quantitative relationships.
11. What are the typical diameters of a SPT sampler, Shelby tube sampler, and modified California sampler in the United States?
12. The SPT blow count (*N* value) at a depth of 8.0 – 8.46 m (25 – 26.5 ft) was obtained as 25. An automatic trip hammer is used. The SPT sampler complied with the ASTM D1586 and had no liner. Grain size distribution of the soil retrieved at this depth finds $D_{50} = 0.6$ mm. The soil in the top 10 m is uniform sandy soil with a unit weight of 19 kN/m^3. The groundwater table was not encountered in the subsoil exploration.
    1. Determine the standard penetration number that is corrected based on the field conditions, N_{60}.
    2. Determine the relative density of the soil at 8.0 m.
    3. Determine the effective friction angle at 8.0 m.

Chapter 2

13. If a relatively undisturbed soil sample is desired, what sampler(s) should be used?

14. How should a soil sample be stored after it is sampled from the subsurface?

15. Compare and state the advantages and disadvantages of SPT and CPT.

16. What is a dilatometer in geotechnical engineering? What does it measure?

17. What is an inclinometer in geotechnical engineering? What does it measure?

18. What does a vane shear test measure?

19. What is a geotechnical engineering report?

20. What is a geotechnical boring log? What information does it commonly include?

21. For each of the following multiple-choice questions, select all of the correct answers.

   (1) Which of the following aspects are included in subsurface exploration?

      A. Determination of the subsurface conditions, such as soil strata (depth, thickness, and types of soil).

      B. Recovery of soil and rock samples, so that further laboratory testing can be performed.

      C. Completion of field tests to obtain the in-situ properties of the soil or rock.

      D. Determination of the depth of the groundwater table.

      E. Identification of the existence of any problematic soils, such as soft soils.

   (2) Before a drilling is conducted, the following site information should be known:

      A. Approximate groundwater elevation, particularly confined aquifer elevation.

      B. Location of underground utility lines.

      C. Location of overhead utility lines.

      D. Stratification of the subsoil.

      E. Water content of the subsoil.

   (3) Augers are used to:

      A. Drill boreholes.

      B. Sample soils.

      C. Perform field tests.

      D. Store soil samples.

      E. Measure the groundwater table.

   (4) The purposes of using bentonite slurry in drilling are to:

      A. Pick up the soil cuttings from the bottom of the borehole and carry them to the top.

B. Keep the hole from collapsing.

C. Maintain the drilling equipment.

D. Retrieve soil samples below the groundwater table.

E. Loosen and cut the soil at the bottom of the borehole during mud rotary drilling.

(5) Mud rotary drilling can be used in the following types of soil:

A. Coarse sand and cobbles, in which the slurry can drain freely.

B. Fine sand.

C. Silt.

D. Clay.

E. Collapsible soils below the groundwater table.

(6) Which of the following sampler can retrieve relatively undisturbed soils?

A. Standard penetration test sampler.

B. Modified California sampler.

C. Shelby tube sampler (thin-walled sampler).

D. Piston sampler.

E. Pitcher barrel sampler.

(7) In a US practice, an SPT is performed and the blow counts in the first, second, and third 6-inch penetrations are 8, 9, and 10, respectively. What is the SPT blow count (N value)?

A. 17

B. 18

C. 19

D. 27

(8) The piston sampler is used to prevent:

A. The soil from entering the sampler before the sampling starts.

B. The soil sample from falling out of the tube when the sampler is pulled out.

C. The borehole from collapsing.

D. Excessive disturbance of the soil sample.

E. Drilling fluid from entering the sampler.

(9) The Pitcher sampler is mostly suitable for sampling:

A. Alternating hard and soft layers or soils of variable hardness.

B. Soft rocks.

C. Loose sand below the groundwater table.

D. Collapsible clayey soil.

E. Gravels.

(10) Compared with the SPT, the CPT has the following advantages:

A. It can delineate the entire subsoil profile.

B. It is fast, economical, and productive.

C. It can eliminate operator errors and can provide reliable and repeatable results.

D. It is particularly suitable for soft soils.

E. It can obtain representative soil samples for further laboratory testing.

(11) Compared with the SPT, the CPT has the following disadvantages:

   A. It cannot obtain soil samples for further laboratory testing and verification.

   B. It cannot penetrate dense, granular soils, such as cobbles.

   C. It usually has high capital investment.

   D. It requires a skilled operator to run.

   E. Data results may contain electronic noise and require calibration.

(12) A flat-plate dilatometer in the context of geotechnical engineering

   A. Employs a flat steel plate and measures the friction resistance when the plate is pushed into the subsoil.

   B. Employs a flat steel plate with a sharp cutting edge, pushes the plate into the subsoil, and measures the strain induced by inflating the plate.

   C. Employs the SPT sampler and measures the in situ soil pressure when the sampler is driven into subsoil.

   D. Measures the rotational resistance when the flat plate dilatometer is pushed into the soil and rotated.

   E. Is particularly designed to be embedded in cracks of bedrocks and measures the shrinking and expansion of rock mass due to temperature change.

(13) An inclinometer in the context of geotechnical subsoil exploration

   A. Measures the deformation normal to the axis of a pipe by passing a probe along the pipe.

   B. Measures the angle of a slope and embankment using ground penetration radar (GPS) technology.

   C. Can determine characteristics of landslides, slope movements, and deflections in retaining walls.

   D. Often involves drilling a borehole and inserting the inclinometer probe into the borehole.

   E. Is a nondestructive technique, therefore, a borehole is never needed.

(14) A vane shear test

   A. Can measure the in situ soil's undrained shear strength.

   B. Employs a vane shear sampler and can retrieve undisturbed soil.

   C. Pushes the vane shear device into subsoil and measures the friction resistance when the plate is pushed into the subsoil.

D. Pushes the vane shear device into the subsoil and measures the strain induced by inflating the device.

E. Cannot be used below the groundwater table due to possible collapsing of the borehole.

(15) In the ground penetration radar (GPR) technique, the resolution of the GPR is affected by:

A. Frequency of antenna.

B. Depth in the subsoil.

C. Subsurface material type.

D. Shape of the cable on the ground surface.

E. Number of electrodes used.

(16) Ground penetration radar:

A. Sends electromagnetic wave of high frequency into the ground and detects the wave that is bounced back from objects in the soil.

B. Uses dropping of weight, an explosive charge, or a hammer onto a plate to generate the seismic energy from the ground surface.

C. Sends electrical direct current into the subsoil and measures the electrical resistivity of the encountered materials.

D. Includes the time-domain and frequency-domain methods.

E. Has limited capabilities in highly conductive materials such as clayey soils and salt-contaminated soils.

(17) Which of the following geophysical techniques are affected by the electrical conductivity of subsurface materials?

A. Ground penetration radar.

B. Electromagnetics in frequency domain and in time domain.

C. Electrical resistivity imaging.

D. Microgravity.

E. Seismic refraction and seismic reflection.

22. Briefly describe (1) ground penetration radar, (2) electromagnetics in time domain and in frequency domain, (3) electrical resistivity imaging, (4) the microgravity method, and (5) seismic refraction and seismic reflection methods, respectively. For each method, describe:

- How each technique works (a diagram can be used).
- The applications of each technique (i.e., what each technique determines).

23. Use published case studies to illustrate (1) ground penetration radar (GPR), (2) electromagnetics in time domain and in frequency domain, (3) electrical resistivity imaging, (4) the microgravity method, and (5) seismic refraction and seismic reflection methods, respectively. Include graphs to show the image results of these geophysical techniques in each of the case studies.

Chapter 2

References

Clayton, C.R.I. (1990). "SPT Energy Transmission: Theory, Measurement, and Significance." *Ground Engineering*, Vol. 23, No. 10, pp. 35–43.

Cubrinovski, M., and Ishihara, K. (1999). "Empirical Correlations between SPT N-values and Relative Density for Sandy Soils." *Soils and Foundations*, Vol. 39, No.5, pp. 61–92.

Day, R.W. (1999). *Geotechnical and Foundation Engineering: Design and Construction*. McGraw-Hill, New York, NY.

Kulhawy, F.H., and Mayne, P.W. (1990). *Manual on Estimating Soil Properties for Foundation Design*, Electric Power Research Institute, Palo Alto, California.

Kumar, S. (2005). *"Geotechnical Field Investigation using Test Borings: Bringing Field Exploration Closer to You" (DVD)*. ASCE Continuing Education, Reston, VA.

Mayne, P.W., Christopher, B.R., DeJong, J. (2002). "Subsurface Investigations – Geotechnical Site Characterization Reference Manual." Publication No. FHWA NHI-01-031. NHI Course No. 132031. National Highway Institute, Federal Highway Administration, U.S. Department of Transportation, Washington, D.C. May 2002.

Naval Facilities Engineering Command (NAVFAC), *Soil Mechanics Design Manual 7.01*. 200 Stovall Street, Alexandria, Virginia 22322. 1986.

Seed, H.B., Tokimatsu, K., Harder, L.F., and Chung, R.M. (1985). "Influence of SPT Procedures in Soil Liquefaction Resistance Evaluations." *ASCE Journal of Geotechnical Engineering*, Vol. 111, No. 12, pp. 1425–1445.

Schmertmann, J.H. (1975). "Measurement of In Situ Shear Strength." *Proceedings, Specialty Conference on In Situ Measurement of Soil Properties*, ASCE, Vol. 2, pp. 57–138.

Skempton, A.W. (1986). "Standard Penetration Test Procedures and the Effect in Sands of Overburden Pressure, Relative Density, Particle Size, Aging and Overconsolidation." *Geotechnique*, Vol. 36, No. 2, pp. 425–447.

US Federal Highway Administration. (2003). Checklist and Guidelines for Review of Geotechnical Reports and Preliminary Plans and Specifications Publication No. FHWA ED-88-053, Federal Highway Administration, U.S. Department of Transportation, Washington, D.C. August 1988. Revised February 2003.

Chapter 3

Shallow Foundation Design

3.1 Introduction to foundation design

A foundation is the structural element that is generally embedded underground and connects the superstructure to the ground. Its function is to transmit the load from the superstructure to the underlying soil or rock such that the superstructure can be safely supported by the soil or rock.

As shown in Figures 3.1 and 3.2, foundations are divided into two major categories: shallow foundations and deep foundations, depending on the depth of embedment of the foundation. Generally, if the ratio of the embedment to the width of a foundation (D/B) is larger than 3, then the foundation is considered a deep foundation; otherwise, the foundation is considered a shallow foundation.

- Shallow foundations include spread footings and mat foundations. A spread footing is an enlargement at the bottom of a bearing wall or a column that distributes the structural load onto a sufficiently large area of soil or rock. A mat foundation, also known as *raft foundation*, covers the entire footprint of the superstructure.
- Deep foundations typically include piles and drilled shafts. Piles are driven into the subsoil, while drilled shafts (also known as *cast-in-drilled-hole*) are of larger diameter and are cast in predrilled holes with reinforcement casing.

Choosing a shallow foundation or a deep foundation depends on many factors, including:

- Loads (or actions) acting on the superstructure, such as horizontal, vertical, and seismic loads.
- Subsurface conditions.
- Performance (serviceability) requirements, such as required bearing capacity and settlement.
- Budget.
- Available materials and contractor's capabilities.

The design of foundations mainly includes three primary aspects:

- Bearing capacity (ultimate limit state).
- Settlement (serviceability limit state).
- Structural design (ultimate and serviceability limit states).

Geotechnical Engineering Design, First Edition. Ming Xiao.
© 2015 John Wiley & Sons, Ltd. Published 2015 by John Wiley & Sons, Ltd.
Companion Website: www.wiley.com/go/Xiao

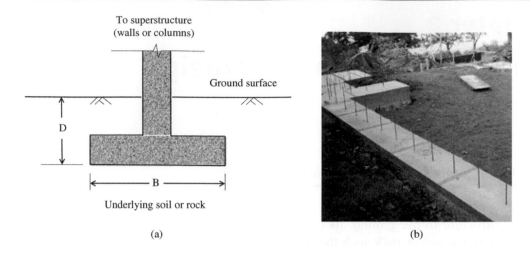

Fig. 3.1   Example of shallow foundation. (a) Sectional schematic of a spread footing, (b) wall footing.

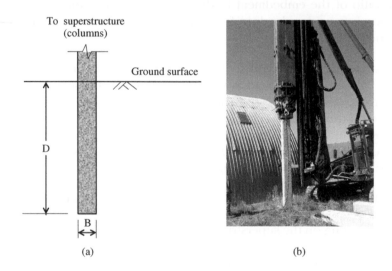

Fig. 3.2   Example of deep foundation. (a) Sectional schematic of a single pile, (b) a concrete pile being driven into subsoil.

The following parameters are required as input values in a foundation design:

- Loads from superstructure.
- Factor of safety for bearing capacity (or partial factors of safety for geotechnical properties and structural loads).
- Required total settlement and differential settlement.
- Subsoil conditions, such as density, cohesion, internal friction angle, corrosion potential, and depth to groundwater table.

A foundation design should determine the following:

- Foundation type: deep or shallow foundation.
- Material: steel, timber, concrete, or masonry.
- Embedment depth.
- Dimensions.
- Bearing capacity and factor of safety (or limit state verification if partial factors of safety are used for design).
- Total and differential settlements.
- Reinforcement, if using reinforced concrete, and structural stability.

Two major types of design approaches are discussed in this chapter and in the following chapters where applicable. They are working stress design and limit states design. The working stress design, or allowable stress design, is based on the concept of factor of safety and has been used in geotechnical engineering for more than a decade. The limit state design considers the limit states (or a set of conditions) that should be avoided. The limit states include serviceability state and ultimate state. The serviceability state refers to a set of conditions, under which the structure can no longer meet its required service or functions such as large settlement and deformation. The ultimate state refers to a set of conditions of partial or total failure of structures.

3.2   Bearing capacity of shallow foundations

Bearing capacity is the capability of the soil beneath a foundation to support a superstructure load. The maximum load-bearing capability of the soil, that is the maximum stress the soil can carry without failure, is called *ultimate bearing capacity*, q_{ult}. Determination of ultimate bearing capacity depends on the foundation shape (square, rectangular, circular), size, embedment depth, subsoil conditions, and the failure mode. In foundation design, a global factor of safety for bearing capacity is commonly used to account for the approximation of design methodologies and uncertainty of the subsoil parameters and to provide sufficient safety margin. Therefore, the *allowable bearing capacity* is used to compare with the maximum stress due to the superstructure.

$$q_{all} = \frac{q_{ult}}{FS} \tag{3.1}$$

where:

q_{ult} = ultimate bearing capacity,
q_{all} = allowable bearing capacity,
FS = factor of safety. In foundation design, the acceptable factor of safety is generally not less than 3.0.

$$FS = \frac{q_{ult}}{q_{all}} \geq 3.0 \tag{3.2}$$

Alternatively, a limit state design approach combined with partial factors of safety can be used (as required in European standards – Eurocodes), in which the uncertainty related to subsoil parameters and design methodologies is accounted for within the partial factors of safety, and these may be different for all subsoil properties. The design resistance

$V_d(= q_{ult}A$, where A is the area of the foundation) is used to compare with the design effect of the actions E_d due to loads imposed on the foundation (which are also affected by various partial factors of safety).

$$V_d \geq E_d \tag{3.3}$$

3.2.1 Failure modes of shallow foundations

When the ultimate bearing capacity of the soil beneath a foundation is exceeded by the stress caused by the superstructure (or when the design effect of the actions is larger than the design resistance), the soil may compress and slide (shear), and a sliding (shear) surface may develop in the soil. This is referred to as bearing capacity failure. There are three commonly identified modes of the bearing capacity failure, depending on the soil's density. They are shown in Figure 3.3 and explained subsequently.

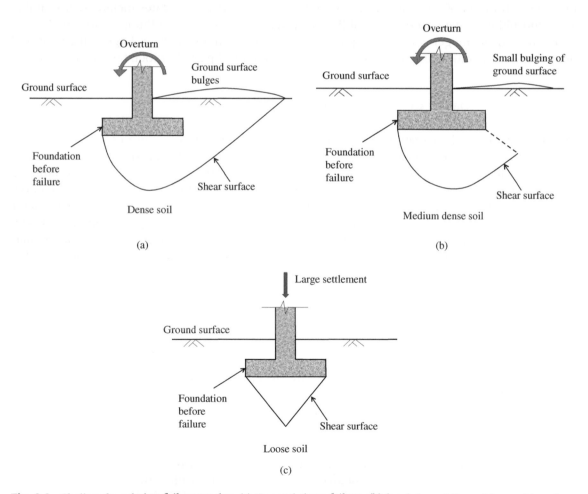

Fig. 3.3 Shallow foundation failure modes. (a) General shear failure, (b) local shear failure, (c) punching shear failure.

- *General shear failure*: General shear failure occurs in dense soil with relative density $D_r > 67\%$ (Coduto 2001). A complete shear failure surface develops from a corner of the foundation and extends to the ground surface; as the foundation overturns to one side, the soil on the other side is pushed up at the ground surface. General shear failure occurs suddenly and is the most catastrophic to the structure. It is the most common failure mode.
- *Local shear failure*: Local shear failure occurs in medium dense soil with $30\% < D_r < 67\%$ (Coduto 2001). The shear failure surface develops from a corner of the foundation and extends locally to the adjacent area beneath or beside the foundation. The failure surface does not extend to the ground surface. As the foundation overturns to one side, a small bulge of the ground surface occurs on the opposite side of the foundation. Sudden failure does not occur; instead, the foundation gradually settles and tilts in the subsoil.
- *Punching shear failure*: Punching shear failure occurs in loose soil with relative density $D_r < 30\%$ (Coduto 2001). The shear failure surface only develops beneath the foundation. Little or no bulging occurs at the ground surface. The foundation settles gradually into the ground. Overturning of the foundation does not occur in punching failure mode; however, the structure no longer meets the service requirement due to excessive settlement.

In practical foundation design, it is only necessary to analyze the bearing capacity for the general failure mode; settlement is then determined to verify that the foundation will not settle excessively. These settlement analyses implicitly protect against local and punching shear failures (Coduto 2001).

3.2.2 Terzaghi's bearing capacity theory

Terzaghi (1943) was the first to present a comprehensive theory of the ultimate bearing capacity of shallow foundations. The geometry of the general shear failure surface for Terzaghi's bearing capacity theory is shown in Figure 3.4. The shear failure surface extends from one corner of the foundation to the ground surface, and the failure surface can develop on either side of the foundation. Hence, symmetrical failure surfaces are shown in the figure. The failure surfaces and the horizontal line at the bottom of the foundation form three zones: zone I, the wedge zone, moves downward with the foundation; zone II, the radial shear zone, generally moves in the lateral direction and pushes zone III, the passive zone.

Terzaghi's bearing capacity theory was based on the following assumptions:

- The embedment of the foundation, D, is less than or equal to the foundation width, B. In the current practice, Terzaghi's bearing capacity theory can still be used if $D \leq 3B$.
- The soil beneath the foundation is a homogeneous, infinite half-space.
- The load on the foundation is concentric and vertical.
- The foundation has a horizontal base on a level ground surface.
- General shear failure is the failure mode for the foundation.

On the basis of the failure surface shown in Figure 3.4, Terzaghi (1943) provided the formulae of ultimate bearing capacity, q_{ult}, for three types of shallow foundations.

For strip foundations (also known as *continuous foundations* or *wall footings*):

$$q_{ult} = c'N_c + qN_q + \frac{1}{2}\gamma BN_r \tag{3.4}$$

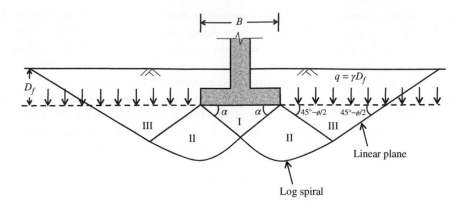

Fig. 3.4   Geometry of failure surface for Terzaghi's bearing capacity theory.

For square foundations:

$$q_{\text{ult}} = 1.3c'N_c + qN_q + 0.4\gamma BN_r \tag{3.5}$$

For circular foundations:

$$q_{\text{ult}} = 1.3c'N_c + qN_q + 0.3\gamma BN_r \tag{3.6}$$

In the aforementioned equations:

c' = effective cohesion of the soil beneath the foundation,
γ = unit weight of the homogeneous soil,
q = effective stress at the bottom of the foundation due to the soil above the foundation and surcharge (if any) at the ground surface. If no surcharge is at the ground surface, then

$$q = \gamma D_{\text{f}} \tag{3.7}$$

where:

D_{f} = embedment depth of the foundation,
B = width of square foundation or diameter of circular foundation.

In the aforementioned equations, $N_c, N_q,$ and N_γ are the dimensionless bearing capacity factors that account for the contributions of cohesion c, surcharge q, and the soil's unit weight γ to the bearing capacity. They are dependent on the effective internal frictional angle of the foundation soil, ϕ'. The calculations of N_c and N_q are expressed in Equations (3.8)–(3.11).

$$N_q = \frac{a^2}{2\cos^2\left(45 + \frac{\phi'}{2}\right)} \tag{3.8}$$

$$a = e^{\pi\left(0.75 - \frac{\phi'}{365}\right)\tan\phi'} \tag{3.9}$$

$$N_c = \frac{N_q - 1}{\tan\varphi'} \quad (\text{for } \varphi' > 0) \tag{3.10}$$

Table 3.1 Terzaghi's bearing capacity factors using equations (3.8)–(3.11), for *general* failure mode.

ϕ'(deg)	N_c	N_q	N_γ	ϕ'(deg)	N_c	N_q	N_γ
0	5.700	1.000	0	27	29.235	15.896	11.602
1	6.015	1.105	0.014	28	31.611	17.808	13.693
2	6.300	1.220	0.035	29	34.243	19.981	16.175
3	6.621	1.347	0.063	30	37.163	22.456	19.129
4	6.964	1.487	0.099	31	40.412	25.282	22.653
5	7.338	1.642	0.144	32	44.036	28.517	26.871
6	7.726	1.812	0.200	33	48.090	32.230	31.935
7	8.152	2.001	0.267	34	52.637	36.504	38.035
8	8.602	2.209	0.348	35	57.754	41.440	45.410
9	9.085	2.439	0.444	36	63.528	47.156	54.360
10	9.607	2.694	0.559	37	70.067	53.799	65.266
11	10.160	2.975	0.694	38	77.495	61.546	78.614
12	10.764	3.288	0.854	39	85.966	70.614	95.028
13	11.409	3.634	1.041	40	95.663	81.271	115.311
14	12.109	4.019	1.262	41	106.807	93.846	140.509
15	12.861	4.446	1.520	42	119.668	108.750	171.990
16	13.678	4.922	1.822	43	134.580	126.498	211.556
17	14.559	5.451	2.175	44	151.950	147.736	261.603
18	15.518	6.042	2.589	45	172.285	173.285	325.342
19	16.557	6.701	3.074	46	196.219	204.191	407.113
20	17.691	7.439	3.641	47	224.550	241.800	512.836
21	18.923	8.264	4.305	48	258.285	287.855	650.673
22	20.271	9.190	5.085	49	298.718	344.636	831.990
23	21.747	10.231	6.000	50	347.510	415.146	1072.797
24	23.361	11.401	7.076	51	406.821	503.382	1395.915
25	25.134	12.720	8.342	52	479.489	614.718	1834.301
26	27.085	14.210	9.836	53	569.275	756.453	2436.199

N_γ values are from Kumbhojkar (1993).

$$N_c = 5.70 \ (\text{for } \phi' = 0) \tag{3.11}$$

The original expression of N_γ by Terzaghi (1943) is:

$$N_\gamma = \frac{4p_{\gamma\min}}{\gamma B^2} - \frac{\tan\phi'}{2} \tag{3.12}$$

where: $P_{\gamma\min}$ is the minimum passive earth force exerting on zone II (Figure 3.4). Terzaghi used trial and error to graphically determine $P_{\gamma\min}$ for various ϕ' and subsequently developed a graphical relationship between ϕ' and N_γ. Kumbhojkar (1993) developed analytical expressions to obtain N_γ and provided the N_γ values for ϕ' from 0 to 53°, which are now widely used. The values of the bearing capacity factors are listed in Table 3.1.

For foundations with local shear failure, Terzaghi (1943) recommended the ultimate bearing capacity calculations follow the same Equations (3.4)–(3.6); meanwhile, the shear strength parameters (c', ϕ') should be reduced using Equations (3.13) and (3.14).

$$c' = \frac{2}{3}c' \tag{3.13}$$

Chapter 3

Table 3.2 Terzaghi's bearing capacity factors for *local* failure mode.

Original ϕ'(deg)	N_c	N_q	N_γ	Original ϕ'(deg)	N_c	N_q	N_γ
0	5.7	1.00	0	27	16.302	6.538	2.430
1	5.900	1.07	0.016	28	17.132	7.073	2.738
2	6.096	1.14	0.022	29	18.027	7.662	3.793
3	6.301	1.22	0.033	30	18.991	8.310	4.315
4	6.514	1.30	0.050	31	20.034	9.025	4.915
5	6.738	1.39	0.071	32	21.164	9.816	5.606
6	6.971	1.49	0.097	33	22.390	10.693	6.403
7	7.216	1.59	0.127	34	23.724	11.668	7.324
8	7.472	1.70	0.161	35	25.178	12.753	8.389
9	7.741	1.82	0.200	36	26.768	13.965	9.624
10	8.024	1.94	0.243	37	28.510	15.323	11.058
11	8.321	2.08	0.290	38	30.425	16.847	12.727
12	8.633	2.22	0.343	39	32.535	18.564	14.672
13	8.962	2.38	0.402	40	34.866	20.504	16.943
14	9.308	2.55	0.466	41	37.451	22.704	19.602
15	9.674	2.73	0.538	42	40.326	25.207	22.720
16	10.061	2.92	0.617	43	43.535	28.065	26.384
17	10.470	3.13	0.705	44	47.129	31.341	30.699
18	10.903	3.36	0.804	45	51.171	35.114	35.792
19	11.362	3.61	0.913	46	55.734	39.476	41.817
20	11.850	3.88	1.036	47	60.909	44.544	50.266
21	12.368	4.17	1.173	48	66.803	50.461	59.769
22	12.920	4.48	1.327	49	73.550	57.406	71.248
23	13.509	4.82	1.500	50	81.313	65.604	85.569
24	14.137	5.20	1.694	51	90.295	75.337	103.668
25	14.809	5.60	1.911	52	100.749	86.968	126.559
26	15.529	6.05	2.156	53	112.992	100.964	155.334

$$\phi' = \tan^{-1}\left(\frac{2}{3}\tan\phi'\right) \tag{3.14}$$

The reduced $\bar{c}'$ should be used in Equations (3.4)–(3.6). The reduced $\bar{\phi}'$ is used in Equations (3.8)–(3.11) to derive the new bearing capacity factors N_c, N_q. The derivation of N_γ is based on the analytical solutions of Kumbhojkar (1993), using the reduced $\bar{\phi}'$. The bearing capacity factors for local shear failure mode are listed in Table 3.2. When looking up the bearing capacity factors in Table 3.2, the original ϕ' is used as the index.

Terzaghi's ultimate bearing capacity calculations are expressed in terms of effective strength parameters (c', ϕ'); however, they can also be used in the total stress analysis using undrained cohesion c_u and friction angle $\phi(=0)$, which simply replace c' and ϕ' in the aforementioned calculations.

3.2.3   The general bearing capacity theory

Terzaghi's bearing capacity theory did not consider rectangular footings, inclined loads, or the shear resistance due to the soil above the footing (i.e., foundation embedment). These factors were taken into account in the extensive research and numerous methods proposed by researchers including Meyerhof (1963), DeBeer (1970), Hansen (1970), Vesic (1973, 1975), and Hanna and Meyerhof (1981). These researchers contributed to the general bearing capacity theory that was originally proposed by Meyerhof (1963). The ultimate bearing capacity can be expressed as

$$q_{ult} = c'N_cF_{cs}F_{cd}F_{ci} + qN_qF_{qs}F_{qd}F_{qi} + \frac{1}{2}\gamma BN_\gamma F_{\gamma s}F_{\gamma d}F_{\gamma i} \tag{3.15}$$

where:

c' = effective cohesion of the soil beneath the foundation,
γ = unit weight of the homogeneous soil,
q = effective stress at the bottom of the foundation due to the soil above the foundation and surcharge (if any) at the ground surface. If no surcharge is at the ground surface, then:

$$q = \gamma D_f \tag{3.16}$$

where:

D_f = embedment depth of foundation,
B = width of square foundation or diameter of circular foundation.

The bearing capacity factors, N_c, N_q, and N_γ, are expressed in Equations (3.17)–(3.20) and are listed in Table 3.3.

$$N_q = \tan^2\left(45 + \frac{\phi'}{2}\right)e^{\pi \tan \phi'} \text{ (from Prandtl 1921)} \tag{3.17}$$

$$N_c = \frac{N_q - 1}{\tan \phi'} \quad (\text{for } \phi' > 0) \text{ (from Reissner 1924)} \tag{3.18}$$

$$N_c = 5.14 \text{ (for } \phi' = 0) \tag{3.19}$$

$$N_\gamma = 2(N_q + 1)\tan \phi' \text{ (from Vesic 1973)} \tag{3.20}$$

F_{cs}, F_{qs}, and $F_{\gamma s}$ are the *shape factors* that take into account the broad range of footing shapes. There are various forms of equations available; they are summarized by Salgado (2008) and Bowles (1996). Equations (3.21)–(3.23) were proposed by DeBeer (1970).

$$F_{cs} = 1 + \left(\frac{B}{L}\right)\left(\frac{N_q}{N_c}\right) \tag{3.21}$$

$$F_{qs} = 1 + \left(\frac{B}{L}\right)\tan \phi' \tag{3.22}$$

$$F_{\gamma s} = 1 - 0.4\left(\frac{B}{L}\right) \tag{3.23}$$

Note that EN 1997-1:2004 (Eurocode 7) suggests the use of alternative shape factors, which may also depend on the loading conditions as follows:

For drained loading:

$$F_{\gamma s} = 1 - 0.3\left(\frac{B}{L}\right) \tag{3.24}$$

Table 3.3   Bearing capacity factors using equations (3.17)–(3.20), for *general* failure mode.

ϕ'(deg)	N_c	N_q	N_γ	ϕ'(deg)	N_c	N_q	N_γ
0	5.140	1.000	0.000	27	23.942	13.199	14.470
1	5.379	1.094	0.073	28	25.803	14.720	16.717
2	5.632	1.197	0.153	29	27.860	16.443	19.338
3	5.900	1.309	0.242	30	30.140	18.401	22.402
4	6.185	1.433	0.340	31	32.671	20.631	25.994
5	6.489	1.568	0.449	32	35.490	23.177	30.215
6	6.813	1.716	0.571	33	38.638	26.092	35.188
7	7.158	1.879	0.707	34	42.164	29.440	41.064
8	7.527	2.058	0.860	35	46.124	33.296	48.029
9	7.922	2.255	1.031	36	50.585	37.752	56.311
10	8.345	2.471	1.224	37	55.630	42.920	66.192
11	8.798	2.710	1.442	38	61.352	48.933	78.024
12	9.285	2.974	1.689	39	67.867	55.957	92.246
13	9.807	3.264	1.969	40	75.313	64.195	109.411
14	10.370	3.586	2.287	41	83.858	73.897	130.214
15	10.977	3.941	2.648	42	93.706	85.374	155.542
16	11.631	4.335	3.060	43	105.107	99.014	186.530
17	12.338	4.772	3.529	44	118.369	115.308	224.634
18	13.104	5.258	4.066	45	133.874	134.874	271.748
19	13.934	5.798	4.681	46	152.098	158.502	330.338
20	14.835	6.399	5.386	47	173.640	187.206	403.652
21	15.815	7.071	6.196	48	199.259	222.300	495.999
22	16.883	7.821	7.128	49	229.924	265.497	613.140
23	18.049	8.661	8.202	50	266.882	319.057	762.859
24	19.324	9.603	9.442	51	311.752	385.982	955.766
25	20.721	10.662	10.876	52	366.660	470.304	1206.482
26	22.254	11.854	12.539	53	434.421	577.496	1535.380

$$F_{qs} = 1 + \left(\frac{B}{L}\right) \sin \phi' \tag{3.25}$$

$$F_{cs} = \frac{(F_{qs} N_q - 1)}{(N_q - 1)} \tag{3.26}$$

For undrained loading:

$$F_{cs} = 1 + 0.2 \left(\frac{B}{L}\right) \tag{3.27}$$

F_{cd}, F_{qd}, and $F_{\gamma d}$ are the *depth factors* that take into account the contribution of foundation embedment to the bearing capacity. There are also various forms of equations available; they are summarized by Salgado (2008) and Bowles (1996). Equations (3.28)–(3.32) (Hansen 1970; Vesic 1973) are examples of commonly used equations.

$$F_{cd} = 1 + 0.4 \left(\frac{D_f}{B}\right) \quad \left(\text{for } \frac{D_f}{B} \leq 1\right) \tag{3.28}$$

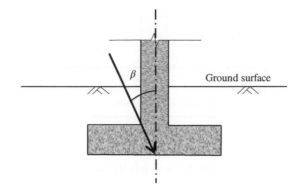

Fig. 3.5 Illustration of inclined load on a shallow foundation.

$$F_{cd} = 1 + 0.4 \tan^{-1}\left(\frac{D_f}{B}\right) \quad \left(\text{for}\frac{D_f}{B} > 1\right) \tag{3.29}$$

$$F_{qd} = 1 + 2\tan\phi'(1 - \sin\phi')^2\left(\frac{D_f}{B}\right) \quad \left(\text{for}\frac{D_f}{B} \leq 1\right) \tag{3.30}$$

$$F_{qd} = 1 + 2\tan\phi'(1 - \sin\phi')^2\tan^{-1}\left(\frac{D_f}{B}\right) \quad \left(\text{for}\frac{D_f}{B} > 1\right) \tag{3.31}$$

$$F_{\gamma d} = 1 \tag{3.32}$$

No recommendation is made in the Eurocodes with regard to the use of depth factors. However, it is clearly stated that a recognized analytical method should be used in the verification of ultimate limit states involving bearing resistance (or capacity).

F_{ci}, F_{qi}, $F_{\gamma i}$ are the *load inclination factors* that take into account the reduction of bearing capacity due to inclined loads, as shown in Figure 3.5. They are expressed in Equations (3.33)–(3.34) (Meyerhof 1963; Hanna and Meyerhof 1981). Other expressions of the inclinations factors are summarized by Salgado (2008) and Bowles (1996).

$$F_{ci} = F_{qi} = \left(1 - \frac{\beta^\circ}{90^\circ}\right)^2 \tag{3.33}$$

$$F_{\gamma i} = \left(1 - \frac{\beta^\circ}{\phi'}\right)^2 \tag{3.34}$$

Alternative expressions are also suggested in Eurocode 7. They are, however, discussed in the context of Sample Problem 3.3.

3.2.4 Effect of groundwater on ultimate bearing capacity

If the groundwater table is near the ground surface, it may affect the ultimate bearing capacity. When considering the effect of the groundwater, the ultimate bearing capacity Equations (3.4)–(3.6), and (3.15) still apply, and the bearing capacity factors still follow the same approach as described earlier. The only changes are the effective stress, q, and the unit weight, γ, in these equations, depending on the elevation of the groundwater table.

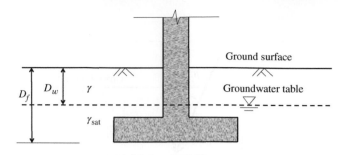

Fig. 3.6　Effect of groundwater table, case I.

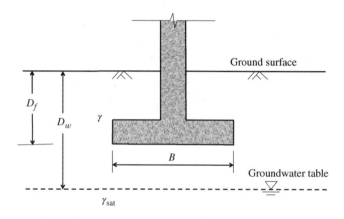

Fig. 3.7　Effect of groundwater table, case II.

Case I:　The groundwater table is at or above the bottom of the foundation, Figure 3.6.
　　　　The effective stress at the bottom of the foundation is as follows:

$$q = \gamma D_w + (\gamma_{sat} - \gamma_w)(D_f - D_w) \tag{3.35}$$

If the soil beneath the foundation is fully saturated, use γ' to replace γ in Equations (3.4)–(3.6), and (3.15). The submerged unit weight is:

$$\gamma' = \gamma_{sat} - \gamma_w \tag{3.36}$$

where:

γ_w = unit weight of water.

Case II:　The groundwater table is below the bottom of the foundation, and $D_f \le D_w \le D_f + B$, Figure 3.7.
　　　　When the groundwater table is not far below the foundation, there is still some effect on the ultimate bearing capacity. The effective stress at the bottom of the foundation still follows Equation (3.7): $q = \gamma D_f$. The unit weight in Equations (3.4)–(3.6), and

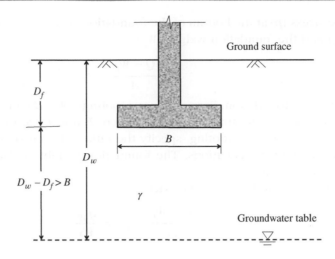

Fig. 3.8 Effect of groundwater table, case III.

(3.15) should use a weighted average:

$$\overline{\gamma} = \gamma' + \frac{D_w - D_f}{B}(\gamma - \gamma') \qquad (3.37)$$

$\overline{\gamma}$ should replace γ in Equations (3.4)–(3.6), and (3.15).

Case III: The groundwater table is below the bottom of the foundation, and $D_w > D_f + B$, Figure 3.8.

When the groundwater table is far below the foundation, i.e., $D_w > D_f + B$, it is assumed that the groundwater does not affect the bearing capacity. Therefore, no revision is made to Equations (3.4)–(3.6), and (3.15).

It is noted that in the Eurocode, the groundwater is only considered at the most critical level as shown in case I. Therefore, case II is not considered.

3.2.5 Foundation design approach based on allowable bearing capacity and the global factor of safety approach

Step 1: Determine the loads from the superstructure and the subsoil conditions (unit weight, cohesion, and internal friction angle).

Step 2: Determine the foundation embedment depth on the basis of the data in Tables 3.4 and 3.5.

Step 3: Determine the groundwater table elevation and whether it affects the ultimate bearing capacity. If the foundation width is unknown, appropriate assumptions can be made on the basis of the groundwater table elevation.

Step 4: Determine the factor of safety; the minimum value is generally 3.0.

Step 5: Express the ultimate bearing capacity using Terzaghi's or the general bearing capacity equation, depending on the foundation type. In the equations, the foundation width, B, is unknown and will be determined.

Step 6: Calculate the stress (p) at the bottom of the foundation that is caused by the superstructure load (Q) and the foundation weight (W_f):

$$p = \frac{Q + W_f}{A} \tag{3.38}$$

where: A = area of foundation. For continuous footings: $A = B \times$ unit length; for rectangular (including square) footings: $A = B \times L$, where B and L are the width and length, respectively. For rectangular footing, specify the ratio between B and L such that only one variable (B) is in the equations. The foundation weight is also a function of the foundation's dimensions.

Step 7: Equate p with the allowable bearing capacity:

$$p = \frac{Q + W_f}{A} = q_{all} = \frac{q_{ult}}{FS} \tag{3.39}$$

Using trial and error, solve for B. The value of B is rounded up to the nearest value that is conventionally accepted in practice.

Table 3.4   Minimum embedment depth (D_f) for square and rectangular footings (from Coduto 2001).

Load, P (kip)	Minimum D_f (inch)	Load, P (kN)	Minimum D_f (cm)
0–65	12	0–300	30
65–140	18	300–500	40
140–260	24	500–800	50
260–420	30	800–1100	60
420–650	36	1100–1500	70
		1500–2000	80
		2000–2700	90
		2700–3500	100

Table 3.5   Minimum embedment depth (D_f) for continuous footings (from Coduto 2001).

Load, P (kip/ft)	Minimum D_f (inch)	Load, P (kN/m)	Minimum D_f (cm)
0–10	12	0–170	30
10–20	18	170–250	40
20–28	24	250–330	50
28–36	30	330–410	60
36–44	36	410–490	70
		490–570	80
		570–650	90
		650–740	100

If a structure has varying loads on different walls or columns, the largest load is used in Equation (3.38) to obtain the largest width B, which can be adopted as the width for all other foundations of the same type. Alternatively, for a more economical design, various loads on the same type of foundation can be arranged into several load groups, and the foundation width B can be obtained for each load group.

3.2.6 Foundation design approach based on allowable bearing capacity and the partial factor of safety approach

Step 1: Determine the characteristic values of the actions (loads), F_k, from the superstructure and classify them according to their nature (i.e., unfavorable, favorable, permanent, variable, accidental, etc.). Characteristic values are defined as cautious estimates of the actions being considered. Characteristic vertical, permanent loads are denoted as $V_{G,k}$; and vertical, variable loads are denoted as $V_{Q,k}$.

Step 2: Determine the partial factors of safety corresponding to each of the action types in the previous step. In the examples that follow, partial factors of safety on permanent and variable unfavorable loads are denoted as $\gamma_{F,G}$ and $\gamma_{F,Q}$, respectively.

Step 3: Calculate the representative values of the actions by multiplying the characteristic actions by the corresponding factor ψ where appropriate (i.e., $F_{rep} = \psi F_k$)

Step 4: Calculate the design values of the actions by multiplying the representative actions by their corresponding partial factors of safety (i.e., $F_d = \gamma_F F_{rep}$, where γ_F is the appropriate partial factor of safety). Design values of permanent and variable actions are denoted as $V_{G,d}$ and $V_{Q,d}$, respectively.

Step 5: Determine the design effect of the actions V_d by summing up all the design actions.

Step 6: Determine the characteristic values (cautious estimates) of the geotechnical parameters (X_k) such as unit weight, cohesion, and internal friction angle (e.g., γ_k, c'_k, ϕ'_k).

Step 7: Divide the characteristic values of geotechnical parameters by their corresponding partial factors of safety to obtain design values of geotechnical parameters X_d $(= X_k/\gamma_M$, where γ_M is a partial factor of safety for material properties). On the basis of the previous step, partial factors of safety for the unit weight, cohesion, and internal friction correspond to $\gamma_\gamma, \gamma_{c'}, and\, \gamma_{\phi'}$.

Step 8: Using a recognized analytical method, calculate the characteristic resistance R_k $(= q_{ult}A$, where A is the area of the foundation).

Step 9: Determine the design resistance R_d by dividing the characteristic resistance by its corresponding partial factor of safety (γ_R).

Step 10: Verify that the design resistance is greater than the design effect of the actions (i.e., $V_d \leq R_d$)

Note that in contrast to the global factor of safety approach in which the foundation size is estimated directly, when designing using partial factors of safety, the foundation size is determined iteratively to ensure that the condition in Step 10 is satisfied. Hence, foundation sizes required for the calculation of q_{ult} require initial estimates.

Chapter 3

Sample Problem 3.1: Shallow foundation design using Terzaghi's bearing capacity theory

As shown in Figure 3.9, a shallow spread footing is designed to support a column (Figure 3.9). The load on the column, including the column weight, is $Q = 1000$ kN. The foundation rests in homogeneous silty sand. Subsurface exploration and laboratory testing found that the soil's effective cohesion is 25 kN/m^2 and the effective friction angle is 32°. The groundwater table is 1.5 m below the ground surface. The bulk unit weight above the groundwater table is 17.5 kN/m^3; the saturated unit weight below the groundwater table is 18.5 kN/m^3. Determine the foundation embedment and design the dimensions of the spread footing. The factor of safety for bearing capacity is 3.0. Use Terzaghi's bearing capacity theory. Assume general shear failure.

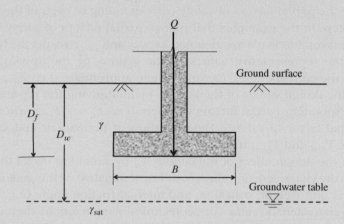

Fig. 3.9   Shallow foundation design sample problem.

Solution: Using global factor of safety

A spread column footing is to be designed. A square footing is chosen.

Step 1: The loads from the superstructure: $Q = 1000$ kN.
The subsoil condition: $c' = 25$ kN/m^2, $\phi' = 32°$, $\gamma = 17.5$ kN/m^3, $\gamma_{sat} = 18.5$ kN/m^3.

Step 2: On the basis of the data in Table 3.4, choose $D_f = 0.6$ m ($< D_w = 1.5$ m).

Step 3: Groundwater table (GWT) elevation: $D_w = 1.5$ m. It is assumed the GWT affects the ultimate bearing capacity.

Step 4: The minimum factor of safety is 3.0.

Chapter 3

Step 5: The Terzaghi's ultimate bearing capacity equation for square footing is:

$$q_{ult} = 1.3c'N_c + qN_q + 0.4\gamma BN_r$$

Using $\phi' = 32°$, find the bearing capacity factors for general failure in Table 3.1:

$$N_c = 44.036, \ N_q = 28.517, \ \text{and} \ N_\gamma = 26.871.$$

The effective stress at the bottom of the foundation is: $q = 0.6 \times 17.5 = 10.5 \ \text{kN/m}^2$.

Use $\overline{\gamma} = \gamma' + D_w - D_f/B(\gamma - \gamma')$ to substitute γ in the bearing capacity equation,
where

$$\gamma' = \gamma_{sat} - \gamma_w = 18.5 - 9.8 = 8.7 \ \text{kN/m}^3 \ \text{and}$$

$$\overline{\gamma} = 8.7 + \frac{1.5 - 0.6}{B}(17.5 - 8.7) = 8.7 + \frac{7.9}{B}$$

Therefore,

$$q_{ult} = 1.3c'N_c + qN_q + 0.4\gamma BN_\gamma$$

$$= 1.3 \times 25 \times 44.036 + 10.5 \times 28.517 + 0.4 \times \left(8.7 + \frac{7.9}{B}\right) B$$

$$\times 26.871$$

$$= 1815.5 + 93.5B$$

Step 6: The foundation slab thickness should be determined by structural analysis. Assume the thickness is 0.3 m. Then, the foundation slab's weight is:

$$W_f = 23.56 \times 0.3B^2 = 7.1B^2 = 7.1B^2 \ (\text{kN})$$

$$(\text{unit weight of concrete is } 23.56 \ \text{kN/m}^3)$$

The stress at the bottom of the foundation is as follows:

$$p = \frac{Q + W_f}{A} = \frac{1000 + 7.1B^2}{B^2} = \frac{1000}{B^2} + 7.1$$

Step 7: The allowable bearing capacity should be at least p.

$$q_{all} = \frac{q_{ult}}{FS} = p$$

$$\frac{1815.5 + 93.5B}{3} = \frac{1000}{B^2} + 7.1$$

Solve for B and find $B = 1.26$ m. Use $B = 1.5$ m.
The stress at the bottom of the foundation is:

$$p = \frac{Q + W_f}{A} = \frac{1000 + 7.1B^2}{B^2} = 451.5 \ \text{kN/m}^2$$

The ultimate bearing capacity is:

$$q_{ult} = 1955.7 \ \text{kN/m}^2$$

The allowable bearing capacity is:

$$q_{all} = \frac{q_{ult}}{FS} = \frac{1955.7}{3} = 651.9 \ \text{kN/m}^2$$

Using partial factors of safety

A spread column footing is to be designed. A square footing is chosen, and $B = 1.5$ m and $D_f = 0.6$ m are the initial estimates.

Step 1: For ease of comparison with the previous solution, it is assumed that the foundation will be subject to a characteristic unfavorable, permanent, and vertical load $V_{G,k} = 408$ kN and a characteristic unfavorable, variable vertical load $V_{Q,k} = 300$ kN.

Step 2: Partial factor of safety for unfavorable permanent loads → $\gamma_{F,G} = 1.35$
Partial factor of safety for unfavorable variable loads → $\gamma_{F,Q} = 1.50$

Step 3: $\psi = 1.0$ (Changes according to building category)

$$V_{G,rep} = \psi V_{G,k} = 1.0(408) = 408 \ \text{kN}$$

$$V_{Q,rep} = \psi V_{Q,k} = 1.0(300) = 300 \ \text{kN}$$

Step 4: Design values of actions

$$F_{G,d} = 408(1.35) = 550 \ \text{kN}$$

$$F_{Q,d} = 30(1.50) = 450 \ \text{kN}$$

Step 5: Design effect of the actions

$$V_d = F_{G,d} + F_{Q,d} = 550 + 450 = 1000 \ \text{kN}$$

Step 6: Characteristic values of geotechnical parameters

$$c'_k = 25 \text{ kN/m}^2$$

$$f'_k = 32°$$

$$\gamma_k \text{(above groundwater level)} = 17.5 \text{ kN/m}^3$$

$$\gamma_k \text{(below groundwater level)} = 18.5 \text{ kN/m}^3$$

Step 7: Partial factors of safety for geotechnical parameters:

$$\gamma'_c = 1.00$$

$$\gamma'_f = 1.00$$

$$\gamma_\gamma = 1.00$$

Design values of geotechnical parameters

$$c'_d = c'_k/\gamma_{c'} = 25/1.00 = 25 \text{ kN/m}^2$$

$$\phi'_d = a\tan(\tan\phi'_k/\gamma_{\phi'}) = a\tan(\tan 32/1.00) = 32°$$

$$\gamma'_d = \gamma'_k/\gamma_\gamma = 17.5/1.00 = 17.5 \text{ kN/m}^3$$

(above ground water level)

$$\gamma'_d = \gamma'_k/\gamma_\gamma = 18.5/1.00 = 18.5 \text{ kN/m}^3$$

(below ground water level)

Step 8: The Terzaghi's ultimate bearing capacity equation for square footing is:

$$R_k = q_{ult}A = (1.3c'_dN_c + qN_q + 0.4\gamma N_\gamma)A$$

Using $\phi'_d = 32°$, find the bearing capacity factors for general failure in Table 3.1:

$$N_c = 44.036, \ N_q = 28.517, \ \text{and} \ N_\gamma = 26.871$$

The effective stress at the bottom of the foundation is: $q = 0.6 \times 17.5 = 10.5$ kN/m^2.
Use $\bar{\gamma} = \gamma\prime + D_w - D_f/B(\gamma - \gamma\prime)$ to substitute γ in the bearing capacity equation,
where:

$$\gamma' = \gamma_{sat} - \gamma_w = 18.5 - 9.8 = 8.7 \text{ kN/m}^3 \ \text{and}$$

$$\bar{\gamma} = 8.7 + \frac{1.5 - 0.6}{1.5}(17.5 - 8.7) = 13.98 \text{ kN/m}^3$$

Therefore,

$$R_k = q_{ult}A = (1.3c'_dN_c + qN_q + 0.4\gamma BN_\gamma)A$$

$$= (1.3 \times 25/44.036 + 10.5 \times 28.517 + 0.4 \times 13.98$$

$$\times 1.5 \times 26.871)(1.5 \times 1.5)$$

$$= 4400 \text{ kN}$$

Step 9: Design resistance

$$R_d = R_k/\gamma_R = 4400/1.40 = 3143 \text{ } kN$$

Step 10: Verification of the ultimate limit state

$$V_d = 1000 \text{ kN} \le R_d = 3143 \text{ kN}$$

Conclusion: Design is satisfactory.

Note that according to Eurocodes, the partial factors of safety may be changed in different countries according to their local experience and requirements. In fact, Eurocode 7 provides three different design approaches, which only differ in the values of these partial factors of safety and the corresponding countries have a choice on what design approach to use.

Sample Problem 3.2: Shallow foundation design using the general bearing capacity theory.

The problem statement is the same as in Sample Problem 3.1. Use the general bearing capacity theory to design the shallow column foundation.

Solution: Using global factor of safety

Step 1: The loads from the superstructure: $Q = 1000$ kN.
The subsoil condition: $c' = 25$ kN/m^2, $\phi' = 32°$, $\gamma = 17.5$ kN/m^3, $\gamma_{sat} = 18.5$ kN/m^3.

Step 2: On the basis of the data in Table 3.4, choose $D_f = 0.6$ m ($< D_w = 1.5$ m).

Step 3: GWT elevation: $D_w = 1.5$ m. It is assumed the GWT affects the ultimate bearing capacity.

Step 4: The minimum factor of safety is 3.0.
Step 5: The general ultimate bearing capacity equation for square footings is:

$$q_{ult} = c'N_cF_{cs}F_{cd}F_{ci} + qN_qF_{qs}F_{qd}F_{qi} + \frac{1}{2}\gamma BN_rF_{\gamma s}F_{\gamma d}F_{\gamma i}$$

Using Table 3.3 and given $\phi' = 32°$, the bearing capacity factors
are: $N_c = 35.490$, $N_q = 23.177$, and $N_\gamma = 30.215$.
The foundation is square, i.e., $B = L$. The shape factors are as
follows:

$$F_{cs} = 1 + \left(\frac{B}{L}\right)\left(\frac{N_q}{N_c}\right) = 1 + 1 \times \frac{23.177}{35.490} = 1.653$$

$$F_{qs} = 1 + \left(\frac{B}{L}\right)\tan\phi' = 1 + 1 \times \tan 32° = 1.625$$

$$F_{\gamma s} = 1 - 0.4\left(\frac{B}{L}\right) = 0.6$$

Assuming $D_f/B \leq 1$, the depth factors are as follows:

$$F_{cd} = 1 + 0.4\left(\frac{D_f}{B}\right) = 1 + \frac{0.24}{B}$$

$$F_{qd} = 1 + 2\tan\phi'(1 - \sin\phi')^2\left(\frac{D_f}{B}\right)$$

$$= 1 + 2 \times \tan 32°(1 - \sin 32°)^2\left(\frac{0.6}{B}\right)$$

$$= 1 + \frac{0.166}{B}$$

$$F_{\gamma d} = 1$$

Because the load is vertical, the load inclination factors F_{ci}, F_{qi}, $F_{\gamma i}$
are all 1.0.
The effective stress at the bottom of the foundation is: $q = 0.6 \times 17.5 = 10.5$ kN/m^2.

Use $\bar{\gamma} = \gamma' + \frac{D_w - D_f}{B}(\gamma - \gamma')$ to substitute γ in the bearing capacity equation,
where:

$$\gamma' = \gamma_{sat} - \gamma_w = 18.5 - 9.8 = 8.7 \text{ kN/m}^3 \quad \text{and}$$

$$\bar{\gamma} = 8.7 + \frac{1.5 - 0.6}{B}(17.5 - 8.7) = 8.7 + \frac{7.9}{B}$$

Therefore,

$$q_{ult} = c'N_cF_{cs}F_{cd}F_{ci} + qN_qF_{qs}F_{qd}F_{qi} + \frac{1}{2}\gamma BN_rF_{\gamma s}F_{\gamma d}F_{\gamma i}$$

$$= 25 \times 35.490 \times 1.653 \times \left(1 + \frac{0.24}{B}\right) \times 1 + 10.5 \times 23.177$$

$$\times 1.625 \times \left(1 + \frac{0.166}{B}\right) \times 1 + \frac{1}{2} \times \left(8.7 + \frac{7.9}{B}\right) B \times 30.215$$

$$\times 0.6 \times 1 \times 1$$

$$= 2145.42 + \frac{352.0}{B} + 78.86B$$

Step 6: The foundation slab thickness should be determined by structural analysis. Assume the thickness is 0.3 m. Then, the foundation slab's weight is:

$$W_f = 23.56 \times 0.3B^2 = 7.1B^2 (kN)$$

(unit weight of concrete is 23.56 kN/m^3)

The stress at the bottom of the foundation is:

$$p = \frac{Q + W_f}{A} = \frac{1000 + 7.1B^2}{B^2} = \frac{1000}{B^2} + 7.1$$

Step 7: The allowable bearing capacity should be at least p.

$$q_{all} = \frac{q_{ult}}{FS} = p$$

$$\frac{2145.42 + \frac{352.0}{B} + 78.86B}{3} = \frac{1000}{B^2} + 7.1$$

Solve for B and find $B = 1.12$ m. Use $B = 1.20$ m.
The stress at the bottom of the foundation is:

$$p = \frac{Q + W_f}{A} = \frac{1000 + 7.1B^2}{B^2} = 701.5 \ kN/m^2$$

The ultimate bearing capacity is:

$$q_{ult} = 2533.4 \ kN/m^2$$

The allowable bearing capacity is:

$$q_{all} = \frac{q_{ult}}{FS} = \frac{2533.4}{3} = 844.4 \, kN/m^2 > p$$

Using partial factors of safety

A spread column footing is to be designed. A square footing is chosen, and $B = 1.20$ m and $D_f = 0.6$ m are taken as an initial estimates.

Step 1: For ease of comparison with the previous solution, it is assumed that the foundation will be subject to a characteristic unfavorable, permanent vertical load $V_{G,k} = 408$ kN and a characteristic unfavorable, variable vertical load $V_{Q,k} = 300$ kN.

Step 2: Partial factor of safety for unfavorable permanent loads → $\gamma_{F,G} = 1.35$

Partial factor of safety for unfavorable variable loads → $\gamma_{F,Q} = 1.50$

Step 3: $\psi = 1.0$ (Changes according to building category)

$$V_{G,rep} = \psi V_{G,k} = 1.0(408) = 408 \text{ kN}$$

$$V_{Q,rep} = \psi V_{Q,k} = 1.0(300) = 300 \text{ kN}$$

Step 4: Design values of actions

$$F_{G,d} = 408(1.35) = 550 \text{ kN}$$

$$F_{Q,d} = 30(1.50) = 450 \text{ kN}$$

Step 5: Design effect of the actions

$$V_d = F_{G,d} + F_{Q,d} = 550 + 450 = 1000 \text{ kN}$$

Step 6: Characteristic values of geotechnical parameters

$$c\prime_k = 25 \text{ kN/m}^2$$

$$\phi'_k = 32°$$

$$\gamma_k(\text{above ground water level}) = 17.5 \text{ kN/m}^3$$

$$\gamma_k(\text{below ground water level}) = 18.5 \text{ kN/m}^3$$

Step 7: Partial factors of safety for geotechnical parameters

$$\gamma_{c'} = 1.00$$

$$\gamma_{\phi'} = 1.00$$

$$\gamma_\gamma = 1.00$$

Design values of geotechnical parameters

$$c'_d = c'_k/\gamma_{c'} = 25/1.00 = 25 \text{ kN/m}^2$$

$$\phi'_d = a\tan(\tan\phi'_k/\gamma_{\phi'}) = a\tan(\tan 32/1.00) = 32°$$

$$\gamma'_d = \gamma'_k/\gamma_\gamma = 17.5/1.00 = 17.5 \text{ kN/m}^3$$

(above ground water level)

$$\gamma'_d = \gamma'_k/\gamma_\gamma = 18.5/1.00 = 18.5 \text{ kN/m}^3$$

(below ground water level)

Step 8: The general ultimate bearing capacity equation for square footings is:

$$q_{ult} = c'N_c F_{cs}F_{cd}F_{ci} + qN_q F_{qs}F_{qd}F_{qi} + \frac{1}{2}\gamma BN_r F_{\gamma s}F_{\gamma d}F_{\gamma i}$$

Using Table 3.3 and given $\phi'_d = 32°$, the bearing capacity factors are: $N_c = 35.490$, $N_q = 23.177$, and $N_\gamma = 30.215$.
The foundation is square, $B = L$. The shape factors as suggested in Eurocodes:

$$F_{\gamma s} = 1 - 0.3\left(\frac{B}{L}\right) = 0.7$$

$$F_{qs} = 1 + \left(\frac{B}{L}\right)\sin\phi' = 1 + \sin 32 = 1.530$$

$$F_{cs} = \frac{(F_{qs}N_q - 1)}{(N_q - 1)} = \frac{1.530 \times 23.177 - 1}{23.177 - 1} = 1.554$$

Assuming $D_f/B \le 1$, the depth factors are as follows:

$$F_{cd} = 1 + 0.4\left(\frac{D_f}{B}\right) = 1 + 0.4\left(\frac{0.6}{1.20}\right) = 1.2$$

$$F_{qd} = 1 + 2\left(\frac{B}{L}\right)\tan\phi'(1 - \sin\phi')$$

$$= 1 + 2 \times 1 \times \tan 32 \times (1 - \sin 32) = 1.587$$

$$F_{\gamma d} = 1$$

Because the load is vertical, the load inclination factors $F_{ci}, F_{qi}, F_{\gamma i}$ are all 1.0.
The effective stress at the bottom of the foundation is: $q = 0.6 \times 17.5 = 10.5 \text{ kN/m}^2$.
Use $\bar{\gamma} = \gamma' + \frac{D_w - D_f}{B}(\gamma - \gamma')$ to substitute γ in the bearing capacity equation,

where:

$$\gamma' = \gamma_{sat} - \gamma_w = 18.5 - 9.8 = 8.7\,kN/m^3 \quad and$$

$$\gamma' = 8.7 + \frac{1.5 - 0.6}{1.2}(17.5 - 8.7) = 15.3\,kN/m^3$$

Therefore,

$$R_k = (c'_d N_c F_{cs} F_{cd} F_{ci} + q N_q F_{qs} F_{qd} F_{qi} + 0.5\gamma B N_\gamma F_{\gamma s} F_{\gamma d} F_{\gamma i})A$$

$$= (25 \times 35.490 \times 1.554 \times 1.2 \times 1 + 10.5 \times 23.177 \times 1.530$$

$$\times 1.587 \times 1 + 0.5 \times 15.3 \times 1.2 \times 30.215 \times 0.7 \times 1 \times 1)$$

$$\times (1.2 \times 1.2)$$

$$= 3513.0\,kN$$

Step 9: Design resistance

$$R_d = R_k/\gamma_R = 3513.0/1.40 = 2509.3\,kN$$

Step 10: Verification of the ultimate limit state

$$V_d = 1000\,kN \le R_d = 2509.3\,kN$$

Design is satisfactory

3.2.7   Bearing capacity of eccentrically loaded shallow foundations

For structures that are subjected to horizontal forces, such as retaining walls under horizontal Earth pressure and high-rise buildings under wind load, there is a rotational moment acting on the structures (Figure 3.10). The moment can cause uneven pressure distribution at the bottom of the foundation. Similarly, for structures that are subjected to eccentric loading, the eccentric load is equivalent to a concentric load plus a rotational moment. In this section, one-way eccentricity is covered, i.e., the load is eccentric in one direction, but concentric in the other direction, as shown in Figure 3.11.

The point of load application is off the centerline by a distance, e, referred to as *eccentricity*.

$$e = \frac{M}{Q} \tag{3.40}$$

The maximum and minimum stresses due to the eccentric loading are:

$$q_{max} = \frac{Q}{BL} + \frac{6M}{B^2 L} = \frac{Q}{BL}\left(1 + \frac{6e}{B}\right) \tag{3.41}$$

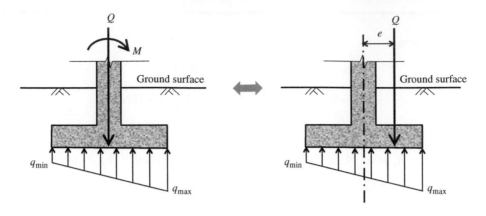

Fig. 3.10　Eccentrically loaded foundation, section view.

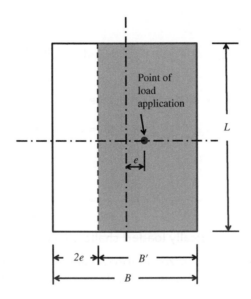

Fig. 3.11　Eccentrically loaded foundation of one-way eccentricity, plan view.

$$q_{min} = \frac{Q}{BL} - \frac{6M}{B^2 L} = \frac{Q}{BL} \left(1 - \frac{6e}{B} \right) \tag{3.42}$$

where:

Q = vertical load from superstructure,
M = moment,
B = foundation width,
L = foundation length,
e = eccentricity.

When $e > B/6$, q_{min} becomes negative. It means that tension between the soil and the foundation is developed on the side where negative q_{min} occurs. Because soil can withstand little or no tension, the foundation and soil become separated and not all the foundation is supported by soil. To avoid this condition, foundation design requires:

$$e > \frac{B}{6} \tag{3.43}$$

Design approach of eccentrically loaded foundation with one-way eccentricity using a global factor of safety:

The following approach is based on the "effective area method" proposed by Meyerhof (1953). The effective area of the foundation is the gray area in Figure 3.11.

Step 1: Calculate the eccentricity: $e = M/Q \leq B/6$ or $L/6$, depending on the eccentricity direction. If $e > B/6$ or $L/6$, the foundation dimension should be increased.

Step 2: Determine the effective dimensions. The effective width is $B' = B - 2e$; the effective length is $L' = L$, unchanged. If the eccentricity is in L direction, then the effective width B is unchanged ($B' = B$), and the effective length is $L' = L - 2e$.

Step 3: Use the general bearing capacity Equation (3.15) to determine the ultimate bearing capacity. Use effective dimensions in the general bearing capacity equation. In evaluating the shape factors, use the effective dimensions. In evaluating the depth factors, use the original dimensions, instead of the effective dimensions.

Step 4: Calculate the ultimate load that the foundation can sustain.

$$Q_{ult} = q_{ult}(L' \times B') \tag{3.44}$$

Step 5: Calculate the factor of safety to check whether the foundation, as a whole entity, can support the load.

$$FS = \frac{Q_{ult}}{Q} \geq 3.0 \tag{3.45}$$

where:
Q includes the load from the superstructure and the weight of the foundation.

Step 6: Calculate the factor of safety to ensure local failure does not occur.

$$FS = \frac{q_{ult}}{q_{max}} \geq 3.0 \tag{3.46}$$

The maximum stress at the bottom of the foundation is:

$$q_{max} = \frac{Q}{BL} + \frac{6M}{B^2L} \tag{3.47}$$

The smaller FS in Equations (3.45) and (3.46) is the FS of the design.

Design approach of eccentrically loaded foundation with one-way eccentricity using a partial factor of safety approach:

Step 1: Determine the characteristic values of the actions (loads), F_k, from the superstructure and classify them according to their nature (i.e., unfavorable, favorable, permanent, variable, accidental, etc.). Characteristic values are defined as cautious estimates of the actions being considered. These values are denoted as $V_{G,k}$, $V_{Q,k}$ for the characteristic vertical permanent and variable loads, respectively, while $M_{Q,k}$ represents a characteristic variable moment.

Step 2: Determine the partial factors of safety corresponding to each of the action types in the previous step (e.g. $\gamma_{F,G}$ and $\gamma_{F,Q}$ for unfavorable permanent and variable loads, respectively).

Step 3: Calculate the representative values of the actions by multiplying the characteristic actions by the corresponding factor ψ where appropriate (i.e., $F_{rep} = \psi F_k$)

Step 4: Calculate the design values of the actions by multiplying the representative actions by their corresponding partial factors of safety (i.e., $F_d = \gamma_F F_{rep}$, where γ_F is the appropriate partial factor of safety). These are denoted as $V_{G,d}$, $V_{Q,d}$ and $M_{Q,d}$ for the permanent and variable vertical actions and the variable moment, respectively)

Step 5: Determine the design effect of the actions V_d by summing up all the vertical design actions.

Step 6: Calculate the design eccentricity and verify that it does not exceed 1/3 of the width of a rectangular footing or 0.6 of the radius of a circular footing. Note that this is a lower requirement (in EN 1997–1:2004) as compared to that described for the global factor of safety approach described earlier.

Step 7: Calculate the design values of the geometrical data (e.g., $B\prime$ and $L\prime$ accounting for the design eccentricity.

Step 8: Determine the characteristic values (cautious estimates) of the geotechnical parameters (X_k) such as unit weight, cohesion, and internal friction angle.

Step 9: Divide the characteristic values of geotechnical parameters by their corresponding partial factors of safety to obtain the design values of geotechnical parameters X_d ($= X_k/\gamma_M$, where γ_M is a partial factor of safety for material properties).

Step 10: Using a recognized analytical method, calculate the characteristic resistance R_k ($= q_{ult}A$, where A is the area of the foundation).

Step 11: Determine the design resistance R_d by dividing the characteristic resistance by its corresponding partial factor of safety (γ_R).

Step 12: Verify that the design resistance is greater than the design effect of the actions (i.e., $V_d \le R_d$)

Sample Problem 3.3: Eccentrically loaded foundation design, with one-way eccentricity.

As shown in Figure 3.12, a shallow spread footing is designed to support a column. The load on the column, including the weight of the foundation, is $Q = 1000$ kN. The foundation is also subjected to a horizontal load that causes an overturning moment of 100 kN × m. The foundation rests in homogeneous silty sand. Subsurface exploration and laboratory testing found that the soil's effective cohesion is 25 kN/m² and the effective friction angle is 32°. The groundwater table is 1.5 m below the ground surface. The bulk unit weight above the groundwater table is 17.5 kN/m³; the saturated unit weight below the groundwater table is 18.5 kN/m³. The foundation embedment is 0.6m, and dimension is $B \times B = 1.2m \times 1.2m$. Calculate the factor of safety for the bearing capacity.

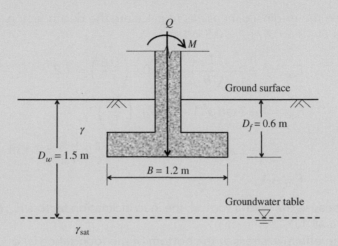

Fig. 3.12 Sample problem for eccentrically loaded foundation design.

Solution: Using a global factor of safety

Use the Meyerhof's effective area method.

Step 1: Eccentricity: $e = M/Q = 100/1000 = 0.1\,m < B/6 = 0.2\,m$, OK. No tension is developed at the bottom of the foundation.

Step 2: Determine the effective dimensions. Effective width: $B' = B - 2e = 1.2 - 0.2 = 1.0$ m; effective length: $L' = L = 1.2$ m.

Step 3: Use the general bearing capacity Equation (3.15) to determine the ultimate bearing capacity.

$$q_{ult} = c'N_c F_{cs}F_{cd}F_{ci} + qN_q F_{qs}F_{qd}F_{qi} + \frac{1}{2}\gamma BN_\gamma F_{\gamma s}F_{\gamma d}F_{\gamma i}$$

Using Table 3.3 and given $\phi' = 32°$, the bearing capacity factors are:

$$N_c = 35.490,\ N_q = 23.177,\ \text{and}\ N_\gamma = 30.215.$$

Use effective dimensions to calculate the shape factors:

$$F_{cs} = 1 + \left(\frac{B'}{L}\right)\left(\frac{N_q}{N_c}\right) = 1 + \frac{1}{1.2} \times \frac{23.177}{35.490} = 1.544$$

$$F_{qs} = 1 + \left(\frac{B'}{L}\right)\tan\phi' = 1 + \frac{1}{1.2} \times \tan 32° = 1.521$$

$$F_{\gamma s} = 1 - 0.4\left(\frac{B'}{L}\right) = 1 - 0.4 \times \frac{1}{1.2} = 0.667$$

Use the original dimensions to calculate the depth factors:
As $D_f/B = 0.6/1.2 = 0.5 < 1$:

$$F_{cd} = 1 + 0.4\left(\frac{D_f}{B}\right) = 1 + 0.4\left(\frac{0.6}{1.2}\right) = 1.2$$

$$F_{qd} = 1 + 2\tan\phi'(1 - \sin\phi')^2\left(\frac{D_f}{B}\right)$$

$$= 1 + 2 \times \tan 32° \times (1 - \sin 32°)^2 \times \frac{0.6}{1.2} = 1.138$$

$$F_{\gamma d} = 1$$

Because the load is vertical, the load inclination factors $F_{ci}, F_{qi}, F_{\gamma i}$ are all 1.0
The effective stress at the bottom of the foundation is: $q = 0.6 \times 17.5 = 10.5 \text{ kN/m}^2$.
Use $\bar{\gamma} = \gamma' + \frac{D_w - D_f}{B}(\gamma - \gamma')$ to substitute γ in the bearing capacity equation,
where:

$$\gamma' = \gamma_{sat} - \gamma_w = 18.5 - 9.8 = 8.7 \text{ kN/m}^3 \quad \text{and}$$

$$\bar{\gamma} = 8.7 + \frac{1.5 - 0.6}{1.2}(17.5 - 8.7) = 15.3 \text{ kN/m}^3$$

Therefore,

$$q_{ult} = c'N_cF_{cs}F_{cd}F_{ci} + qN_qF_{qs}F_{qd}F_{qi} + \frac{1}{2}\bar{\gamma}B'N_\gamma F_{\gamma s}F_{\gamma d}F_{\gamma i}$$

$$= 25 \times 35.490 \times 1.544 \times 1.2 \times 1 + 10.5 \times 23.177$$

$$\times 1.521 \times 1.138 \times 1 + \frac{1}{2} \times 15.3 \times 1.0 \times 30.215$$

$$\times 0.667 \times 1 \times 1$$

$$= 2219.3 \text{ kN/m}^3$$

Step 4: Calculate the ultimate load that the foundation can sustain:

$$Q_{ult} = q_{ult}(L'B') = 2219.3 \times 1.0 \times 1.2 = 2663.1 \text{ kN}$$

Step 5: The factor of safety for the entire foundation:

$$FS = \frac{Q_{ult}}{Q} = \frac{2663.1}{1000} = 2.66$$

Step 6: Calculate the factor of safety for local failure.

$$q_{max} = \frac{Q}{BL} + \frac{6M}{B^2 L} = \frac{1000}{1.2 \times 1.2} + \frac{6 \times 100}{1.2^2 \times 1.2} = 1041.7\,kN/m^2$$

$$FS = \frac{q_{ult}}{q_{max}} = \frac{2219.3}{1041.7} = 2.13$$

The smaller FS is the FS of the design. Therefore, FS = 2.13 < 3.0. The dimension of the foundation should be increased to satisfy FS ≥ 3.0.

Using partial factors of safety

A spread column footing is to be designed. A square footing is chosen, and $B = 1.2$ m and $D_f = 0.6$ m are taken as initial estimates.

Step 1: For ease of comparison with the previous solution, it is assumed that the foundation will be subjected to a characteristic unfavorable, permanent vertical load $V_{G,k} = 408$ kN and a characteristic unfavorable, variable vertical load $V_{Q,k} = 300$ kN, as well as an unfavorable variable overturning moment $M_{Q,k} = 16$ kN · m

Step 2: Partial factor of safety for unfavorable permanent loads
→ $\gamma_{F,G} = 1.35$
Partial factor of safety for unfavorable variable loads
→ $\gamma_{F,Q} = 1.50$

Step 3: $\psi = 1.0$ (Changes according to building category)

$$V_{G,rep} = \psi V_{G,k} = 1.0(408) = 408\,kN$$

$$V_{Q,rep} = \psi V_{Q,k} = 1.0(300) = 300\,kN$$

$$M_{Q,rep} = \psi M_{Q,k} = 1.0(16) = 16\,kN \cdot m$$

Step 4: Design values of actions

$$F_{G,d} = 408(1.35) = 550\,kN$$

$$F_{Q,d} = 30(1.50) = 450\,kN$$

$$M_{Q,d} = 16(1.50) = 24\,kN \cdot m$$

Chapter 3

Step 5: Design effect of the actions

$$V_d = F_{G,d} + F_{Q,d} = 550 + 450 = 1000 \text{ kN}$$

Step 6: Design eccentricity

$$e_d = M_d/V_d = 100/1000 = 0.1 \text{ m} < 1.2/3 = 0.4 \text{ m}$$

Design eccentricity is acceptable
Note that this is a lower requirement (in EN 1997–1:2004) as compared to that described for the global factor of safety approach described earlier.

Step 7: Design values of geometrical data

$$B' = B - 2e_d = 1.2 - 0.2 = 1.0 \text{ m}$$

$$L' = L = 1.2 \text{ m}$$

Step 8: Characteristic values of geotechnical parameters

$$c'_k = 25 \text{ kN/m}^2$$

$$f'_k = 32°$$

$$\gamma_k(\text{above ground water level}) = 17.5 \text{ kN/m}^3$$

$$\gamma_k(\text{below ground water level}) = 18.5 \text{ kN/m}^3$$

Step 9: Partial factors of safety for geotechnical parameters

$$\gamma_{c'} = 1.00$$

$$\gamma_{\phi'} = 1.00$$

$$\gamma_\gamma = 1.00$$

Design values of geotechnical parameters

$$c'_d = c'_k/\gamma_{c'} = 25/1.00 = 25 \text{ kN/m}^2$$

$$\phi'_d = a\tan(\tan\phi'_k/\gamma_{\phi'}) = a\tan(\tan 32/1.00) = 32°$$

$$\gamma'_d = \gamma'_k/\gamma_\gamma = 17.5/1.00 = 17.5 \text{ kN/m}^3$$

(above ground water level)

$$\gamma'_d = \gamma'_k/\gamma_\gamma = 18.5/1.00 = 18.5 \text{ kN/m}^3$$

(below ground water level)

Step 10: The general ultimate bearing capacity equation for square foot-ings is:

$$q_{ult} = c'N_cF_{cs}F_{cd}F_{ci} + qN_qF_{qs}F_{qd}F_{qi} + \frac{1}{2}\gamma B'N_rF_{\gamma s}F_{\gamma d}F_{\gamma i}$$

Using Table 3.3 and given $\phi\prime_d = 32°$, the bearing capacity factors are: $N_c = 35.490$, $N_q = 23.177$, and $N_\gamma = 30.215$.
Use effective dimensions to calculate the shape factors:

$$F_{\gamma s} = 1 - 0.3\left(\frac{B'}{L'}\right) = 1 - 0.3\left(\frac{1.0}{1.2}\right) = 0.75$$

$$F_{qs} = 1 + \left(\frac{B}{L}\right)\sin\phi' = 1 + \left(\frac{1.0}{1.2}\right)\sin 32 = 1.441$$

$$F_{cs} = \frac{(F_{qs}N_q - 1)}{(N_q - 1)} = \frac{1.521 \times 23.177 - 1}{23.177 - 1} = 1.489$$

Use the original dimensions to calculate the depth factors:
Since $D_f/B = 0.6/1.2 = 0.5 < 1$:

$$F_{cd} = 1 + 0.4\left(\frac{D_f}{B}\right) = 1 + 0.4\left(\frac{0.6}{1.2}\right) = 1.2$$

$$F_{qd} = 1 + 2\tan\phi'(1 - \sin\phi')^2\left(\frac{D_f}{B}\right)$$

$$= 1 + 2 \times \tan 32° \times (1 - \sin 32°)^2 \times \frac{0.6}{1.2} = 1.138$$

$$F_{\gamma d} = 1.0$$

Because the load is vertical, the load inclination factors F_{ci}, F_{qi}, $F_{\gamma i}$ are all 1.0.
The effective stress at the bottom of the foundation is: $q = 0.6 \times 17.5 = 10.5$ kN/m^2.
Use $\bar{\gamma} = \gamma' + \frac{D_w - D_f}{B}(\gamma - \gamma')$ to substitute γ in the bearing capacity equation,
where:

$$\gamma' = \gamma_{sat} - \gamma_w = 18.5 - 9.8 = 8.7 \text{kN/m}^3 \quad \text{and}$$

$$\bar{\gamma} = 8.7 + \frac{1.5 + 0.6}{1.2}(17.5 - 8.7) = 15.3 \text{kN/m}^3$$

Therefore,

$$R_k = (c'_d N_c F_{cs}F_{cd}F_{ci} + qN_qF_{qs}F_{qd}F_{qi} + 0.5\gamma BN_\gamma F_{\gamma s}F_{\gamma d}F_{\gamma i})A'$$

$$= (25 \times 35.490 \times 1.489 \times 1.2 \times 1 + 10.5 \times 23.177 \times 1.441$$

$$\times 1.138 \times 1 + 0.5 \times 15.3 \times 1.0 \times 30.215 \times 0.75 \times 1 \times 1)$$

$$(1.2 \times 1.0)$$

$$= 2589.3 \text{ kN}$$

Step 11: :Design resistance

$$R_d = R_k/\gamma_R = 2589.3/1.40 = 1849.5 \text{ kN}$$

Step 12: Verification of the ultimate limit state

$$V_d = 1000 \text{ kN} \leq R_d = 1849.5 \text{ kN}$$

Design is satisfactory

3.2.8   Mat foundations

Individual spread footings can be combined to form one concrete slab supporting multiple columns and walls in order to increase the resistance to differential settlement and distortions of the slab. A combined footing, as shown in Figure 3.13, is a structural unit or assembly of units supporting more than one column load. Conventionally, combined footings refer to foundations supporting only one row of columns.

A mat foundation, also known as a "raft foundation," covers nearly the entire footprint of the superstructure. As defined by the American Concrete Institute (ACI), committee 336 (footings, mats, and drilled piers), a mat foundation is "a continuous footing supporting an array of columns in several rows in each direction, having a slab-like shape with or without depressions or openings, covering an area of at least 75% of the total area within the outer limits of the assembly." A mat is generally reinforced concrete, i.e., 1–2 m (3–6 ft) thick. Figure 3.14 illustrates the concept of mat foundations, and Figure 3.15 shows a mat foundation during construction.

A mat foundation is often the choice under the following conditions.

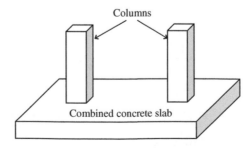

Fig. 3.13   Combined footing.

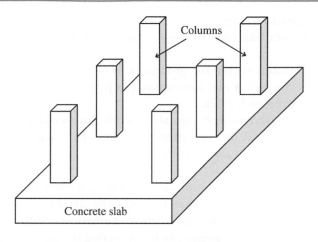

Fig. 3.14 Mat foundation.

Fig. 3.15 Mat foundation of the Four Seasons Hotel, San Francisco, California. (Photo courtesy of Prof. Ross W. Boulanger, University of California, Davis).

- When the superstructure load is high or the subsoil is erratic or weak, the foundation size required to provide adequate bearing capacity is very large. It will be economical and convenient to construct a footing that covers the entire footprint of the structure. Often a mat foundation is used when spread footings cover more than one-half of the foundation area (ACI 2002).
- If the vertical loads at different locations vary significantly or the soil's strengths at different locations are significantly different, excessive differential settlement is likely to occur. Large horizontal forces such as wind load and lateral soil pressure can also cause the stress distribution at the bottom of the foundation to be nonuniform, causing differential settlements.

To reduce the differential settlement, a mat foundation is preferred because the structural continuity and the flexural strength of the foundation will compensate for the differential settlements.

- If the foundation elevation is within the groundwater fluctuation zone, it can be subjected to hydraulic uplift forces (heave). A mat foundation can provide sufficient resistance to heave; it can also provide a waterproof barrier.
- If the structure is subjected to nonuniform lateral loads, uneven deformation and subsequent damage to the foundation and the superstructure can occur. A mat foundation can provide the structural continuity to resist the uneven deformation.

The ultimate bearing capacity of mat foundations can be determined using the same Equations (3.5), (3.6), and (3.15) that are used for spread footings. The design of a mat foundation includes complex structural analyses that require the realistic evaluation of the soil pressure exerted at the bottom of the foundation. Soil response can be estimated by modeling the soil as coupled or uncoupled "soil springs." Structural engineers generally use finite element methods, finite grid methods, finite difference methods, and approximate flexible methods to analyze and design mat foundations. The detailed design can be referenced in ACI 336.2R-88: Suggested Analysis and Design Procedures for Combined Footings and Mats (Reapproved 2002) (ACI 2002).

3.3   Settlements of shallow foundations

Often times, a foundation failure is not due to inadequate bearing capacity, but due to excessive settlement. Therefore, settlement is another important factor that controls shallow foundation design. The settlement of a shallow foundation includes soil's elastic deformation and plastic deformation (consolidation), and these deformations are caused by vertical stress increase due to external load. In this section, determinations of vertical stress increase due to external load are first presented, followed by the methods to determine elastic and consolidation settlements.

3.3.1   Vertical stress increase due to external load

Boussinesq (1883) developed a method to determine the vertical stress increase under a point load in a homogeneous, elastic, isotropic, and semi-infinite medium, as expressed in Equation (3.48):

$$\Delta\sigma_z = \frac{3P}{2\pi z^2 \left[1 + \left(\frac{r}{2}\right)^2\right]^{\frac{5}{2}}} \tag{3.48}$$

where:
$\Delta\sigma_z$ = vertical normal stress increase,
P = point load,
r = planar radius from the point of application,
z = depth.

Equation (3.48) became the basis for calculating the vertical stress increases under various types of loads, such as strip loads, uniformly distributed loads on rectangular or circular footings, and loads of embankments of different shapes. Newmark (1935), by integrating Equation (3.48), derived the vertical stress at any point below a uniformly loaded flexible area of soil of any shape; the mathematical expressions were plotted on the stress increase contour lines (pressure isobars), popularly known as Newmark's Influence Charts. Some analytical formulas for the vertical stress increases in a homogeneous, elastic, isotropic, and semi-infinite medium due to various types of loads are summarized in Table 3.6, based on NAVFAC (1986) and Das (2011).

Poulos and Davis (1974) provided simplified formulas, as expressed in Equations (3.49)–(3.52), to compute the vertical stress increase beneath the *center* of a shallow foundation. The equations produce answers that are within 5% of the Boussinesq values (Coduto 2001).

For a circular foundation with diameter B, the vertical stress increase at depth z below the foundation is:

$$\Delta\sigma_z = (q - \sigma_0') \left[1 - \left(\frac{1}{1 + \left(\frac{B}{2z} \right)^2} \right)^{1.50} \right] \tag{3.49}$$

For a square foundation with length B, the vertical stress increase at depth z below the foundation is:

$$\Delta\sigma_z = (q - \sigma_0') \left[1 - \left(\frac{1}{1 + \left(\frac{B}{2z} \right)^2} \right)^{1.76} \right] \tag{3.50}$$

For a continuous foundation with width B, the vertical stress increase at depth z below the foundation is:

$$\Delta\sigma_z = (q - \sigma_0') \left[1 - \left(\frac{1}{1 + \left(\frac{B}{2z} \right)^2} \right)^{2.60} \right] \tag{3.51}$$

For a rectangular foundation with width B and length L, the vertical stress increase at depth z below the foundation is:

$$\Delta\sigma_z = (q - \sigma_0') \left[1 - \left(\frac{1}{1 + \left(\frac{B}{2z} \right)^{1.38+0.62B/L}} \right)^{2.60-0.84\,B/L} \right] \tag{3.52}$$

In Equations (3.49)–(3.52):

q = uniform stress at the bottom of the foundation,
σ'_0 = effective stress due to self-weight of soil at the bottom of the foundation, and
z = depth, starting from the bottom of the foundation.

Another approximate but popular method to compute vertical stress increases is the "2:1 method," as illustrated in Figure 3.16. In this method, the vertical stress is assumed to propagate and diminish downward at a 2:1 (vertical to horizontal) slope, and the total load at any depth beneath the foundation is equal to the total load at the bottom of the foundation.

Table 3.6 Vertical stress increases due to various types of loads using Boussinesq method.

Load type	Stress diagram	Equation	Note
Point load		$$\Delta\sigma_z = \frac{3P}{2\pi z^2\left[1+\left(\frac{r}{z}\right)^2\right]^{\frac{5}{2}}}$$	P = force
Line load of infinite length		$$\Delta\sigma_z = \frac{2pz^3}{\pi(x^2+z^2)^2}$$	p = force/length
Uniform strip load	Vertical view	$$\Delta\sigma_z = \frac{p}{\pi}[\alpha + \sin\alpha \cdot \cos(\alpha + 2\gamma)]$$	p = force/area
Uniformly loaded rectangular area		$$\Delta\sigma_z = \frac{p}{4\pi}\left[\frac{2mn\sqrt{m^2+n^2+1}}{m^2+n^2+m^2n^2+1}\left(\frac{m^2+n^2+2}{m^2+n^2+1}\right) + \tan^{-1}\left(\frac{2mn\sqrt{m^2+n^2+1}}{m^2+n^2-m^2n^2+1}\right)\right]$$	Only applies to vertical stress increase below the *corner*. $m = \frac{B}{z}, n = \frac{L}{z}$
Uniformly loaded circular area		$$\Delta\sigma_z = p\left[1-\left(\frac{1}{1+\left(\frac{R}{z}\right)^{1.5}}\right)^{1.5}\right]$$	Only applies to vertical stress increase below the *center*
Slope load	Vertical view	$$\Delta\sigma_z = \frac{p_0}{\pi D}[x\beta + z]$$	p_0 = force/area

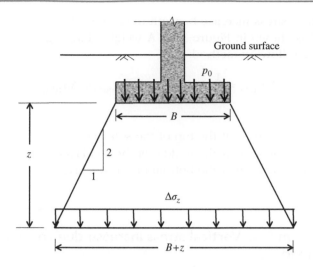

Fig. 3.16   Illustration of 2:1 method.

For a rectangular foundation of dimensions $B \times L$, the vertical stress increase at depth z below the foundation is:

$$\Delta \sigma_z = \frac{p_0 \times B \times L}{(B+z) \times (L+z)} \qquad (3.53)$$

For a strip (continuous) foundation of width B, the vertical stress increase at depth z below the foundation is:

$$\Delta \sigma_z = \frac{p_0 \times B}{(B+z)} \qquad (3.54)$$

When calculating the settlement of a soil layer beneath a foundation, the *average* vertical stress increase in the soil layer is used. The vertical stress increase beneath a foundation is nonlinear.

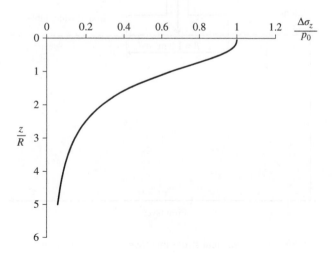

Fig. 3.17   Vertical stress increase beneath the center of uniformly loaded circular area.

For example, the vertical stress increase beneath the center of a uniformly loaded circular area (radius R, pressure p) is shown in Figure 3.17. A weighted average vertical stress increase can be calculated using the following method.

$$\Delta\sigma_{z(av)} = \frac{1}{6}(\Delta\sigma_{z(top)}) + 4\Delta\sigma_{z(mid)} + \Delta\sigma_{z(bot)} \qquad (3.55)$$

where:

$\Delta\sigma_{z(top)}$ = vertical stress increase at the top of the soil layer,
$\Delta\sigma_{z(mid)}$ = vertical stress increase at the middle of the soil layer,
$\Delta\sigma_{z(bot)}$ = vertical stress increase at the bottom of the soil layer.

Sample Problem 3.4: Vertical stress increase due to shallow foundation load

As shown in Figure 3.18, a square spread footing supports a column. The load on the column, including the weight of the foundation, is $Q = 1000$ kN. The foundation rests in homogeneous silty clay soil. The groundwater table is at the bottom of the foundation. Below the foundation is a 6-m thick homogeneous layer of silty clay. The soil's effective cohesion is 30 kN/m², and the effective friction angle is 15°. The bulk unit weight above the groundwater table is 17.5 kN/m³; the saturated unit weight below the groundwater table is 18.5 kN/m³. The foundation embedment is 0.6 m, and the foundation dimensions are $B \times B = 1.2$ m $\times$ 1.2 m. Determine the average vertical stress increase in the soil layer due to the foundation loading.

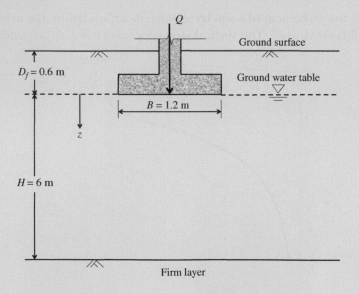

Fig. 3.18   Vertical stress increase sample problem.

Solution:

Use the simplified Equation (3.50) given by Poulos and Davis (1974). For square foundations with length B, the vertical stress increase at depth z below the center of the foundation is:

$$\Delta\sigma_z = (q - \sigma_0')\left[1 - \left(\frac{1}{1 + \left(\frac{B}{2z}\right)^2}\right)^{1.76}\right]$$

where:

q = uniform stress at the bottom of the foundation,

q = $Q/B^2 = 1000/1.2^2 = 694.4\,\text{kN/m}^2$,

σ_0' = effective stress due to self-weight of soil at the bottom of the foundation,

σ_0' = $0.6 \times 17.5 = 10.5\,\text{kN/m}^2$.

The vertical stress increases are calculated at the top, middle, and bottom of the soil layer and listed in the following table. For comparison, the vertical stress increases using Boussinesq equation (Table 3.6) and the 2:1 method are also listed.

The Boussinesq equation for vertical stress increase below the corner of a rectangular foundation is:

$$\Delta\sigma_z = \frac{p}{4\pi}\left[\frac{2mn\sqrt{m^2 + n^2 + 1}}{m^2 + n^2 + m^2n^2 + 1}\left(\frac{m^2 + n^2 + 2}{m^2 + n^2 + 1}\right) + \tan^{-1}\left(\frac{2mn\sqrt{m^2 + n^2 + 1}}{m^2 + n^2 - m^2n^2 + 1}\right)\right]$$

where: $m = B/z$, and $n = L/z$, and $B = L = 0.6$ m (divide the foundation into four sections). The result should be multiplied by 4 to obtain the stress increase beneath the center of the foundation.

The 2:1 method for vertical stress increase below a rectangular foundation is:

$$\Delta\sigma_z = \frac{p_0 \times B \times L}{(B + z) \times (L + z)}$$

where $B = L = 1.2$ m.

The average vertical increase uses Equation (3.55):

$$\Delta\sigma_{z(av)} = \frac{1}{6}(\Delta\sigma_{z(top)} + 4\Delta\sigma_{z(mid)} + \Delta\sigma_{z(bot)})$$

Chapter 3

z(m)	$\Delta\sigma_z$ (Poulos and Davis, 1974) (kN/m^2)	$\Delta\sigma_z$ (Boussinesq) (kN/m^2)	2:1 method (kN/m^2)
0	683.9	694.4	694.4
3	45.6	69.7	56.7
6	11.9	13.0	19.3
Average vertical stress increase (kN/m^2)	146.4	164.4	156.7

3.3.2　Elastic settlement

Elastic settlement is due to the elastic deformation of soils. It occurs within a short period of time after the initial loading; therefore, it is also called "immediate settlement." A number of solutions for calculation of elastic settlement exist in the literature for different theories, initial governing assumptions, foundation geometries, and specific situations. Most of the solutions provide similar results. Mayne and Poulos (1999) provided the following settlement equation for shallow spread footings and mat foundations that account for homogeneous to linearly varying soil modulus of elasticity profiles, finite to infinite soil layer thickness, foundation flexibility, undrained and drained loading conditions, and foundation embedment depth:

$$S_e = q_0 B_e I_G I_F I_E \frac{(1-\mu^2)}{E_0} \tag{3.56}$$

where:

S_e　=　elastic settlement,
B_e　=　equivalent diameter.

For circular footings of diameter D:

$$B_e = D \tag{3.57}$$

For rectangular footings with dimensions L and B:

$$B_e = \sqrt{\frac{4BL}{\pi}} \tag{3.58}$$

μ　　　　=　Poisson's ratio,
E_0　　　=　the soil's modulus of elasticity at depth $z = 0$ at the bottom of the foundation, and
I_G, I_F, I_E　=　displacement influence factors.

A more generalized situation, known as the "Gibson" case, is that a footing rests on a non-homogeneous elastic medium whose modulus of elasticity increases linearly with depth. The relationship is expressed as follows and illustrated in Figure 3.19.

$$E_s = E_0 + k_E z \tag{3.59}$$

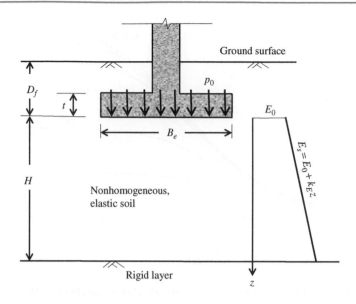

Fig. 3.19   Shallow foundation on nonhomogeneous, elastic soil.

where:

E_s = soil's modulus of elasticity at depth $z \geq 0$,
k_E = rate of increase of modulus of elasticity with depth,
z = depth starting from the bottom of the foundation.

The displacement influence factor, I_G, accounts for the effects of a soil's elastic modulus and a foundation's relative distance to an underlying rigid layer. Mayne and Poulos (1999) provided a chart to determine the I_G values, based on the "normalized Gibson modulus," β. The definition of β is expressed in Equation (3.60). Figure 3.20 shows the relationship between I_G and β.

$$\beta = \frac{E_0}{K_E B_e} \tag{3.60}$$

The displacement influence factor, I_F, also known as the rigidity correction factor, accounts for the effect of a foundation's rigidity.

$$I_F = \frac{\pi}{4} + \frac{1}{4.6 + 10\left(\dfrac{E_f}{E_{s(av)}}\right)\left(\dfrac{2t}{B_e}\right)^3} = \frac{\pi}{4} + \frac{1}{4.6 + 10K_F} \tag{3.61}$$

where:

E_f = elastic modulus of the foundation slab,
$E_{s(av)}$ = average elastic modulus of the soil. According to Equation (3.59), the soil's elastic modulus increases linearly in the soil layer of thickness of H. Therefore,

$$E_{s(av)} = E_0 + \frac{H}{2}k_E \tag{3.62}$$

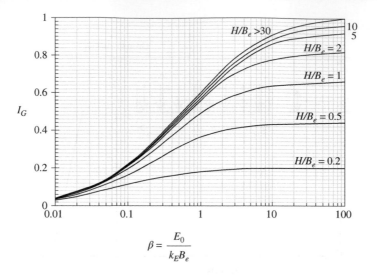

$$\beta = \frac{E_0}{k_E B_e}$$

Fig. 3.20   Influence factor, I_G, for nonhomegeneous, elastic soil (Mayne and Poulos 1999). (Used with permission of ASCE.)

K_F = foundation flexibility factor:

$$K_F = \left(\frac{E_f}{E_{s(av)}} \right) \left(\frac{2t}{B_e} \right)^3 \tag{3.63}$$

Figure 3.21 shows the rigidity correction factor variation with K_F. The line divides the foundation into three categories: (1) perfectly rigid with $K_F > 10$; (2) intermediate flexibility with $0.01 < K_F < 10$; and (3) perfectly flexible with $K_F < 0.01$.

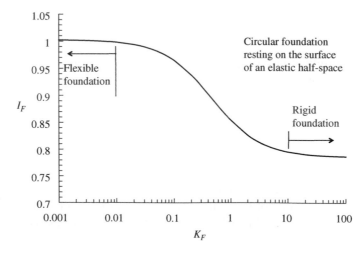

Fig. 3.21   Variation of rigidity correction factor I_F with foundation flexibility factor K_F (Mayne and Poulos 1999). (Used with permission of ASCE.)

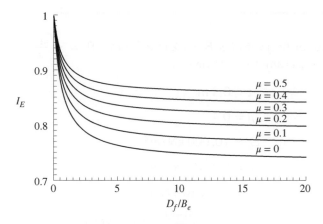

Fig. 3.22   Settlement correction factor for shallow foundation embedment (Mayne and Poulos 1999). (Used with permission of ASCE.)

The displacement influence factor, I_E, also known as the "settlement correction factor," accounts for the effects of a foundation's embedment and a soil's Poisson's ratio. I_E is expressed in the following equation and shown in Figure 3.22.

$$I_E = 1 - \frac{1}{3.5^{1.22\mu-0.4}\left(1.6 + \frac{B_e}{D_f}\right)} \tag{3.64}$$

where:

μ = Poisson's ratio,
D_f = foundation embedment depth.

Sample Problem 3.5: Elastic settlement using the Mayne and Poulos method

The problem statement is the same as in Sample Problem 3.4 and is shown in Figure 3.18. The thickness of the foundation slab is 0.3 m. The Poisson's ratio is 0.4. The elastic modulus increases linearly with the depth: $E_s = 10,000(kN/m^2) + 500(kN/m^2/m) \times z$. The thickness of the foundation slab is 0.3 m, and the elastic modulus of the foundation is $1.5 \times 10^7 kN/m^2$. Determine the elastic settlement of the soil layer beneath the foundation.

Solution:

Use the Mayne and Poulos (1999) method. The elastic settlement is:

$$S_e = q_0 B_e I_G I_F I_E \frac{(1-\mu^2)}{E_0}$$

For a square footing with $L \times B = 1.2 \text{ m} \times 1.2 \text{ m}$: $B_e = \sqrt{\frac{4B^2}{\pi}} = 1.35 \text{ m}$
The following parameters are given:

$$t = 0.3 \text{ m}$$

$$\mu = 0.4$$

$$E_0 = 10,000 \text{ kN/m}^2$$

$$k_E = 500 \text{ kN/m}^2/\text{m}$$

$$E_f = 1.2 \times 10^7 \text{kN/m}^2$$

$$q_0 = \frac{Q}{B^2} = \frac{1000}{1.2^2} = 694.4 \text{ kN/m}^2$$

The displacement influence factor, I_G, is determined using Figure 3.20.
The normalized Gibson modulus is: $\beta = E_0/(k_E B_e) = 10,000/(500 \times 1.35) = 14.8$
In addition, $\dfrac{H}{B_e} = \dfrac{6}{1.35} = 4.44$

From Figure 3.20, find $I_G \approx 0.85$.
The rigidity correction factor, I_F, is:

$$I_F = \frac{\pi}{4} + \frac{1}{4.6 + 10 \left(\dfrac{E_f}{E_{s(av)}} \right) \left(\dfrac{2t}{B_e} \right)^3} = \frac{\pi}{4} + \frac{1}{4.6 + 10K_F}$$

$$E_{s(av)} = E_0 + \frac{H}{2}k_E = 10,000 + \frac{6}{2} \times 500 = 11500 \text{ kN/m}^2$$

$$K_F = \left(\frac{E_f}{E_{s(av)}} \right) \left(\frac{2t}{B_e} \right)^3 = \left(\frac{1.2 \times 10^7}{11500} \right) \left(\frac{2 \times 0.3}{1.35} \right)^3 = 91.6$$

$$I_F = \frac{\pi}{4} + \frac{1}{4.6 + 10 \left(\dfrac{E_f}{E_{s(av)}} \right) \left(\dfrac{2t}{B_e} \right)^3} = \frac{\pi}{4} + \frac{1}{4.6 + 10 \times 91.6} = 0.786$$

The settlement correction factor, I_E, is:

$$I_E = 1 - \frac{1}{3.5e^{1.22\mu - 0.4} \times \left(1.6 + \dfrac{B_e}{D_f} \right)} = 1 - \frac{1}{3.5e^{1.22 \times 0.4 - 0.4} \times \left(1.6 + \dfrac{1.35}{0.6} \right)} = 0.932$$

Therefore, the elastic settlement is:

$$S_e = q_0 B_e I_G I_F I_E \frac{(1 - \mu^2)}{E_0}$$

$$= 694.4 \times 1.35 \times 0.85 \times 0.786 \times 0.932 \times \left(\frac{1 - 0.4^2}{10,000} \right)$$

$$= 0.049 \text{ m} = 4.9 \text{ cm}$$

3.3.3   Consolidation settlement

When saturated soil is subjected to an external load, the pore water pressure increases imme-diately on the application of the external load. With time, the increase of pore water pressure gradually decreases, and the effective stress gradually increases; as pore water drains from the soil, the pore volume and total volume of the soil gradually decrease. This entire process is called "consolidation" or "primary consolidation." The soil volume decrease in the vertical direction due to primary consolidation is the *primary consolidation settlement*. After the completion of the primary consolidation, the soil grains may rotate or slide against other grains or may be crushed, causing a permanent plastic deformation. This type of deformation is called "secondary consol-idation settlement." Typically, the secondary consolidation settlement is very small for mineral soils (sand or clay); therefore, it is often ignored in the practical design. For organic soils such as peat, however, the secondary consolidation settlement is the highest among geotechnical mate-rials (Mesri and Ajlouni 2007) and should be evaluated if a foundation is to rest on organic soils.

Theoretically, "consolidation is any process which involves a decrease in water content of a saturated soil without replacement of water by air" (Terzaghi 1943). In practice, consolidation settlement of granular soil (sand and gravel) is small and is usually ignored. Consolidation settlement is generally calculated only for clayey and silty soils, where drainage is slow and consolidation can be significant.

Determination of primary consolidation settlement is based on the consolidation test (ASTM D 2435, ASTM D 4186, and BS 1377–6:1990). An example of a typical consolidation curve (also referred to as $e \sim \log p$ curve) is shown in Figure 3.23. Three important parameters can be obtained from a consolidation test: compression index c_c, swell index c_s, and preconsolidation pressure $\sigma\prime_c$. The preconsolidation pressure is the maximum pressure that the soil mass has ever been subjected to during its entire history. Consolidation settlement, S_c, depends on the following two consolidation conditions:

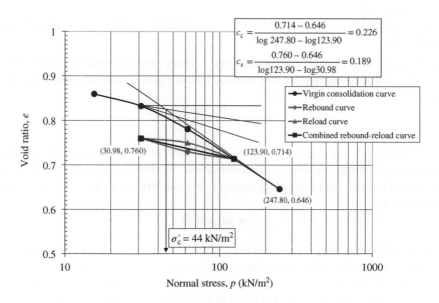

Fig. 3.23   Example of consolidation test curve ($e \sim \log p$ curve).

(a) Normally consolidated soils ($\sigma'_c = \sigma'_0$, or overconsolidation ratio OCR $= \sigma'_c/\sigma'_0 = 1$):

$$S_c = \frac{c_c H}{1 + e_0} \log \frac{\sigma'_0 + \Delta\sigma'_{av}}{\sigma'_0} \qquad (3.65)$$

(b) Overconsolidated soils ($\sigma'_c > \sigma'_0$, or overconsolidation ratio OCR $= \sigma'_c/\sigma'_0 > 1$). Under this condition, two subcategories exist:

- If $\sigma'_c \geq \sigma'_0 + \Delta\sigma'_{av}$:

$$S_c = \frac{c_s H}{1 + e_0} \log \frac{\sigma'_0 + \Delta\sigma'_{av}}{\sigma'_0} \qquad (3.66)$$

- If $\sigma'_c < \sigma'_0 + \Delta\sigma'_{av}$:

$$S_c = \frac{c_s H}{1 + e_0} \log \frac{\sigma'_c}{\sigma'_0} + \frac{c_c H}{1 + e_0} \log \frac{\sigma'_0 + \Delta\sigma'_{av}}{\sigma'_c} \qquad (3.67)$$

In the aforementioned equations:

σ'_0 = average effective stress of the soil stratum for which the consolidation settlement is calculated;

σ'_c = preconsolidation pressure, derived from $e \sim \log p$ curve using the Casagrande method, refer to the example in Figure 3.23;

$\Delta\sigma'_{av}$ = average vertical stress increase in the soil stratum due to external load, and it can be calculated using Equation (3.49);

c_c = compression index;

c_s = swell index;

H = thickness of the soil stratum for which the consolidation settlement is calculated; and

e_0 = initial void ratio of the soil stratum for which the consolidation settlement is calculated.

An alternative approach that produces the same or very similar results requires the coefficient of volume compressibility m_v, which can also be found from odometer results such as those in Figure 3.23:

$$m_v = \frac{1}{1 + e_0} \frac{e_0 - e_1}{\Delta\sigma'_{av}} \qquad (3.68)$$

As it can be expected, the value of m_v is stress and state dependent; therefore, it can be calculated under both normally and overconsolidated conditions, provided the laboratory data are available. Knowing the value of coefficient compressibility (m_v), the consolidation settlement can be calculated as:

$$S_c = \Delta\sigma'_{av} m_v H \qquad (3.69)$$

Consolidation settlement is caused by vertical stress increase. When the vertical stress increase, $\Delta\sigma'_z$, at a certain depth is less than one-tenth of the *in situ* effective stress σ'_0, it is assumed that consolidation settlement is negligible. Therefore, consolidation settlement should be considered either for the depth $\Delta\sigma'_z \geq 0.1\sigma'_0$, or when there is an underlying firm layer whose compressibility is negligible.

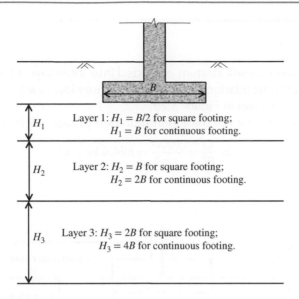

Fig. 3.24 Classical method of dividing soil layer beneath a shallow foundation for consolidation settlement calculation. (After Coduto 2001.)

Because the vertical stress increase is nonlinear with depth, taking an average of the vertical stress increase over a thick soil stratum will produce a large error. A thick soil layer should be divided into a number of thin layers, for which consolidation settlement is calculated and then summed. Approximate thicknesses of soil layers for manual computation of consolidation settlement of shallow foundations are shown in Figure 3.24.

Sample Problem 3.6: **Primary consolidation settlement**

The problem statement is the same as in Sample Problem 3.4 and is illustrated in Figure 3.25. A square spread footing supports a column. The load on the column, including the weight of the foundation, is $Q = 1000$ kN. The foundation rests in homogeneous silty clay soil. The groundwater table is at the bottom of the foundation. Below the foundation is a 6-m thick homogeneous layer of silty clay. The soil's effective cohesion is 60 kN/m^2, and the effective friction angle is 15°. The bulk unit weight above the groundwater table is 17.5 kN/m^3; the saturated unit weight below the groundwater table is 18.5 kN/m^3. The foundation embedment is 0.6 m, and the foundation dimensions are $B \times B = 1.2$ m $\times$ 1.2 m.

Laboratory testing on soil samples retrieved at three different depths produced the soil properties that are shown in Figure 3.25. Determine the primary consolidation settlement of the soil layer of 6 m in thickness beneath the foundation.

Solution:

The 6-m saturated clay soil stratum is divided into three layers to obtain better prediction of the consolidation settlement. The three layers with their respective characteristics are shown in Figure 3.25.

The stress at the bottom of the foundation due to external load is:

$$p_0 = \frac{Q}{B^2} = \frac{1000}{1.2^2} = 694.4 \text{ kN/m}^2$$

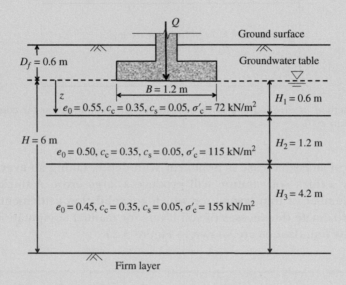

Fig. 3.25   Sample problem for primary consolidation settlement.

To simplify the calculation and to be conservative, the 2:1 method is used to calculate the vertical stress increases in each layer.

$$\Delta\sigma_z = \frac{p_0 \times B \times L}{(B + z) \times (L + z)}$$

The average vertical stress increase in each layer is determined using Equation (3.55).

$$\Delta\sigma_{s(av)} = \frac{1}{6}(\Delta\sigma_{z(top)} + 4\Delta\sigma_{s(mid)} + 4\Delta\sigma_{s(bot)})$$

The average effective stress of each layer is the effective stress at the middle of the soil layer. The results are summarized in the following tables.

z(m)	Vertical stress increase (kN/m²)
0	694.4
0.3	444.4
0.6	308.6
1.2	173.6
1.8	111.1
3.9	38.4
6.0	19.2

Layer	Average vertical stress increase, $\Delta\sigma'_{z(av)}$(kN/m²)	Average *in situ* effective stress, $\Delta\sigma'_{0(av)}$(kN/m²)	Preconsolidation pressure, σ'_c(kN/m²)
1	463.3	2.6	72
2	185.7	10.4	115
3	47.4	33.9	155

In layer 1: $\sigma'_c > \sigma'_{0(av)}$, the soil is overconsolidated.
In addition, $\sigma'_c < \sigma'_{0(av)} + \Delta\sigma'_{z(av)} = 465.9 \text{ kN/m}^2$.

$$S_c = \frac{c_s H}{1 + e_0} \log \frac{\sigma'_c}{\sigma'_{0(av)}} + \frac{c_c H}{1 + e_0} \log \frac{\sigma'_{0(av)} + \Delta\sigma'_{(z)av}}{\sigma'_c}$$

$$= \frac{0.05 \times 0.6}{1 + 0.55} \log \frac{72}{2.6} + \frac{0.35 \times 0.6}{1 + 0.55} \log \frac{2.6 + 463.3}{72}$$

$$= 0.137 \text{ m}$$

$$= 13.7 \text{ cm}$$

In layer 2: $\sigma'_c > \sigma'_{0(av)}$, the soil is overconsolidated.
In addition, $\sigma'_c < \sigma'_{0(av)} + \Delta\sigma'_{z(av)} = 196.1 \text{ kN/m}^2$.

$$S_c = \frac{c_s H}{1 + e_0} \log \frac{\sigma'_c}{\sigma'_{0(av)}} + \frac{c_c H}{1 + e_0} \log \frac{\sigma'_{0(av)} + \Delta\sigma'_{(z)av}}{\sigma'_c}$$

$$= \frac{0.05 \times 1.2}{1 + 0.5} \log \frac{115}{10.4} + \frac{0.35 \times 1.2}{1 + 0.5} \log \frac{10.4 + 185.7}{115}$$

$$= 0.106 \text{ m}$$

$$= 10.6 \text{ cm}$$

In layer 3: $\sigma'_c > \sigma'_{0(av)}$, the soil is overconsolidated.

Chapter 3

In addition, $\sigma_c' < \sigma_{0(av)}' + \Delta\sigma_{z(av)}' = 81.3 \ \text{kN/m}^2$.

$$S_c = \frac{c_s H}{1 + e_0} \log \frac{\sigma_{0(av)}' + \Delta\sigma_{(z)av}'}{\sigma_{0(av)}'}$$

$$= \frac{0.05 \times 4.2}{1 + 0.45} \log \frac{33.9 + 47.4}{33.9}$$

$$= 0.055 \ \text{m}$$

$$= 5.5 \ \text{cm}$$

Therefore, total consolidation settlement of the 6-m clay layer is:

$$S_c = 13.7 + 10.6 + 5.5 = 29.8 \ \text{cm}$$

It is noted that the settlement is rather large. Preloading and drainage may be required before construction of the foundation, or the foundation dimension should be increased.

Homework Problems

1. *Bearing capacity problem.* Figure 3.26 shows that a *square* shallow spread footing is designed to support a column. The load on the column, including the column weight, is 650 kN. The foundation rests in homogeneous sandy clay. Subsurface exploration and laboratory testing found that the soil's effective cohesion is 55 kN/m² and the effective friction angle is 15°. The groundwater table is 1.2 m below the ground surface. The bulk unit weight above the groundwater table is 18 kN/m³; the saturated unit weight below the groundwater table is 18.8 kN/m³. Use *Terzaghi's bearing capacity theory*, and assume the general shear failure.
   (1) Determine the foundation embedment and design the dimensions of the spread footing. The factor of safety for bearing capacity is 3.0.
   (2) Choose the foundation dimensions that are acceptable in the construction, and then calculate the allowable bearing capacity on the basis of these dimensions.
2. *Bearing capacity problem.* The problem statement is the same as in Problem 1 and shown in Figure 3.26. Use the *general bearing capacity theory.*

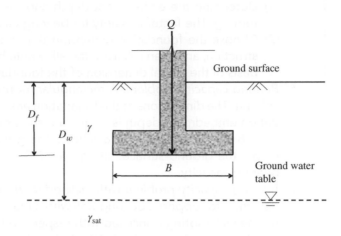

Fig. 3.26   Shallow foundation design sample problem.

(1) Determine the foundation embedment and design the dimension of the spread footing. The factor of safety for bearing capacity is 3.0.

(2) Choose the foundation dimensions that are acceptable in the construction, and then calculate the allowable bearing capacity on the basis of these dimensions.

3. *Bearing capacity problem.* A *circular* foundation is designed to support a water tank. The diameter of the tank is 8 m. The total load from the water tank is approximately 2500 kN. The foundation embedment is 1.0 m. The subsoil is silty sand with $c' = 0$ and $\phi' = 35°$. The groundwater table is far below the foundation. The foundation soil is compacted to 100% of the maximum dry unit weight of 18.8 kN/m³. Determine the factor of safety for the bearing capacity. Any method can be used.

4. *Bearing capacity problem.* A *continuous* footing is designed to support a bearing wall with a load of 300 kN/m. The groundwater table is 1.0 m below the ground surface. The subsoil is silty sand with effective cohesion of 15 kN/m² and effective internal friction angle of 35°. The bulk unit weight above the groundwater table is 17.5 kN/m³; the saturated unit weight below the groundwater table is 18.5 kN/m³. Use *Terzaghi's bearing capacity theory* and assume *local* shear failure.

   (1) Determine the embedment depth and the dimensions of the wall footing. The factor of safety for bearing capacity is 3.0.

   (2) Choose the foundation dimensions that are acceptable in the construction, and then calculate the allowable bearing capacity on the basis of these dimensions.

5. *Bearing capacity problem.* The problem statement is the same as in Problem 4. Use the *general bearing capacity theory.* Assume general failure mode.

(1) Determine the embedment depth and the dimension of the wall footing. The factor of safety for bearing capacity is 3.0.
(2) Choose the foundation dimension that is acceptable in the construction, and then calculate the allowable bearing capacity on the basis of the actual dimension of the foundation.

6. *Bearing capacity problem.* A *rectangular* footing is used to support a column. The dimensions of the foundation are 3 m × 4 m, and the foundation embedment depth is 1.5 m. The soil's strength parameters are $c = 38$ kN/m^2 $\phi = 25°$ $\gamma = 18$ kN/m^3. The groundwater table is 25 m below the ground surface. Using FS = 3.0, determine the allowable bearing capacity.

7. *Bearing capacity problem with inclined load.* A *continuous* footing is designed to support a bearing wall with a load of 300 kN/m. The load on the wall footing is inclined with respect to the vertical direction at 15°. The groundwater table is 10 m below the ground surface. The subsoil is silty sand with effective cohesion of 15 kN/m^2 and effective internal friction angle of 35°. The bulk unit weight above the groundwater table is 17.5 kN/m^3; the saturated unit weight below the groundwater table is 18.5 kN/m^3. Use FS = 3.0. Determine the foundation's embedment depth, dimensions, and use the actual dimensions that are generally accepted in field construction to determine the allowable bearing capacity.

8. *Bearing capacity problem with one-way eccentricity.* A *rectangular* footing is to support a column. The load on the column, including the weight of the foundation, is $Q = 1200$ kN. The foundation is also subjected to an overturning moment of 150 kN·m. The foundation rests in homogeneous silty sand with $c' = 0$ and $\phi' = 35°$. The foundation embedment depth is 1 m, and the dimensions are $L \times B = 2$ m × 1.5 m. The one-way eccentricity is in the L (longer dimension) direction. The groundwater table is at the same level at the bottom of the foundation. The bulk unit weight above the groundwater table is 17 kN/m^3; the saturated unit weight below the groundwater table is 18 kN/m^3. Determine the factor of safety for the bearing capacity.

9. *Bearing capacity problem with one-way eccentricity.* A *continuous* footing is designed to support a bearing wall with a vertical load of 340 kN/m. In addition, the foundation is subjected to an overturning moment of 170 kN·m per unit length (meter) of the wall. The groundwater table is 15 m below the ground surface. The bulk unit weight above the groundwater table is 17.5 kN/m^3; the saturated unit weight below the groundwater table is 18.5 kN/m^3. The subsoil is silty sand with effective cohesion of 25 kN/m^2 and effective internal friction angle of 35°. Use FS of at least 3.0. Determine the foundation's embedment depth, dimensions, and use these dimensions to

determine the allowable bearing capacity. The trial-and-error method may be used to obtain the dimensions.

10. *Concepts of mat foundations.* Select the correct answer(s) in the following multiple-choice questions.
    (1) A mat foundation is:
        A. A type of shallow foundation.
        B. A type of deep foundation.
        C. Neither a shallow nor a deep foundation.
        D. A footing supporting an array of columns in several rows in each direction.
        E. Also known as a raft foundation.
        F. Required to cover at least 75% of the total footprint of the structure.
    (2) The following conditions may warrant the consideration of a mat foundation:
        A. The superstructure load is high or the subsoil is erratic or weak.
        B. Excessive differential settlement is likely to occur.
        C. The foundation elevation is at the groundwater fluctuation zone.
        D. The structure is subjected to nonuniform lateral loads.
        E. A shallow firm stratum is present in the subsoil.

11. *Vertical stress increase problem.* As shown in Figure 3.27, a *continuous* footing is designed to support a bearing wall with load of 300 kN/m. The groundwater table is far below the ground surface. The subsoil is homogeneous silty sand with bulk unit weight of 17.5 kN/m^3. The width of the wall foundation is $B = 1$ m, the foundation embedment depth is $D_f = 0.5$ m. Use the following methods to determine the vertical stress increases at $z = 0$, $z = 2.5B$, and $z = 5B$, and calculate the average vertical stress increase in the soil layer below the center of the footing from $z = 0$ to $5B$. Note, z starts from the bottom of the foundation.
    (1) Boussinesq method.
    (2) Poulos and Davis method.
    (3) 2:1 method.

12. *Vertical stress increase problem.* As shown in Figure 3.27, a *square* spread footing is designed to support a column. The load on the column, including the foundation weight, is 600 kN. The dimensions of the foundation are $B \times B = 1.2$ m $\times 1.2$ m, and the foundation embedment is $D_f = 0.6$ m. The subsoil is homogeneous silty sand with bulk unit weight of 18 kN/m^3. Use the following methods to determine the vertical stress increases at $z = 0$, $z = 2.5B$, and $z = 5B$, and calculate the average vertical stress increase in the soil layer below the center

of the footing from $z = 0$ to $5B$. Note that z starts from the bottom of the foundation.
(1) Boussinesq method.
(2) Poulos and Davis method.
(3) 2:1 method.

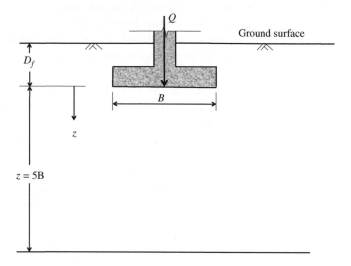

Fig. 3.27   Illustration of shallow foundation with subsoil stratum.

13. *Vertical stress increase problem.* The problem statement is the same as in Problem 12, except that the foundation is *rectangular* with $L \times B$ $= 1.5\,\text{m} \times 1.2\,\text{m}$. Use the following methods to determine the vertical stress increases at $z = 0$, $z = 2.5B$, and $z = 5B$, and calculate the average vertical stress increase in the soil layer from $z = 0$ to $5B$. Note that z starts from the bottom of the foundation (Figure 3.27).
    (1) Boussinesq method.
    (2) Poulos and Davis method.
    (3) 2:1 method.

14. *Vertical stress increase problem.* The problem statement is the same as in Problem 12, except that the foundation is *circular* with diameter $B=1.5\,\text{m}$. Use Poulos and Davis method to determine the depth at which the vertical stress increase is 10% of the *in situ* effective stress.

15. *Elastic settlement problem.* As shown in Figure 3.27, a *square* spread footing is designed to support a column. The load on the column, including the foundation weight, is 600 kN. The dimensions of the foundation are $B \times B = 1.2\,\text{m} \times 1.2\,\text{m}$, and the foundation embedment depth is $D_f = 0.6\text{m}$. The subsoil is heterogeneous silty sand with bulk unit weight of 18 kN/m³. The soil's Poisson's ratio is 0.5. The soil's elastic modulus increases linearly with depth: $E_s = 10{,}000\ (\text{kN/m}^2)$

+ 40 $(kN/m^2/m) \times z$, and z starts from the bottom of the foundation. The thickness of the foundation slab is 0.3m, the elastic modulus of the foundation is 1.2×10^7 kN/m². Determine the elastic settlement of the soil layer from $z = 0$ to 5B, using the Mayne and Poulos method.

16. *Elastic settlement problem.* A *circular* spread footing is designed to support a water tank. The load of the tank is 1800 kN. The diameter of the foundation is 6 m, the foundation embedment depth is $D_f = 1.5$ m. The subsoil is homogeneous silty sand with bulk unit weight of 18 kN/m³. The soil's Poisson's ratio is 0.4. The soil's elastic modulus increases linearly with depth: E_s = 30,000 (kN/m²) + 300 $(kN/m^2/m) \times z$, and z starts from the bottom of the foundation. The thickness of the foundation slab is 1 m; the elastic modulus of the foundation is 1.5×10^7 kN/m². Bedrock is at 10 m below the ground surface. Determine the elastic settlement of the soil layer beneath the foundation, using the Mayne and Poulos method.

17. *Consolidation settlement problem.* As shown in Figure 3.28, a *square* shallow foundation is built in a clayey soil. The saturated clay layer is 4 m thick. The preconsolidation stress (σ_c') is determined to be 70 kN/m². Other parameters are given in the figure. Determine the primary consolidation settlement in the clay layer beneath the foundation.

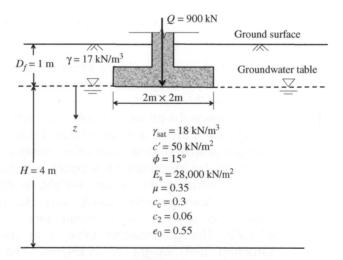

Fig. 3.28   Diagram for problem 17.

18. *Consolidation settlement problem.* A *rectangular* shallow foundation and the subsoil condition are shown in Figure 3.29. The foundation

rests on a saturated clayey soil. Above the groundwater table (GWT): $\gamma = 18 \text{ kN/m}^3$; below the GWT: $\gamma_{\text{sat}} = 19 \text{ kN/m}^3$.

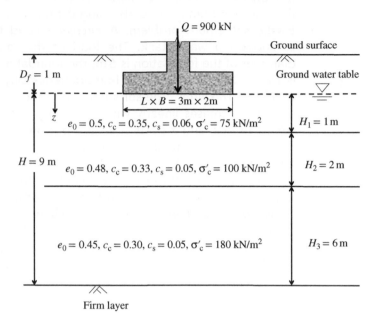

$Q = 900 \text{ kN}$

Ground surface

$D_f = 1 \text{ m}$

Ground water table

$L \times B = 3\text{m} \times 2\text{m}$

z

$e_0 = 0.5, c_c = 0.35, c_s = 0.06, \sigma'_c = 75 \text{ kN/m}^2$

$H_1 = 1 \text{ m}$

$H = 9 \text{ m}$

$e_0 = 0.48, c_c = 0.33, c_s = 0.05, \sigma'_c = 100 \text{ kN/m}^2$

$H_2 = 2 \text{ m}$

$e_0 = 0.45, c_c = 0.30, c_s = 0.05, \sigma'_c = 180 \text{ kN/m}^2$

$H_3 = 6 \text{ m}$

Firm layer

Fig. 3.29   Diagram for problem 18.

(1) Divide the soil layer into sublayers based on Figure 3.24; then calculate the average vertical stress increase in each of the sublayers. Use the 2:1 method.
(2) Determine the primary consolidation settlement in each of the sublayers and then the *entire* saturated clay layer.

19. *Comprehensive shallow foundation design.* A *square* spread footing is to be designed to support a column. The load on the column, including the column weight, is 800 kN. The foundation rests in homogeneous sandy clay. Geotechnical investigation found the soil's strength parameters are $c' = 38 \text{ kN/m}^2$ and $\phi' = 25°$. The groundwater table is at the ground surface. The saturated unit weight is 19 kN/m^3. The soil's Poisson's ratio is 0.5. The soil's elastic modulus increases linearly with depth: $E_s = 4,000(\text{kN/m}^2) + 230(\text{kN/m}^2/\text{m}) \times z$, and z starts from the bottom of the foundation. The thickness of the foundation slab is chosen to be 0.5m, the elastic modulus of the foundation is $1.2 \times 10^7 \text{kN/m}^2$. The initial void ratio, compression index, swell index, and preconsolidation pressure are assumed to be consistent throughout the depth. Their values are $e_0 = 0.45$, $c_c = 0.33$, $c_s = 0.05$, $\sigma_c' = 100 \text{ kN/m}^2$.

Geotechnical investigation also found the homogeneous sandy clay extends to significant depth. Design the shallow foundation by performing the following tasks:

(1) Determine the foundation embedment depth. Use the general bearing capacity theory to determine foundation dimensions to satisfy a minimum factor of safety of 3.0.

(2) Calculate the allowable bearing capacity based on the actual dimensions of the foundation that are generally accepted in field construction.

(3) Determine the depth at which the vertical stress increase is 10% of the *in situ* effective stress. Any method can be used to calculate the vertical stress increase.

(4) Determine the elastic settlement of the soil layer until the depth where vertical stress increase is 10% of the *in situ* effective stress. Use the Mayne and Poulos method.

(5) Determine primary consolidation settlement of the saturated clay layer until the depth where vertical stress increase is 10% of the *in situ* effective stress. When calculating the settlement, first divide the soil layer into sublayers based on Figure 3.24; then calculate the average vertical stress increase in each of the sublayers; then primary consolidation settlement of each layer can be calculated. If any parameter is needed but not provided in the problem statement, make appropriate assumptions and explicitly state them.

20. *Comprehensive mat foundation design.* A mat foundation is to be designed. The dimensions of the mat are 30 m by 30 m. The stress caused by the superstructure and the mat at the bottom of the foundation is 480 kN/m². The embedment depth of the mat foundation is determined to be 10 m below the ground surface, and the thickness of the mat is 3.0 m. The groundwater table (GWT) is at the bottom of the foundation. The subsoil is homogeneous clayey soil with the following parameters.

Soil strength parameters: $c' = 75$ kN/m², $\phi' = 25°$.

Above GWT : $\gamma = 17.5$ kN/m³; below GWT : $\gamma_{sat} = 18.5$ kN/m³.

Poisson's ratio $\mu = 0.4$.

The soil's elastic modulus increases linearly with depth: E_s(kN/m²) = 15,000 (kN/m²) + 750 (kN/m²/m) × z and z starts from the bottom of the foundation.

The elastic modulus of the foundation is $E_f = 1.25 \times 10^7$ kN/m².

Compressibility parameters: $e_0 = 0.55, c_c = 0.3, c_s = 0.06, \sigma'_c = 100$ kN/m².

Geotechnical investigation also found the homogeneous sandy clay extends to significant depth. Design the mat foundation by performing the following tasks:

Chapter 3

(1) Calculate the factor of safety for bearing capacity.
(2) Determine the depth at which the vertical stress increase is 10% of the *in situ* effective stress. Any method can be used to calculate the vertical stress increase.
(3) Determine the elastic settlement of the soil layer until the depth where the vertical stress increase is 10% of the *in situ* effective stress. Use the Mayne and Poulos method.
(4) Determine primary consolidation settlement of the saturated clay layer until the depth where the vertical stress increase is 10% of the *in situ* effective stress. To do so, first divide the soil layer into sublayers based on Figure 3.24; then calculate the average vertical stress increase in each of the sublayers; then primary consolidation settlement of each layer can then be calculated.

If any parameter is needed but not provided in the problem statement, make appropriate assumptions and explicitly state them.

References

American Concrete Institute (ACI), Committee 336 (Footings, Mats, and Drilled Piers). (2002) ACI 336.2R-88: Suggested Analysis and Design Procedures for Combined Footings and Mats (Reapproved 2002). *ACI,* Farmington Hills, MI.

Boussinesq, J. (1883). *Application des Potentials á L'tude de L'Équilibre et du Mouvement des Solides Élastiques,* Gauthier-Villars, Paris.

Bowles, J.E. (1996). *Foundation Analysis and Design,* 5th ed. The McGraw-Hill Companies, Inc., New York, NY.

Coduto, D.P. (2001). *Foundation Design, Principles and Practices.* 2nd edition. Prentice Hall, Upper Saddle River, NJ.

Das, B.M. (2011). *Principles of Foundation Engineering,* 7th Edition, Cengage Learning, Stamford, CT, USA.

DeBeer, E.E. (1970). "Experimental Determination of the shape factors and bearing capacity factors of sand." *Geotechnique,* Vol. 20, No. 4, pp. 387–411.

Hanna, A.M., and Meyerhof, G.G. (1981). "Experimental evaluation of bearing capacity of footings subjected to inclined loads." *Canadian Geotechnical Journal,* Vol. 18, No. 4, pp. 599–603.

Hansen, J.B. (1970). *A Revised and Extended Formula for Bearing Capacity.* Bulletin 28, Danish Geotechnical Institute, Copenhagen.

Kumbhojkar, A.S. (1993). "Numerical evaluation of Terzaghi's N_γ." *ASCE Journal of Geotechnical Engineering,* Vol. 119, No. 3, pp. 598–607.

Mayne, P.W., and Poulos, H.G. (1999). "Approximate displacement influence factors for elastic shallow foundations." *ASCE Journal of Geotechnical and Geoenvironmental Engineering,* Vol. 125, No. 6, pp.453–460.

Meyerhof, G.G. (1953). The bearing capacity of foundations under eccentric and inclined loads. *Proceedings, Third International Conference on Soil Mechanics and Foundation Engineering,* Zurich, Vol. 1, pp. 440–445.

Meyerhof, G.G. (1963). "Some recent research on the bearing capacity of foundations." *Canadian Geotechnical Journal,* Vol. 1, No. 1, pp. 16–26.

Mesri, G., and Ajlouni, M. (2007) "Engineering properties of fibrous peats." *ASCE Journal of Geotechnical and Geoenvironmental Engineering,* Vol. 133, No. 7, pp 850–866.

Naval Facilities Engineering Command (NAVFAC) (1986). *Naval Facilities Engineering Command Design Manual 7.01 Soil Mechanics Design Manual,* NAVFAC DM-7.01, Naval Facilities Engineering Command, 200 Stovall Street, Alexandria, Virginia 22322. September 1986.

Newmark, N.M. (1935). *Simplified Computation of Vertical Pressures in Elastic Foundations,* Engineering Experiment Station Circular No. 24, University of Illinois, Urbana.

Prandtl, L. (1921). "Über die Eindringungsfestigkeit (Härte) plastischer Baustoffe und die Festigkeit von Schneiden." (On the penetration strengths (hardness) of plastic construction materials and the strength of cutting edges)," *Zeitschrift für angewandte Mathematik und Mechanik*, Vol. 1, No. 1, pp. 15–20.

Poulos, H.G., and Davis, E.H. (1974). *Elastic Solutions for Soil and Rock Mechanics*, John Wiley, New York.

Reissner, H. (1924). "Zum Erddruckproblem." (Concerning the earth-pressure problem), *Proceedings, First International Congress of Applied Mechanics*, Delft, pp. 295–311.

Salgado, R. (2008). *The Engineering of Foundations*, 1st Edition The McGraw-Hill Companies, Inc., New York, NY.

Terzaghi, K. (1943). *Theoretical Soil Mechanics*, John Wiley & Sons, New York.

Vesic, A. S. (1973). "Analysis of ultimate loads of shallow foundations." *ASCE Journal of the Soil Mechanics and Foundations Division*, Vol. 99, No. SM1, pp. 45–73.

Vesic, A.S. (1975). "Bearing capacity of shallow foundations." *Foundation Engineering Handbook*. First Edition, pp. 121–147. Winterkorn, H.F., and Fang, H.-Y., edit. Van Nostrand Reinhold, New York.

Chapter 3

Chapter 4

Introduction to Deep Foundation Design

4.1 Introduction to deep foundations

4.1.1 Needs for deep foundation

The selection of pile types depends on the subsoil conditions, loading requirements, and the performance requirements. Vesic (1977) summarized the typical situations where piles may be needed, as shown in Figure 4.1.

4.1.2 Foundation types

Deep foundations can be categorized on the basis of construction materials, foundation shapes, installation methods, and load transfer mechanisms.

Construction materials: concrete, steel, timber, and composite piles. An example of composite piles is a timber pile with a precast, concrete tapered pile tip (TPT).
Shapes: H-pile, hollow pile, and pipe pile; all are of steel piles.
Installation methods: precast, cast in place, driven pile, drilled/bored pile.
Load transfer mechanisms: friction pile, toe bearing pile, and combination.

Deep foundations can be broadly categorized into two large groups: driven piles and drilled piles. Driven piles are slender piles that are driven into the subsoil using various types of hammers; driven piles displace soils and significantly affect the soil condition around the piles. Drilled piles, also known as cast-in-drilled-hole (CIDH) piles, usually have large diameters (>1.0 m) and are cast with concrete in predrilled holes in the ground.

4.1.3 Driven pile foundation design and construction process

The design and construction of a driven pile foundation is a complex process. The following design and construction steps are based on the "Design and Construction of Driven Pile Foundation" (Hannigan et al. 1998) (Figure 4.2a,b).

Geotechnical Engineering Design, First Edition. Ming Xiao.
© 2015 John Wiley & Sons, Ltd. Published 2015 by John Wiley & Sons, Ltd.
Companion Website: www.wiley.com/go/Xiao

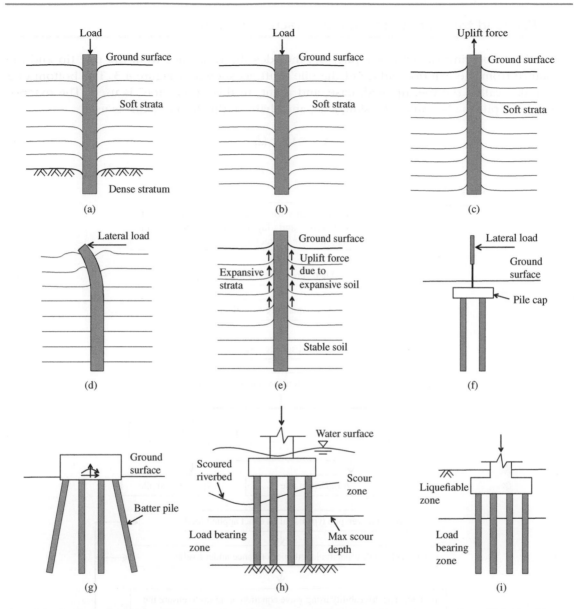

Fig. 4.1 Typical situations where deep foundations are needed. (a) Soft stratum on dense stratum, (b) soft stratum extends to great depth with no underlying dense stratum, (c) foundation is subjected to uplift force because of fluctuating groundwater table and freeze-thaw of pore water, (d) foundation is subjected to horizontal load, (e) the topsoil stratum is expansive or collapsible soil, (f) foundation for highway signs and sound walls that are subjected to horizontal loads such as wind and earthquake, (g) foundation is subjected to large vertical and horizontal loads, (h) bridge foundation at scour zone, (i) the topsoil stratum is susceptible to liquefaction. (Modified after Vesic 1977 and Hannigan et al. 1998.)

4.2   Pile load transfer mechanisms and factor of safety

The ultimate bearing capacity of a pile, Q_u, includes the resistance from the pile tip and the resistance from the exterior surface of the pile shaft, as shown in Figure 4.3. The bottom of a pile is also called the toe, tip, end, base, and point. In this book, "toe" is used. The exterior surface of a pile is also called the skin, side, and shaft. In this book, "skin" is used.

$$Q_u = Q_s + Q_t \tag{4.1}$$

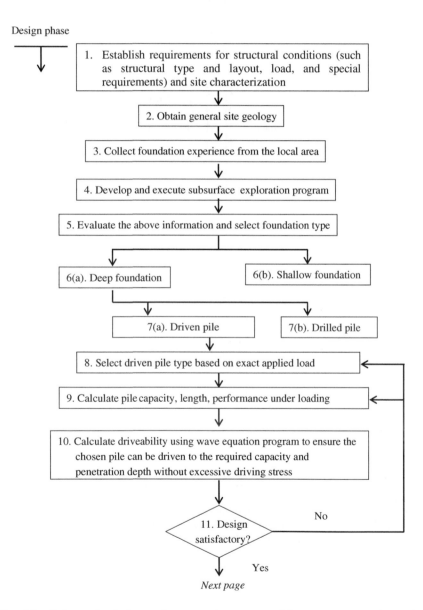

Fig. 4.2   Driven pile design and construction process (after Hannigan et al. 1998).

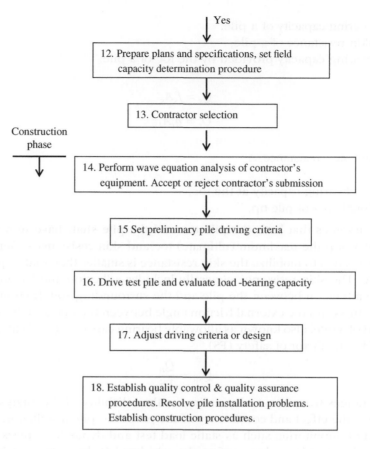

Yes

12. Prepare plans and specifications, set field capacity determination procedure

13. Contractor selection

Construction phase

14. Perform wave equation analysis of contractor's equipment. Accept or reject contractor's submission

15 Set preliminary pile driving criteria

16. Drive test pile and evaluate load-bearing capacity

17. Adjust driving criteria or design

18. Establish quality control & quality assurance procedures. Resolve pile installation problems. Establish construction procedures.

Fig. 4.2 (*continued*)

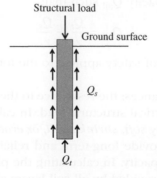

Structural load

Ground surface

Q_s

Q_t

Fig. 4.3 Pile load transfer mechanism.

where

Q_u = ultimate bearing capacity of a pile,
Q_s = ultimate skin resistance of a pile,
Q_t = ultimate bearing capacity provided by the toe of a pile.

$$Q_s = f_s A_s \tag{4.2}$$

$$Q_t = q_t A_t \tag{4.3}$$

where

f_s = ultimate unit skin resistance,
A_s = skin area,
q_t = ultimate unit bearing capacity at the toe,
A_t = cross sectional area of pile tip.

Equation (4.1) assumes that both the pile toe and the pile shaft have moved sufficiently to *simultaneously* develop the maximum (ultimate) toe and skin resistances. Generally, however, the displacement needed to mobilize the skin resistance is smaller than that required to mobilize the toe resistance. The skin resistance generally includes adhesion and friction. Adhesion (c_a) is because of the attraction between the pile and the surrounding soil; friction depends on the vertical effective stress and the external friction angle between the pile and the surrounding soil.

The design load of a pile, also known as allowable bearing capacity, Q_{all}, is the ultimate bearing capacity divided by the factor of safety (FS).

$$Q_{all} = \frac{Q_u}{FS} \tag{4.4}$$

The FS typically ranges from 2 to 4, depending on the reliability of the analysis method, input design parameters, the effect and consistency of the proposed pile installation method, and the level of construction monitoring such as static load test and dynamic analysis. FS = 2.0 can be used if a static loading test is used to confirm the calculated Q_u by a static analysis method. The typical value of FS is 3.0.

An alternative approach that uses partial factors of safety can also be employed for the calculation of the allowable bearing capacity Q_{all} such that:

$$Q_{all} = \frac{Q_t}{\gamma_t} + \frac{Q_s}{\gamma_s} \tag{4.5}$$

where γ_t and γ_s are partial factors of safety applied to the toe resistance Q_t and skin resistance Q_s, respectively.

There are two types of pile resistances: the resistance to the pile driving during the pile installation and the resistance to the vertical structural load. In calculating the load bearing capacity of a pile, the resistance provided by *soft, shrink-swell, or erodible* soil layers should not be considered, as these layers do not provide long-term and reliable resistances, and an FS is used to obtain the allowable bearing capacity. In calculating the pile driving resistance, both the toe and the skin resistances that are provided by all soil layers should be considered, and an FS is not used. The resistance to pile driving will assist contractors in selecting adequate pile driving equipment. This concept is illustrated in Figure 4.4 and Equations (4.6)–(4.8).

For pile driving, all resistances are considered:

$$Q_u = Q_{s1} + Q_{s2} + Q_{s3} + Q_t \tag{4.6}$$

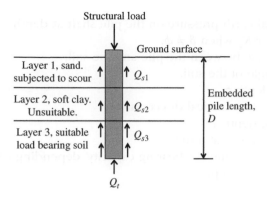

Fig. 4.4   Pile resistances for pile driving and load bearing. (After Hannigan et al. 1998.)

For load bearing, only resistances in suitable load bearing strata are considered:

$$Q_u = Q_{s3} + Q_t \tag{4.7}$$

$$Q_{\text{all}} = \frac{Q_{s3} + Q_t}{FS} \tag{4.8}$$

4.3   Static bearing capacity of a single pile

There are many methods available to determine the static bearing capacity of a driven pile. Some are based on field testing such as the SPT and CPT; some are semiempirical methods that are based on both theories and empirical data. In this book, three semiempirical analysis methods are introduced, as well as other approaches within the framework of limit state design as advocated in the structural Eurocodes (e.g., EN 1997–1:2004). Designers should fully understand the applications and limitations of each method.

4.3.1   Nordlund method, for cohesionless soil

The Nordlund method (Nordlund 1963) is based on field observations and considers tapered pile shape and its soil displacement in calculating the shaft resistance. The method applies only to cohesionless soil ($c = 0$). It follows the results of several pile load test programs including timber piles, H piles, closed-end pipe piles, and tapered piles. These piles had pile widths from 250 to 500 mm (10 to 20 inches). The Nordlund method tends to overpredict pile capacity for piles with widths larger than 600 mm (24 inch).

The Nordlund method for computing the ultimate bearing capacity of a driven pile is:

$$Q_u = Q_s + Q_t = \sum_{Z=0}^{Z=L} \left[K_Z C_K \sigma'_Z \frac{\sin(\delta + \omega)}{\cos \omega} I_Z \Delta Z \right] + \alpha_t N_q A_t \sigma'_t \tag{4.9}$$

where:

z = depth, starting from the ground surface,
L = length of pile,

K_z = coefficient of lateral earth pressure on the pile shaft at depth z,
C_K = correction factor for K_z when $\delta \neq \phi$,
δ = external friction angle between the pile and the soil,
ϕ = internal friction angle of the soil,
σ'_z = effective stress at depth z,
ω = angle of pile taper from vertical direction,
l_z = perimeter of pile at depth z,
Δz = length of pile in each soil stratum,
α_t = dimensionless factor for the toe bearing capacity, depending on the pile's
      depth–width relationship,
N_q = bearing capacity factor,
A_t = cross-sectional area at the pile toe,
σ'_t = effective stress at the pile toe, and the limiting value of σ'_t is 150 kPa.

For uniform pile diameter (no tapering), $\omega = 0$. So Equation (4.9) becomes:

$$Q_u = Q_s + Q_t = \sum_{Z=0}^{Z=L}[(K_Z C_K \sigma'_Z \sin\delta)l\Delta z] + \alpha_t N_q A_t \sigma'_t \qquad (4.10)$$

Equation (4.9) indicates the unit skin resistance is

$$f_s = K_z C_K \sigma'_Z \sin\delta \qquad (4.11)$$

and the unit toe bearing capacity is:

$$q_t = \alpha_t N_q \sigma'_t \qquad (4.12)$$

K_z depends on ω, ϕ, and the displaced soil volume per unit pile length, V. For piles with uniform diameter ($\omega = 0$), K_z can be determined using Figure 4.5. For tapered pile ($\omega \neq 0$), K_z

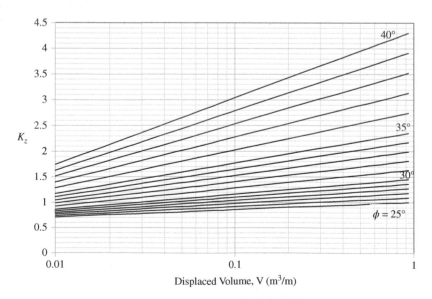

Fig. 4.5 Coefficient of lateral earth pressure, K_z, for $\omega = 0$. (Graph is based on the data in Hannigan et al. (1998).)

Table 4.1 Trend lines for coefficient of lateral earth pressure K_z versus displaced volume V.

ϕ	Trend line equation	R^2
25°	$K_z = 0.150 \log V + 1.005$	0.9987
26°	$K_z = 0.180 \log V + 1.095$	0.9994
27°	$K_z = 0.210 \log V + 1.187$	0.9997
28°	$K_z = 0.241 \log V + 1.278$	0.9996
29°	$K_z = 0.270 \log V + 1.368$	0.9999
30°	$K_z = 0.300 \log V + 1.459$	0.9999
31°	$K_z = 0.360 \log V + 1.641$	0.9998
32°	$K_z = 0.420 \log V + 1.822$	0.9999
33°	$K_z = 0.481 \log V + 2.006$	0.9999
34°	$K_z = 0.541 \log V + 2.190$	0.9999
35°	$K_z = 0.600 \log V + 2.369$	1.0000
36°	$K_z = 0.741 \log V + 2.766$	1.0000
37°	$K_z = 0.880 \log V + 3.156$	1.0000
38°	$K_z = 1.019 \log V + 3.550$	1.0000
39°	$K_z = 1.161 \log V + 3.947$	1.0000
40°	$K_z = 1.300 \log V + 4.340$	1.0000

On the basis of the data in Figure 4.5

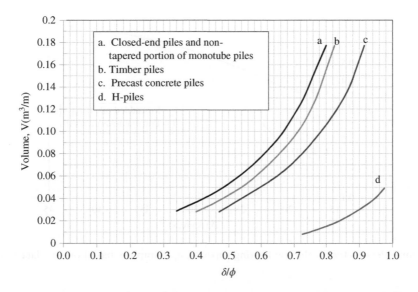

a. Closed-end piles and non-tapered portion of monotube piles
b. Timber piles
c. Precast concrete piles
d. H-piles

Fig. 4.6 Relationship of δ/ϕ with pile displacement volume, V, for various types of piles. (After Nordlund 1979 and Hannigan et al. 1998.)

can be determined by referring to Nordlund (1979) and Hannigan et al. (1998). The trend line expressions of Figure 4.5 are listed in Table 4.1

The external friction angle, δ, can be obtained using Figure 4.6, given the pile type, pile displacement per unit length of pile, V, and the internal friction angle of soil, ϕ. Four types of piles with uniform pile diameters are shown in the figure.

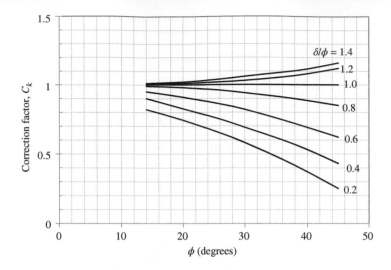

Fig. 4.7 Correction factor C_F. (After Nordlund 1979 and Hannigan et al. 1998.)

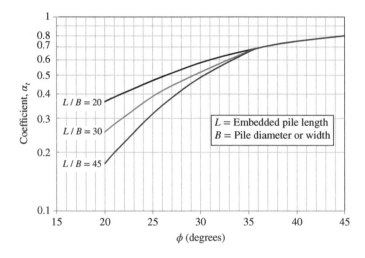

Fig. 4.8 Dimensionless factor for the toe bearing capacity, α_t. (Graph is based on the data in Hannigan et al. (1998).)

The correction factor, C_K, can be obtained from Figure 4.7, using ϕ and δ/ϕ.

The dimensionless factor for the toe bearing capacity, α_t, depends on the soil's internal friction angle at the toe and the length to width (diameter) ratio. It can be approximated using Figure 4.8.

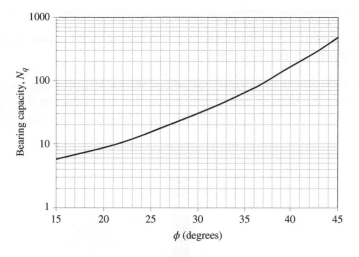

Fig. 4.9   Bearing capacity factor, N_q. (The graph is based on the data in Hannigan et al. (1998).)

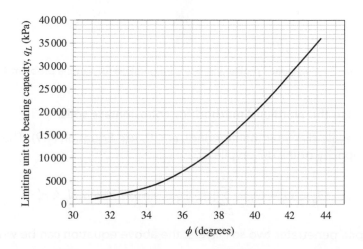

Fig. 4.10   Limiting unit toe bearing capacity. (After Meyerhof 1976 and Hannigan et al. 1998.)

The bearing capacity factor, N_q, depends on the soil's internal friction angle at the toe. It can be approximated using Figure 4.9.

The limiting value for the unit toe bearing capacity (q_t) is:

$$q_t \le q_L \tag{4.13}$$

where the limiting value q_L depends on the soil's internal friction angle at the toe. It can be approximated using Figure 4.10.

Sample Problem 4.1: Ultimate bearing capacity of a pile driven into multiple layers of cohesionless soils.

As shown in Figure 4.11, a concrete pile is driven into the top two layers of subsoil strata. The subsoil profile and properties are shown in the figure. The pile's diameter is 40 cm throughout the pile. Determine the ultimate bearing load of the pile.

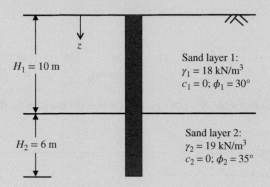

z

$H_1 = 10$ m

Sand layer 1:
$\gamma_1 = 18$ kN/m^3
$c_1 = 0; \phi_1 = 30°$

$H_2 = 6$ m

Sand layer 2:
$\gamma_2 = 19$ kN/m^3
$c_2 = 0; \phi_2 = 35°$

Fig. 4.11   Subsurface profile and pile configuration for sample problem 4.1.

Solution: Use the Nordlund method for cohesionless soils.

For uniform pile diameter (no tapering), $\omega = 0$, use Equation (4.9). The ultimate bearing capacity of the driven pile is:

$$Q_u = Q_s + Q_t = \sum_{Z=0}^{Z=L} [(K_Z C_K \sigma'_Z \sin \delta) l \Delta z] + \alpha_t N_q A_t \sigma'_t$$

As the pile penetrates two soil layers, the above equation can be written as:

$$Q_u = Q_s + Q_t = (K_{z(1)} C_{K(1)} \sigma'_{z(1)} \sin \delta_1) l H_1 + (K_{z(2)} C_{K(2)} \sigma'_{z(2)} \sin \delta_2) l H_2 + \alpha_t N_q A_t \sigma'_t$$

The perimeter of the pile is: $l = \pi B = 3.14 \times 40$ cm $= 125.6$ cm $= 1.256$ m
The cross-sectional area at the pile toe is: $A_t = \frac{1}{4} \pi B^2 = 1256$ cm$^2 = 0.1256$ m^2
The effective stress at the pile toe:

$$\sigma'_t = \gamma_1 H_1 + \gamma_2 H_2 = 18 \times 10 + 19 \times 6 = 294 \text{ kN/m}^2$$

As the limiting value of σ'_t is 150 kPa, choose $\sigma'_t = 150$ kPa
   Table 4.2 is developed to obtain the parameters in the ultimate bearing capacity equation.
   The ultimate skin resistance is:

$$Q_s = (K_{z(1)} C_{z(1)} \sigma'_{z(1)} \sin \delta_1) l H_1 + (K_{z(2)} C_{z(2)} \sigma'_{z(2)} \sin \delta_2) l H_2$$

$$= (1.176 \times 0.94 \times 90 \times \sin 24.6°) \times 1.256 \times 10$$

$$+ (1.803 \times 0.91 \times 237 \times \sin 28.7°) \times 1.256 \times 6$$

$$= 520.1 + 1407.2$$

$$= 1927.3 \text{ kN}$$

The unit toe resistance is:

$$q_t = \alpha_t N_q \sigma'_t = 0.65 \times 79 \times 150 = 7702.5 \text{ kN/m}^2$$

Table 4.2   Parameters in the ultimate bearing capacity equation.

Soil strata	Given parameters	Figures/Tables used	Derived parameters
Layer 1	$\phi_1 = 30°$ Displaced soil volume: $V = 0.114 \text{ m}^3/\text{m}$	Figure 4.5, Table 4.1	$K_{z(1)} = 0.300$ $\log V + 1.459 = 1.176$
	Displaced soil volume: $V = 0.114 \text{ m}^3/\text{m}$, $\phi_1 = 30°$, and precast concrete pile	Figure 4.6, curve (c)	$\delta_1/\phi_1 = 0.82$, so $\delta_1 = 24.6°$,
	$\phi_1 = 30°$, $\delta_1/\phi_1 = 0.82$	Figure 4.7 (use $\delta/\phi = 0.8$ in the chart)	$C_{K(1)} \approx 0.94$
	$\gamma_1 = 18 \text{ kN/m}^3, H_1 = 10 \text{ m}$	N/A	Average $\sigma'_{z(1)} = 18 \times 5$ $= 90 \text{ kN/m}^2$
Layer 2	$\phi_2 = 35°$ Displaced soil volume: $V = 0.114 \text{ m}^3/\text{m}$	Figure 4.5, Table 4.1	$K_{z(2)} = 0.600 \log V +$ $2.369 = 1.803$
	Displaced soil volume: $V = 0.114 \text{ m}^3/\text{m}$, $\phi_2 = 35°$, and precast concrete pile	Figure 4.6, curve (c)	$\delta_2/\phi_2 = 0.82$, so $\delta_2 = 28.7°$,
	$\phi_2 = 35°$, $\delta_2/\phi_2 = 0.82$,	Figure 4.7 (use $\delta_2/\phi_2 = 0.8$ in the chart)	$C_{K(2)} \approx 0.91$
	$\gamma_2 = 19 \text{ kN/m}^3, H_2 = 6 \text{ m}$	N/A	Average $\sigma'_{Z(2)}$ $= 18 \times 10 + 19 \times 3$ $= 237 \text{ kN/m}^2$
	$\phi_2 = 35°$, $L = H_1 + H_2 = 16 \text{ m}$, $B = 0.4 \text{ m}, L/B = 40$	Figure 4.8	$\alpha_t \approx 0.65$
	$\phi_2 = 35°$	Figure 4.9	$N_q \approx 75$

Chapter 4

Using Figure 4.10 and $\phi_2 = 35°$, the limiting value for q_t is:

$$q_L = 5000 \text{ kPa}$$

So, use $q_t = 5000$ kPa.
The ultimate toe resistance is: $Q_t = q_t A_t = 5000 \times 0.1256 = 628$ kN
The total ultimate bearing capacity of the driven concrete pile is:

$$Q_u = Q_s + Q_t = 1927.3 + 628 = \mathbf{2553.3 \text{ kN}}$$

4.3.2 α-method, for undrained cohesive soil

The α-method is a total stress method that uses undrained soil shear strength parameters to calculate the static pile capacity in cohesive soils. It was originally proposed by Tomlinson (1971) and included in adhesion and friction. In undrained saturated clay, it is assumed the internal friction angle of the soil (ϕ) and the external friction angle between the pile and the soil (δ) are zero, so the friction on the skin of the pile is assumed to be zero. The undrained cohesion is expressed as c_u. The unit skin resistance is determined by the adhesion, c_a, which is the attraction between the soil and the pile:

$$f_s = c_a = \alpha s_u \tag{4.14}$$

where $\alpha =$ empirical adhesion factor. Tomlinson (1994) gave an empirical relationship between the pile adhesion (c_a) and the undrained shear strength (s_u). The relationship is also dependent on the pile types and the length to width ratio. The $\alpha \sim s_u$ relationship can be determined using the $c_a \sim s_u$ relationship and is given in Figure 4.12.

Terzaghi et al. (1996) provided a $\alpha \sim c_u$ relationship that is also plotted in Figure 4.12. The relationship is conservative at lower undrained cohesion (e.g., $c_u < 160$ kPa). This relationship did not consider different pile materials and the effect of pile length to width ratio.

Sladen (1992) derived an equation to compute α directly on the basis of the undrained shear strength, s_u, and the effective overburden stress, $\bar{q}$.

$$\alpha = C_1 \left(\frac{\bar{q}}{s_u} \right)^{0.45} \tag{4.15}$$

where: $C_1 = 0.4$ to 0.5 for bored piles, and $C_1 \geq 0.5$ for driven piles.
The unit toe resistance is:

$$q_t = c_u N_c \tag{4.16}$$

where N_c is the bearing capacity factor that depends on the pile diameter and embedment. As $\phi = 0$, N_c usually takes the value of 9 for deep foundations. c_u is the undrained shear strength at the toe of the pile.

It should be noted that the movement required to mobilize the toe resistance is several times greater than that required to mobilize the skin resistance. When the toe resistance is fully mobilized, the skin resistance may have decreased to a residual value. Therefore, the toe resistance

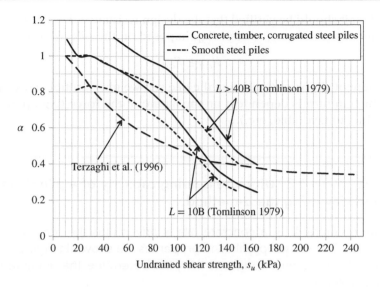

Fig. 4.12 Adhesion factor, α. (Note: D = distance from the ground surface to the bottom of clay layer or pile toe, whichever is less; B = pile diameter.)

contribution to the ultimate pile capacity in cohesive soils is sometimes ignored except in hard cohesive deposits (Hannigan et al. 1998).

The α-method is an example of empirical approach that can be adopted within the context of limit state design while also using partial factors of safety as required in some international standards (e.g. EN 1997–1:2004). Ideally, a set of ground test results should be available (i.e., to estimate values of c_u and from different boreholes). In the case of EN 1997–1:2004 it is recommended to first calculate the characteristic bearing resistance $R_{c;k}$ as the minimum value of:

$$R_{c;k} = \frac{(R_{toe;cal} + R_{skin;cal})_{average}}{\xi_3} \tag{4.17}$$

and

$$R_{c;k} = \frac{(R_{toe;cal} + R_{skin;cal})_{minimum}}{\xi_4} \tag{4.18}$$

where $R_{toe;cal}$ is the calculated toe resistance, $R_{skin;cal}$ is the calculated skin resistance, and ξ_3 and ξ_4 are correlation factors related to the number of ground test results. It is normally specified that the value of the correlation factors reduces as the number of ground test results increases.

Note that the terminology of "resistance" instead of "capacity" basically refers to units of force (kN) instead of pressure (kPa) as suggested in Chapter 3 for shallow foundation design. Once the characteristic resistance $R_{c;k}$ has been estimated, a design value of the bearing resistance $R_{c;d}$ can be calculated as:

$$R_{c;d} = \frac{R_{toe;k}}{\gamma_{toe}} + \frac{R_{skin;k}}{\gamma_{skin}} \tag{4.19}$$

where $R_{toe;k}$ is the characteristic (cautious estimate) of the toe resistance, $R_{skin;k}$ is the characteristic (cautious estimate) of the skin resistance, and γ_{toe} and γ_{skin} are partial factors of safety applied to the toe and skin resistance, respectively.

In a design context it should be verified that the design resistance, $R_{c;d}$, is greater than or equal to the design value of the actions (forces), $F_{c;d}$, imposed on the pile. That is,

$$R_{c;d} \geq F_{c;d} \tag{4.20}$$

where $F_{c;d}$ is the sum of all forces (including the pile self-weight) multiplied by their corresponding partial FS, which in turn is dependent on the nature of the force (e.g., permanent, variable, accidental, favorable, unfavorable, etc.).

Sample Problem 4.2: Ultimate bearing capacity of a pile driven into undrained clay using the α-method.

As shown in Figure 4.13, a concrete pile is driven into the top two layers of subsoil strata. The subsoil profile and properties are shown in Figure 4.13. The pile's diameter is 50 cm throughout the pile. Determine the ultimate bearing capacity of the pile.

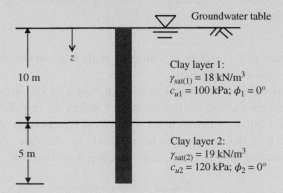

Fig. 4.13   Subsurface profile and pile configuration for sample problem 4.2.

Solution:

As the subsoil is clay and is beneath the groundwater table, use the α-method. Because $\phi = 0$, the undrained shear strength $s_u = c_u$.

The unit skin resistance is:

$$f_s = c_a = \alpha s_u$$

Three methods are used to determine and compare α, as shown in Table 4.3. The α values determined using Tergazhi's method are used.

The perimeter of the pile is: $l = \pi B = 1.57 m$

The cross-sectional area at the pile toe is: $A_t = \frac{1}{4}\pi B^2 = 0.196 m^2$

Layer 1: $f_{s(1)} = \alpha s_u = 0.48 \times 100 = 48 kPa$
Layer 2: $f_{s(2)} = \alpha s_u = 0.42 \times 120 = 50.4 kPa$

Table 4.3   Determination of α.

Soil strata	Methods	Input values	Figure or equation used	α
Layer 1	Tomlinson (1979)	$L/B = 10/0.5 = 20$, $s_u = 100$ kPa. Concrete pile.	Figure 4.12	0.75 (use interpolation)
	Terzaghi et al. (1996)	$s_u = 100$ kPa	Figure 4.12	0.48
	Sladen (1992)	$s_u = 100$ kPa; $C_1 = 0.5$; $\bar{q} = (18 - 9.81) \times 10/2 = 41$ kPa	$\alpha = C_1 \left(\frac{\bar{q}}{s_u}\right)^{0.45}$	0.335
Layer 2	Tomlinson (1979)	$L/B = 5/0.5 = 10$, $s_u = 120$ kPa. Concrete pile.	Figure 4.12	0.47
	Terzaghi et al. (1996)	$s_u = 120$ kPa	Figure 4.12	0.42
	Sladen (1992)	$s_u = 120$ kPa; $C_1 = 0.5$; $\bar{q} = (18 - 9.81) \times 10 + (19 - 9.81) \times 5/2 = 105$ kPa	$\alpha = C_1 \left(\frac{\bar{q}}{s_u}\right)^{0.45}$	0.471

The total skin resistance is:

$$Q_s = f_{s(1)}A_1 + f_{s(2)}A_2$$
$$= 48 \times 1.57 \times 10 + 50.4 \times 1.57 \times 5 = 1149.2\,\text{kN}$$

The unit toe resistance is: $q_t = c_u N_c = 120 \times 9 = 1080\,\text{kPa}$
The total toe resistance is:

$$Q_t = q_t A_t = 1080 \times 0.196 = 212.0\,\text{kN}$$

The total ultimate bearing capacity of the driven concrete pile is:

$$Q_u = Q_s + Q_t = 1149.2 + 212.0 = \mathbf{1361.2\,kN}$$

Alternative solution for limit state verification using partial factors of safety:
The cross-sectional area at the pile toe is: $A_t = \frac{1}{4}\pi B^2 = 0.196\,\text{m}^2$
The characteristic value of the bearing resistance is the minimum value of:

$$R_{c;k} = \frac{(R_{toe;cal} + R_{skin;cal})_{average}}{\xi_3} = \frac{c_u N_c A_t + \alpha s_u \pi Bl}{\xi_3}$$

and

$$R_{c;k} = \frac{(R_{toe;cal} + R_{skin;cal})_{minimum}}{\xi_4} = \frac{c_u N_c A_t + \alpha s_u \pi B l}{\xi_4}$$

Note that as there is only one ground test result, the equations above differ only in terms of the correlation factors. However, if there were more tests available, individual calculations with each of individual values of c_u and s_u would be required (while also being able to use a lower value for the correlation factors ξ_3 and ξ_4). In this case, however, for a single test, ξ_3 and ξ_4 are both equal to 1.4 as suggested in EN-1997-1:2004. Hence, on the basis of the values calculated in the other solution:

$$R_{c;k} = \frac{1361.2 \ kN}{1.4} = 972.3 \ kN$$

and

$$R_{c;d} = \frac{R_{toe;k}}{\gamma_{toe}} + \frac{R_{skin;k}}{\gamma_{skin}} = \frac{212.0}{1.1} + \frac{1149.2}{1.1} = 1237.5 \ kN$$

Although this concludes the sample solution, it is important to emphasize that in a design situation, the verification of the limit state requires the comparison of $R_{c;d}$ against $F_{c;d}$ (e.g., bearing resistance should be higher than the design actions).

4.3.3 β-method, for drained cohesionless and cohesive soils

The β-method is based on effective stress and uses the drained shear strength of the soil. It was proposed by Burland (1973). It can be used to evaluate the long-term bearing capacity of a driven pile. However, cohesion is not considered in the β-method. In this method, the unit skin resistance is:

$$f_s = \beta \cdot \overline{\sigma}' \tag{4.21}$$

where
$\overline{\sigma}'_0$ = average vertical effective stress along the pile shaft,
β = Bjerrum–Burland beta coefficient, and

$$\beta = K_s \tan \delta \tag{4.22}$$

where
K_s = earth pressure coefficient,
δ = external friction angle between the pile and the soil.

Cohesion is not considered in the β-method.

β depends on the soil type and the effective friction angle. Direct evaluation of β can be obtained from Table 4.4 and Figure 4.14 (Fellenius 1991).

Table 4.4   Approximate ranges of β and N_t.

Soil type	ϕ' (degrees)	β	N_t
Clay	25–30	0.23–0.40	3–30
Silt	28–34	0.27–0.50	20–40
Sand	32–40	0.30–0.60	30–150
Gravel	35–45	0.35–0.80	60–300

After Fellenius 1991

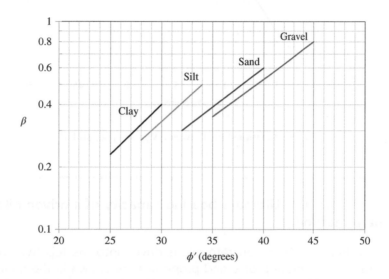

Fig. 4.14   Relationships between β, soil type, and ϕ'.

Bhushan (1982) suggested an equation for β for large-displacement piles (closed-end pipe piles, solid concrete piles).

$$\beta = 0.18 + 0.0065D_r \tag{4.23}$$

where D_r = relative density of surrounding soil.

The unit toe bearing capacity is:

$$q_t = N_t \cdot \sigma_t' \tag{4.24}$$

where

N_t = bearing capacity factor at the toe,
σ_t' = effective overburden stress at the toe.

N_t depends on the soil type and the effective friction angle. The evaluation of N_t can be obtained from Table 4.4 and Figure 4.15 (Fellenius 1991).

Fellenius (1991) recommended the use of a relative low N_t values in clays. In the initial design, conservative values of β and N_t should be used. In the β-method, no limiting values are placed on the skin and the toe resistances.

Provided that the same principles related to the use of partial factors of safety and the calculation of characteristic bearing resistance on the basis of the number of test results used, there is no restriction on whether the β-method can be used within the context of EN 1997–1:2004.

Chapter 4

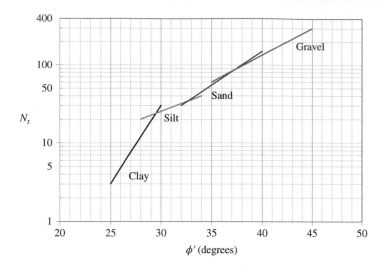

Fig. 4.15 Relationships between N_t, soil type, and ϕ'.

Sample Problem 4.3: Ultimate bearing capacity of a driven pile using the β-method.

As shown in Figure 4.16, a concrete pile is driven into the top two layers of subsoil strata. The subsoil profile and properties are shown in the figure. The pile's diameter is 50 cm throughout the pile. Determine the ultimate bearing capacity of the pile.

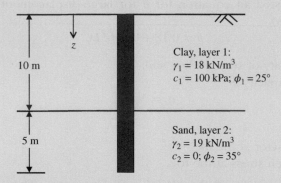

Clay, layer 1:
$\gamma_1 = 18$ kN/m^3
$c_1 = 100$ kPa; $\phi_1 = 25°$

Sand, layer 2:
$\gamma_2 = 19$ kN/m^3
$c_2 = 0$; $\phi_2 = 35°$

10 m

5 m

Fig. 4.16 Subsurface profile and pile configuration for sample problem 4.3.

Solution:

The β -method is used.

The unit skin resistance is: $f_s = \beta \cdot \overline{\sigma}'$

In layer 1 (clay):

$$\overline{\sigma}_0' = 18 \times (10/2) = 90\,kN/m^2.$$

Use Table 4.4, and given clay soil with $\phi = 35°$, select minimum $\beta = 0.23$.

$$f_{s(1)} = \beta \cdot \overline{\sigma}' = 0.23 \times 90 = 20.7\,kN/m^2$$

In layer 2 (sand):

$$\overline{\sigma}_0' = 18 \times 10 + 19 \times (5/2) = 227.5\,kN/m^2.$$

Use Table 4.4, and given sandy soil with $\phi = 25°$, select minimum $\beta = 0.30$.

$$f_{s(2)} = \beta \cdot \overline{\sigma}' = 0.30 \times 227.5 = 68.25\,kN/m^2$$

The perimeter of the pile is: $l = \pi B = 1.57m$
The cross-sectional area at the pile toe is: $A_t = \frac{1}{4}\pi B^2 = 0.196\,m^2$
The ultimate total skin resistance is:

$$Q_s = f_{s(1)}A_1 + f_{s(2)}A_2$$
$$= 20.7 \times 1.57 \times 10 + 68.25 \times 1.57 \times 5 = 860.7\,kN$$

The unit toe bearing capacity is: $q_t = N_t \cdot \sigma_t'$
Using Table 4.4, and given sandy soil with $\phi = 25°$, select minimum $N_t = 30$.
The effective overburden stress at the toe is:

$$\sigma_t' = 18 \times 10 + 19 \times 5 = 275\,kN/m^2$$
$$q_t = N_t \cdot \sigma_t' = 30 \times 275 = 8250\,kN/m^2$$

The ultimate toe bearing capacity is:

$$Q_t = q_t A_t = 8250 \times 0.196 = 1617\,kN$$

The total ultimate bearing capacity of the driven concrete pile is:

$$Q_u = Q_s + Q_t = 860.7 + 1617 = \mathbf{2477.7\,kN}$$

4.3.4   Bearing capacity (resistance) on the basis of the results of static load tests

Using limit state design (e.g., as required for EN 1997–1:2004) several limit states such as the loss of overall stability, structural failure of the pile, combined failure in the ground and the pile, excessive settlement or heave, bearing resistance failure, among others, should be verified. Considering the case of bearing capacity, during design, it should therefore be satisfied that

$$F_{c;d} \leq R_{c;d} \tag{4.25}$$

where $F_{c;d}$ is the design load on a pile and $R_{c;d}$ is the design value of the pile bearing resistance. The fact that these variables are defined as design values implies that they involve the estimation of cautious assessment of the forces (actions), the bearing resistance (characteristic values), and the use of partial factors of safety to multiply the actions and divide the bearing resistance.

The characteristic value of the bearing resistance $(R_{c;k})$ based on static load test in piles is taken from the minimum value of:

$$R_{c;k} = \frac{(R_{c;m})_{average}}{\xi_1} \tag{4.26}$$

and

$$R_{c;k} = \frac{(R_{c;m})_{minimum}}{\xi_2} \tag{4.27}$$

where $(R_{c;m})_{average}$ is the average measured bearing resistance from static load tests, $(R_{c;m})_{minimum}$ is the minimum measured bearing resistance from static load tests, and ξ_1 and ξ_2 are correlation factors related to the number of piles tested. Note that the values of these correlation factors may be changed in design standards according to local practice. As it would be expected, the larger the number of piles tested, the lower their value as they can be related to reliability and confidence obtained by having additional test results.

In certain cases the bearing resistance can be determined from the characteristic values of the toe resistance $R_{t;k}$ and the skin resistance $R_{s;k}$ such that

$$R_{c;k} = R_{t;k} + R_{s;k} \tag{4.28}$$

Hence the design resistance, $R_{c;d}$ may be derived from any of the two following possibilities:

$$R_{c;d} = \frac{R_{c;k}}{\gamma_{total}} \tag{4.29}$$

or

$$R_{c;d} = \frac{R_{t;k}}{\gamma_{toe}} + \frac{R_{s;k}}{\gamma_{skin}} \tag{4.30}$$

where γ_{total}, γ_{toe}, and γ_{skin} are partial factors of safety on the total, toe and skin resistance, respectively.

Sample Problem 4.4: Ultimate bearing capacity of pile from static load tests using partial factors of safety.

Three test piles have been subject to static load test to verify the bearing capacity of a certain soil. The measured axial load for individual tests on each of the piles were 450 kN, 422 kN, and 468 kN. Determine the design bearing resistance assuming that $\xi_1 = 1.2$, $\xi_2 = 1.05$ and $\gamma_{total} = 1.1$

Solution:

The average and minimum measured resistances are estimated as:

$$(R_{c;m})_{average} = \frac{450 + 422 + 468}{3} = 446.7 \ kN$$

$$(R_{c;m})_{minimum} = 422 \ kN$$

Hence,

$$R_{c;k} = \frac{(R_{c;m})_{average}}{\xi_1} = \frac{446.7 \ kN}{1.2} = 372.3 \ kN$$

$$R_{c;k} = \frac{(R_{c;m})_{minimum}}{\xi_2} = \frac{422 \ kN}{1.05} = 401.9 \ kN$$

Therefore, the characteristic bearing resistance $R_{c;k}$ is the minimum of these two values (i.e., $R_{c;k} = 372.3 \ kN$), and the design bearing resistance may be estimated as:

$$R_{c;d} = \frac{R_{c;k}}{\gamma_{total}} = \frac{372.3 \ kN}{1.1} = 338.5 \ kN$$

Within a design context, it must be verified that the design bearing resistance estimated above is greater than the design value of the forces (actions) that need to be supported by the pile. As discussed in the previous chapter, design values of actions are simply the product of characteristic values and partial factors of safety that depend on the nature of the action (i.e., variable, permanent, accidental, etc.).

4.4   Vertical bearing capacity of pile groups

To support a large foundation load and eccentric loading, piles are often installed in groups. A pile cap is constructed over a group of piles to connect the piles together. Because of the close spacing between individual piles in a pile group, the soil zones that are affected by the piles, known as the stress zones, overlap (Figure 4.17). Consequently, the ultimate bearing capacity of a pile group may be different from the sum of the ultimate bearing capacity of individual piles. The efficiency of a pile group is defined as:

$$\eta_g = \frac{Q_{u(g)}}{n Q_u} \tag{4.31}$$

Chapter 4

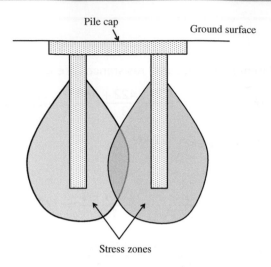

Fig. 4.17   Overlap of stress zones in a pile group.

where

η_g = pile group efficiency,
$Q_{u(g)}$ = ultimate bearing capacity of pile group,
n = number of piles in the pile group,
Q_u = ultimate bearing capacity of an individual pile in the pile group. It can be calculated using the Nordlund method, the α-method, or the β-method, depending on the stress and soil conditions.

In a pile group, the minimum center-to-center spacing of $3b$ ($b =$ individual pile diameter or width) is recommended to optimize the group capacity and minimize installation problems.

When a pile group is in a soft and cohesive soil or in a cohesionless soil that is underlain by a weak, cohesive soil, block shear failure of the pile group can occur: the pile group can be punched into the underlying weak, cohesive soil. The ultimate bearing capacity of a pile group against the block shear failure is:

$$Q_{u(g)} = A_s f_s + A_t q_t \qquad (4.32)$$

where

A_s = skin area of the pile group. Figure 4.18 shows the pile group dimensions.
$A_s = 2(B_1 + B_2)L$, $B_1 =$ length of the pile group base, $B_2 =$ width of the pile group base, $L =$ embedment length of the pile group.
A_t = area of the toe of the pile group, $A_t = B_1 B_2$,
f_s = unit skin resistance along the pile group block,
q_t = unit toe bearing capacity.

It is noted that the soil encompassed in the pile group moves (shears) with the pile group, and the skin resistance along the pile group block is between the soil in the pile group and the soil surrounding the pile group. Therefore, the cohesion and the internal friction angle of the soil should be used to determine f_s.

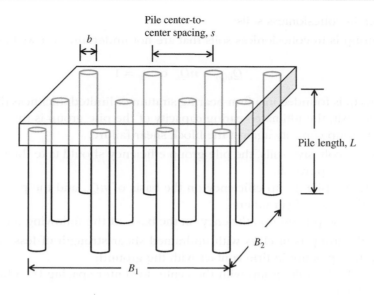

Fig. 4.18   Pile group dimensions.

For undrained clayey soil, $f = 0$, the undrained shear strength, $s_u = c_u, c_u$, is undrained cohesion.

$$f_s = c_{u1} \tag{4.33}$$

$$q_t = c_{u2}N_c \tag{4.34}$$

where

c_{u1} = undrained cohesion of the soil along the pile group shaft,
c_{u2} = undrained cohesion of the soil beneath the pile group toe,
N_c = 9. A limiting value of N_c is given by (Hannigan et al. 1998):

$$N_c = 5\left(1 + \frac{L}{5B_1}\right)\left(1 + \frac{B_1}{5B_2}\right) \leq 9 \tag{4.35}$$

For cohesionless soil, $c = 0$. The β-method can be used to determine the unit skin resistance (f_s) and the unit toe resistance (q_t). As the skin resistance is between the soil in the pile group and the soil surrounding the pile group, the internal friction angle (ϕ) is used to replace the external friction angle (δ) in calculating f_s,.

$$f_s = \beta \cdot \overline{\sigma}' = (K_s \tan \phi)\overline{\sigma}' \tag{4.36}$$

If a pile group penetrates multiple soil layers, the skin resistance should be calculated for each layer, using the appropriate method on the basis of the soil type.

The following design procedures are recommended by Hannigan et al. (1998) in estimating pile group efficiency for driven piles.

(1) For driven piles in cohesionless soils:
- If the pile group is in cohesionless soils that are not underlain by a weak deposit:

$$Q_{u(g)} = nQ_u \text{ or } \eta_g = 1 \tag{4.37}$$

- If the pile group is founded in a firm bearing stratum of limited thickness that is underlain by a weak deposit, the ultimate bearing capacity of the pile group is the smaller value of nQ_u and the group capacity against the block shear failure.

(2) For driven piles in cohesive soils, the pile group efficiency should take the smaller value of the following 2-step approach.
Step 1 Calculate the pile group efficiency on the basis of nQ_u and the pile group capacity against the block shear failure.
Step 2 Determine the pile group efficiency on the basis of the following recommendations.
- If a pile group is in clays with undrained shear strength of less than 95 kPa and the pile cap is not in firm contact with the ground:
  ○ $\eta_g = 0.7$ for pile group with the center-to-center spacing (s) 3 times of the pile diameter (b).
  ○ $\eta_g = 1.0$ for pile group with the center-to-center spacing (s) greater than 6 times of the pile diameter (b).
  ○ For the values of s between $3b$ and $6b$, linear interpolation is used.

$$\eta_g = 0.4 + \frac{0.1}{b}s \tag{4.38}$$

- If a pile group is in clays with undrained shear strength less than 95 kPa and the pile cap is in firm contact with the ground: $\eta_g = 1.0$.
- If a pile group is in clays with undrained shear strength higher than 95 kPa, $\eta_g = 1.0$, regardless of the contact of pile cap with the ground.

Sample Problem 4.5: Ultimate bearing capacity and pile group efficiency of a pile group penetrating two soil strata.

A concrete pile group is driven into multistrata subsoil. The pile group dimensions and the subsoil conditions are shown in Figure 4.19. Determine the ultimate bearing capacity and the pile group efficiency of the pile group.

Solutions:

The pile group is driven into cohesive soils.

Step 1: Calculate nQ_u.
   The ultimate bearing capacity of the individual pile, Q_u, can be determined using the α-method. Each pile's dimensions and the

subsoil conditions are the same as in Sample Problem 4.2. So,

$$Q_u = \mathbf{1361.2\ kN}$$

$$nQ_u = 12 \times 1361.2 = \mathbf{16334.4\ kN}$$

Step 2: Calculate pile group capacity against the block shear failure.
The ultimate bearing capacity of a pile group against the block shear failure is:

$$Q_{u(g)} = \sum (A_s f_s) + A_t q_t$$

Skin area of the pile group in layer 1: $A_s = 2(B_1 + B_2)H_1 = 2 \times (7 + 5) \times 10 = 240\ m^2$.
Unit skin resistance in layer 1: $f_s = c_1 = 100$ kPa.
Skin area of the pile group in layer 2: $A_s = 2(B_1 + B_2)H_2 = 2 \times (7 + 5) \times 5 = 120\ m^2$.
Unit skin resistance in layer 2: $f_s = c_2 = 120$ kPa.
Area of the toe of the pile group: $A_t = B_1 B_2 = 7 \times 5 = 35\ m^2$
Unit toe resistance: $q_t = c_2 N_c$
Limiting bearing capacity factor:

$$N_c = 5\left(1 + \frac{L}{5B_1}\right)\left(1 + \frac{B_1}{5B_2}\right) = 5\left(1 + \frac{15}{5 \times 7}\right)\left(1 + \frac{7}{5 \times 7}\right)$$

$$= 8.93 \leq 9$$

$$q_t = c_2 N_c = 120 \times 8.39 = 1071.6\ kN/m^2$$

$$Q_{u(g)} = \sum (A_s f_s) + A_t q_t$$

$$= 240 \times 100 + 120 \times 120 + 35 \times 1071.6$$

$$= \mathbf{75906\ kN}$$

The pile group efficiency:

$$\eta = \frac{Q_{u(g)}}{nQ_u} = \frac{75906}{16334} = \mathbf{4.65}$$

Step 3: Determine the pile group efficiency on the basis of the following recommendations.

The pile group is in clays with undrained shear strength higher than 95 kPa,
$\eta_g = \mathbf{1.0}$, regardless of the contact of the pile cap with the ground.
The lowest and the most conservative pile group efficiency is chosen from the above calculations. So: $\eta_g = \mathbf{1.0}$.
The ultimate bearing capacity of the pile group is: $nQ_u = \mathbf{16334.4\ kN}$

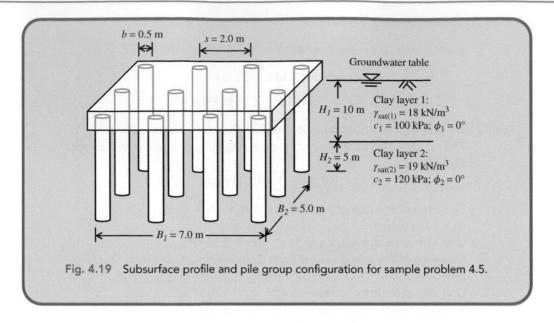

Fig. 4.19   Subsurface profile and pile group configuration for sample problem 4.5.

4.5   Settlement of pile groups

The settlement of a pile group includes the elastic compression of the piles, elastic (immediate) settlement of the soil at the toe of the piles, and consolidation settlement of the soils around and beneath the pile group. Pile groups that are supported in cohesionless soils are subjected only to immediate (elastic) settlement; consolidation settlement is negligible. Pile groups that are supported in cohesive soils are subjected to both immediate (elastic) settlement and consolidation settlement. The settlement of a pile group is much greater than the settlement of a single pile because the soil zone affected by the loading adjacent to and beneath the pile group is much larger, as shown in Figure 4.20.

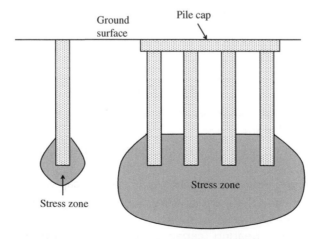

Fig. 4.20   Stress zones in a single pile foundation and a pile group foundation.

4.5.1   Elastic compression of piles

On the basis of Hooke's law, the elastic compressive deformation of a pile itself is:

$$\Delta = \frac{Q_a L}{AE} \tag{4.39}$$

where

Δ = elastic compression of an individual pile,
Q_a = design axial load on an individual pile (kN),
L = pile length,
A = pile cross-sectional area,
E = modulus of elasticity of the pile (MPa). For steel, $E = 207,000$ MPa; for concrete,
  $E = 27800$ MPa.

4.5.2   Empirical equations for pile group settlement using field penetration data.

Meyerhof (1976) gave empirical equations for the settlement of pile groups using standard penetration test (SPT) and cone penetration test (CPT) data.

The settlement of a pile group in granular soils using SPT data is:

$$s_g = \frac{\Delta q \sqrt{B/D}}{2N_{60}D}(mm) \tag{4.40}$$

where

S_g = settlement of a pile group (mm),
Δq = foundation pressure at the toe of the pile group, equal to the total foundation load
  divided by the group area at the toe,
B = width of the pile group; if the pile group's dimensions are B_1 and B_2 as shown in
  Figure 4.15, use the larger dimension as B,
D = individual pile diameter or width,
N_{60} = average SPT blow count in the zone from B to $2B$ below the pile toe.

The settlement of a pile group using CPT data is:

$$S_g = \frac{k_1 B \Delta q}{2q_c} \tag{4.41}$$

where:

$$k_1 = 1 - \frac{L}{8B} \geq 0.5 \tag{4.42}$$

q_c = average static cone tip resistance within the depth of B below the pile toe.

4.5.3   Consolidation settlement of a pile group in saturated cohesive soil

Terzaghi and Peck (1967) proposed that the consolidation settlement of a pile group in saturated clay could be evaluated using an equivalent footing situated at a depth of $L/3$ above the pile toe. It is assumed that the soil in the top two-third of the pile length is not subjected to consolidation settlement. It is also assumed that the vertical stress increase from $L/3$ above the

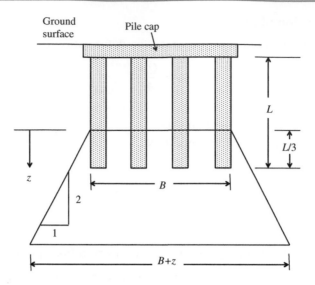

Fig. 4.21   2:1 method in the calculation of consolidation settlement of pile group.

pile toe follows the 2 : 1 method that is described in Chapter 3 (Figure 3.16). The illustration is shown in Figure 4.21.

The consolidation calculation approach follows the same procedure as described in Chapter 3, such as Equations (3.59)–(3.61). Two important notes should be followed when calculating consolidation settlement.

1. The consolidation settlement is caused by vertical stress increase. When the vertical stress increase, $\Delta\sigma_z'$, at a certain depth is less than one-tenth of the in-situ effective stress σ_0', it is assumed the consolidation settlement is negligible. Therefore, consolidation settlement should be considered for a depth that is either $\Delta\sigma_z' < 0.1\sigma_0'$, or there is an underlying firm layer whose compressibility is negligible.
2. A thick soil layer should be divided into a number of thin layers; the consolidation settlement for each thin layer should be calculated; and the individual settlements should be summed. Approximate thicknesses of soil layers for manual computation of consolidation settlement of shallow foundations are shown in Figure 3.24.

Sample Problem 4.6: Settlement of a pile group in saturated clay.

The problem statement is the same as in the Sample Problem 4.5. The pile group configuration and subsoil condition are shown in Figure 4.19. The pile group is subjected to a foundation load of 5425 kN. Assume the clay layer 2 (in Figure 4.19) extends to significant depth. Soil samples were retrieved at depths of 5, 10, 12, 18 m. Laboratory testing on the soil samples yielded the following results:

At 5 m : $e_0 = 0.51, c_c = 0.32, c_s = 0.09, \sigma'_c = 45.5\,\text{kN/m}^2$.

At 10 m : $e_0 = 0.53, c_c = 0.32, c_s = 0.09, \sigma'_c = 85.5\,\text{kN/m}^2$.

At 12 m : $e_0 = 0.53, c_c = 0.32, c_s = 0.09, \sigma'_c = 101.1\,\text{kN/m}^2$.

At 18 m : $e_0 = 0.53, c_c = 0.32, c_s = 0.09, \sigma'_c = 175.0\,\text{kN/m}^2$.

Determine the total settlement of the pile cap.

Solution:

Step 1: Elastic compression of the piles

$$\Delta = \frac{Q_a L}{AE}$$

Design axial load on individual pile: $Q_a = \dfrac{5425\,\text{kN}}{12} = 452.1\,\text{kN}$

Pile length $L = 15$ m

Individual pile cross-sectional area $A = \frac{1}{4}\pi b^2 = 0.196\,\text{m}^2$

Modulus of elasticity of concrete pile $E = 27800$ MPa.

$$\Delta = \frac{Q_a L}{AE} = \frac{452.1(\text{kN}) \times 15(\text{m})}{0.196(\text{m}^2) \times 27800 \times 10^3 (\text{kN/m}^2)}$$

$$= 1.243 \times 10^{-3}\,\text{m} = 1.243\,\text{mm}$$

Step 2: Primary consolidation settlement of the pile group.

The 2:1 method is used to calculate the vertical stress increases because of the pile group loading. The vertical stress increases are assumed to occur starting at the depth of two-thirds of the pile length (Figure 4.21).

(1) Determine the depth of the soil layer for which the consolidation settlement should be calculated. The consolidation settlement should be considered to a depth of $\Delta\sigma'_z \leq 0.1\,\sigma'_0$.

Use 2:1 method: $\Delta\sigma'_z = \dfrac{p B_1 B_2}{(B_1 + z)(B_2 + z)}$

where: $p = \dfrac{Q}{B_1 B_2} = \dfrac{5425}{7 \times 5} = 155\,\text{kN/m}^2$

z starts from the 2/3 of the pile length.

The effective stress: $\sigma'_0 = (18 - 9.81) \times 10 + (19 - 9.81)z$

Let: $\Delta\sigma'_z = 0.1\sigma'_0$, and solve $z = 11.2$ m

(2) As the vertical stress increase is nonlinear with depth, the thick soil layers should be divided into a number of thin layers, and the consolidation settlement for each thin layer then can be

Chapter 4

calculated and the individual settlements summed. Following Figure 3.24 and for easy calculation, the second clay layer of the subsoil is divided into two layers, as shown in Figure 4.22.

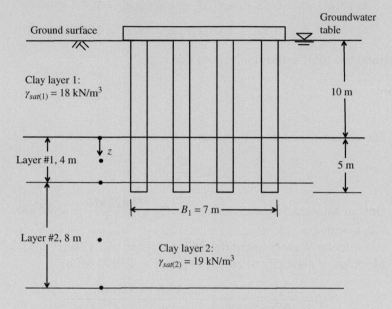

Fig. 4.22   Sample problem 4.6, layer division in the subsoil.

The vertical stress increases, the average vertical stress increase, and the average effective stress in each layer are calculated and listed as follows. Note that the vertical stress increase is calculated from a depth of 10 m, while the effective stress is calculated from the ground surface.

The average vertical stress increase uses Equation (3.55):

$$\Delta\sigma_{z(av)} = \frac{1}{6}(\Delta\sigma_{z(top)} + 4\Delta\sigma_{z(mid)} + \Delta\sigma_{z(bo)})$$

z (m)	Vertical stress increase (kN/m²)
0	155.0
2	86.0
4	54.8
8	27.8
12	16.8

Layer	Average vertical stress increase, $\Delta\sigma'_{z(av)}$ (kN/m^2)	Average in-situ effective stress, $\sigma'_{0(av)}$ (kN/m^2)	Preconsolidation pressure, σ'_c (kN/m^2)	Thickness, H (m)
#1	92.3	$(18-9.81)\times10 + (19-9.81)$ $\times 2 = 100.3$	101.1 (at depth = 12m)	4
#2	30.5	$(18-9.81)\times10 + (19-9.81)\times$ $8 = 155.4$	175.0 (at depth = 18m)	8

In layer #1: $\sigma'_c \approx \sigma'_{0(av)}$, the soil is considered as normally consolidated.

$$S_c = \frac{C_c H}{1+e_0} \log \frac{\sigma'_{0(av)} + \Delta\sigma'_{z(av)}}{\sigma'_{0(av)}}$$

$$= \frac{0.32 \times 4}{1 + 0.53} \log \frac{100.3 + 92.3}{100.3}$$

$$= 0.237\,\text{m}$$

In layer #2: $\sigma'_c > \sigma'_{0(av)}$, the soil is overconsolidated.
  And $\sigma'_c < \sigma'_{0(av)} + \Delta\sigma'_{z(av)} = 185.9\,\text{kN/m}^2$.

$$S_c = \frac{C_s H}{1+e_0} \log \frac{\sigma'_c}{\sigma'_{0(av)}} + \frac{C_c H}{1+e_0} \log \frac{\sigma'_{0(av)} + \Delta\sigma'_{z(av)}}{\sigma'_c}$$

$$= \frac{0.09 \times 8}{1 + 0.55} \log \frac{175.0}{155.4} + \frac{0.32 \times 8}{1 + 0.55} \log \frac{155.4 + 30.5}{175.0}$$

$$= 0.024 + 0.043$$

$$= 0.067\,\text{m}$$

The total consolidation settlement is: 0.304 m or 30.4 cm.
The total settlement of the pile cap is: 30.4 + 0.1 = 30.5 cm.
It is noted that the consolidation settlement is rather large. Ground improvement or foundation redesign may be needed.

Chapter 4

Homework Problems

1. A 20-m-long concrete pile is driven into cohesionless soil of two strata. The topsoil stratum has unit weight of 18.5 kN/m³, internal friction angle of 30°, and thickness of 12 m. The second stratum has unit weight of 19.0 kN/m³, internal friction angle of 35°, and thickness of 50 m. The groundwater table is found to be at 42 m below the ground surface. The concrete pile is circular in cross section with a diameter of 40 cm. Determine the ultimate bearing load of the pile.

2. A concrete pile is driven into a homogeneous cohesionless soil. The soil's unit weight is 18.5 kN/m³, and its internal friction angle is 35°. The groundwater table is not found during the subsoil exploration. The pile is subjected to a load of 800 kN. Using a factor of safety of 3 and pile diameter of 30 cm, determine the required pile length.

3. A 15-m closed-end steel pipe pile is driven into layered undrained clay. The top layer has unit weight of 18.5 kN/m³, undrained cohesion of 90 kN/m², and a thickness of 10 m. The second layer has unit weight of 19.5 kN/m³, undrained cohesion of 120 kN/m², and it extends to a great depth. The groundwater table is at the ground surface. The pile diameter is 40 cm. Determine the ultimate bearing load of the pile.

4. A subsoil profile is shown in Figure 4.23. The concrete pile's diameter is 50 cm. Determine the total length of the concrete pile to take a load of 250 kN with a factor of safety of 3.

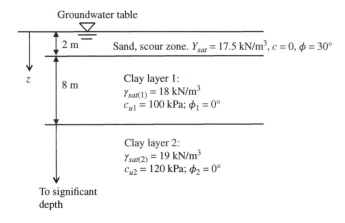

Groundwater table

2 m　　Sand, scour zone. $Y_{sat} = 17.5$ kN/m³, $c = 0$, $\phi = 30°$

z　　8 m　　Clay layer 1:
$\gamma_{sat(1)} = 18$ kN/m³
$c_{u1} = 100$ kPa; $\phi_1 = 0°$

Clay layer 2:
$\gamma_{sat(2)} = 19$ kN/m³
$c_{u2} = 120$ kPa; $\phi_2 = 0°$

To significant depth

Fig. 4.23　Subsoil profile for Problem 4.

5. A concrete pile is designed to support a load of 4600 kN. The pile is driven into a homogeneous drained clayey sand with $c' = 50$ kN/m² and $\phi' = 32°$. The unit weight of the subsoil is 19 kN/m³. The concrete

pile is square in cross section with a width of 30 cm. Use FS = 3. Determine the minimum length of the pile.

6. As shown in Figure 4.24, a concrete pile is driven into the top two layers of subsoil strata. The subsoil profile and properties are shown in the figure. The pile's diameter is 50 cm throughout the pile. Determine the ultimate bearing load of the pile.

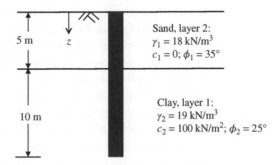

Sand, layer 2:
$\gamma_1 = 18$ kN/m^3
$c_1 = 0; \phi_1 = 35°$

Clay, layer 1:
$\gamma_2 = 19$ kN/m^3
$c_2 = 100$ kN/m^2; $\phi_2 = 25°$

5 m z

10 m

Fig. 4.24   Subsoil profile for Problem 6.

7. A pile group comprises four circular concrete piles. The diameter of each pile is 40 cm. The spacing between two adjacent piles is 120 cm. The pile group is driven into a homogeneous sandy riverbed to support a bridge pier. It is assumed the river flows year-round. The saturated unit weight of subsoil is 19 kN/m^3, the cohesion is zero, and the internal friction angle is 36°. The pile length is determined to be 12 m. Determine the ultimate bearing capacity and pile group efficiency of the pile group.

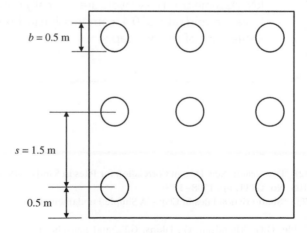

$b = 0.5$ m

$s = 1.5$ m

0.5 m

Fig. 4.25   Layout of pile group for Problem 8.

Chapter 4

8. A 30-m-long closed-end steel pipe pile group is driven into layered undrained clay. The pile cap is square, and the nine piles are evenly spaced. The layout of the pile group is shown in Figure 4.25. The topsoil layer has a unit weight of 18 kN/m^3, undrained cohesion of 100 kN/m^2, and a thickness of 10 m. The second layer has unit weight of 19 kN/m^3, undrained cohesion of 150 kN/m^2, and it extends to great depth. The groundwater table is at the ground surface. Determine the ultimate bearing capacity and pile group efficiency of the pile group.

9. The subsoil profile of a riverbed is shown in Figure 4.23. It is determined that a pile group comprising four piles is needed to support the bridge pier. The four piles are evenly spaced. The center-to-center spacing is three times of the pile diameter, and each pile's outside circumference is assumed to align with the edge of the pile cap. Each concrete pile's diameter is 50 cm, and length is 15 m. Determine the ultimate bearing capacity and pile group efficiency of the pile group.

10. A 15-m-long closed-end steel pipe pile group is driven into a homogeneous clay. The pile cap is square, and the nine piles are evenly spaced. The layout of the pile group is shown in Figure 4.25. The pile group is subjected to a vertical load of 5200 kN. The soil has a unit weight of 18.5 kN/m^3, cohesion of 100 kN/m^2, friction angle of 10 degrees. The clay layer is 100 m deep, and beneath the clay layer is dense sand. The groundwater table is at the ground surface. Preliminary laboratory testing found that the clay's void ratio is 0.45, compression index is 0.3, swell index is 0.08, and the clay is overconsolidated. The preconsolidation pressure is 200 kN/m^2. Determine the primary consolidation settlement of the pile group.

11. The problem statement is the same as in Problem 9, and the subsoil profile is shown in Figure 4.23. The total load on the pile cap is 6000 kN. Assume both clay layers are normally consolidated. Both clay layers have the void ratio of 0.4, compression index of 0.3. Determine the total settlement of the pile cap.

References

Bhushan, K. (1982). "Discussion: New Design Correlation for Piles in Sands." *ASCE Journal of Geotechnical Engineering Division*, Vol. 108, No. GT11, pp. 1508–1510.

Burland, J.B. (1973). "Shaft Friction Piles in Clay – A Simple Fundamental Approach." *Ground Engineering*, Vol. 21, No. 2, pp. 135–142.

Hannigan, P.J., Goble, G.G., Thendean, G., Likins, G.E., and Rausche, F. (1998). Design and Construction of Driven Pile Foundation – Volume I. Publication No. FHWA HI-97-013. Federal Highway Administration, U.S. Department of Transportation, Washington, DC. Revised November 1998.

Meyerhof, G.G. (1976). "Bearing Capacity and Settlement of Pile Foundations." *ASCE Journal of Geotechnical Engineering Division*, Vol. 102, No. GT3, pp. 195–228.

Nordlund, R.L. (1979). Point Bearing and Shaft Friction of Piles in Sand. Proceedings, 5th Annual Fundamentals of Deep Foundation Design, Missouri-Rolla, MO, 1979.

Sladen, J.A. (1992). "The Adhesion Factor: Applications and Limitations." *Canadian Geotechnical Journal*, Vol. 29, No. 2, pp. 322–326.

Terzaghi, K., and Peck, R.B. (1967). Soil Mechanics in Engineering Practice, 2nd edition. John Wiley and Sons, New York.

Terzaghi, K., Peck, R.B., and Mesri, G. (1996). Soil Mechanics in Engineering Practice, 3rd edition. John Wiley and Sons, New York.

Tomlinson, M.J. (1971). "Some Effects of Pile Driving of Skin Friction." Proceedings, Conference on Behaviour of Piles, ICE, London, pp. 107–114.

Tomlinson, M.J. (1994). Pile Design and Construction Practice, 4th edition. E&FN Spon, London.

Vesic, A.S. (1977). Design of Pile Foundations. Synthesis of Highway Practice No. 42, National Cooperative Highway Research Program, Transportation Research Board, National Research Council, Washington, D.C.

Chapter 4

Chapter 5

Slope Stability Analyses and Stabilization Measures

5.1 Introduction

A slope can be a natural slope or one created by excavation or an embankment created by engineering fill. A slope failure is defined as the displacement of a portion of the slope mass downward relative to the mass beneath the sliding surface. The scale of a slope failure varies from less than a meter in slope height to the sliding of a large portion of a mountain. For example, the 1974 Rio Mantaro landslide in Peru involved a sliding mass 6 km long, 2 km high, and 1.5 billion cubic meters in volume (Lee and Duncan 1975). Figure 5.1(a) shows a major slope failure that occurred in Oso, Washington, USA, on March 22, 2014. The failure surface exhibits a rotational sliding surface, as illustrated in Figure 5.1(b).

Slope failures can be categorized into different types on the basis of the shapes of the failure surfaces and the nature of the slope movements.

1. *Surficial (or translational) slope failure* (Figure 5.2a). The sliding surface is parallel to the slope surface. The sliding layer is usually shallow compared to the slope height, and the failure portion slides along a planar surface. Translational slope failure occurs where there is a weak seam (e.g., a thin and smooth clay layer) below the slope surface or where a loose topsoil rests on a hard subsoil.
2. *Rotational slope failure* (Figure 5.2b). A large mass of the slope rotates along a curved failure surface, which is often simplified as a circular or log-spiral curve for simplified analysis.
3. *Landslide*. As shown in Figure 5.2c, a landslide usually involves a large volume of sliding mass and multiple rupture surfaces that may include translational and rotational failures. The failure portion may include various types of soils and rocks and multiple slopes.
4. *Lateral spread*. A lateral spread refers to the lateral movement of a fractured soil mass. This lateral movement can be multidirectional and can occur on a very shallow slope or slightly inclined ground. The movement is similar to translational slope failure and is typically caused by earthquakes, as shown in Figure 9.21(b).
5. *Debris flow, or mudslide*. This type of slope failure involves the relatively rapid movement of soils that are entrained by flowing water or wind; the soil flows like a liquid.

Geotechnical Engineering Design, First Edition. Ming Xiao.
© 2015 John Wiley & Sons, Ltd. Published 2015 by John Wiley & Sons, Ltd.
Companion Website: www.wiley.com/go/Xiao

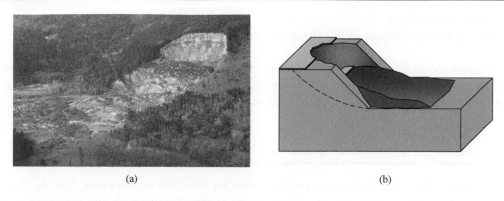

(a)   (b)

Fig. 5.1   Example of slope failure. (a) Landslide in Oso, Washington, USA. March 22, 2014. (photo courtesy of Washington State Department of Transportation, USA). (b) 3-D illustration of the shape of the failure.

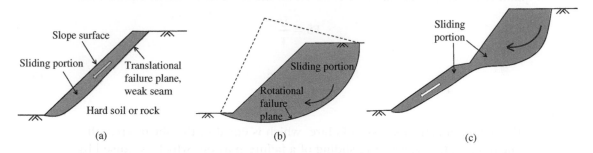

(a)   (b)   (c)

Fig. 5.2   Types of slope failures. (a) Surficial slope failure, (b) rotational slope failure, (c) landslide.

6. *Creep.* Creep is a slow and almost imperceptible movement of the failure portion of a slope. It can be near surface and translational, or deep-seated and rotational. Inclinometers are often used to monitor the creep of a slope.
7. *Rock falls.* Rocks on the upslope can be mobilized by wind, runoff, or gravity and can leap, free-fall, or roll down the slope.

   A slope failure is caused by the imbalance of the external shear stress (or sliding moment) and the internal shear strength (or resisting moment) of the slope:

$$\left. \begin{array}{l} \text{Shear stress} > \text{Shear strength} \\ \text{or: Rotational moment} > \text{Resisting moment} \end{array} \right\} \Rightarrow \text{Slope failure}$$

The following factors may *increase* the shear stress or sliding moment:

- Additional surcharge at the top of a slope.
- Application of lateral force that may be caused by seepage, earthquake, or pile driving.

   The following factors may *decrease* the shear strength or resisting moment:

- Weathering of a rock slope.
- Discontinuities such as weak seams and faults that are developed in the slope.

Chapter 5

- Saturation of the slope.
- Removal of lateral support of the slope, for example, the cut toe of a slope.

This chapter focuses on the stability analyses of *unreinforced soil slopes* under *static loading*. The design and analyses of reinforced soil slopes are presented in Chapter 8, and some basic seismic slope stability analyses are presented in Chapter 9.

5.2   Overview of slope stability analyses

Slope stability analysis methods are typically based on the limit equilibrium. That is, the forces or moments that cause a slope failure (sliding) are at equilibrium with the forces or moments that resist the slope sliding. This is referred to as the critical condition. A factor of safety (FS) is used to quantify the slope stability and is based on the force or moment equilibrium:

$$FS = \frac{\tau_f}{\tau} \tag{5.1}$$

or:

$$FS = \frac{M_{resist}}{M_{slide}} \tag{5.2}$$

where

τ_f = the maximum shear stress at failure, which is equal to the shear strength,
τ = shear stress that causes the sliding of a failure portion, which is caused by external loads such as gravity, foundation loading, seismic force, etc.,
M_{resist} = total resisting moments that resist a rotational sliding,
M_{slide} = total sliding moments that cause a rotational sliding.

In general, a value of 1.3–1.5 is used as an acceptable factor of safety.
Alternatively, within the context of limit state design, it should be mentioned that:

$$E_d \leq R_d \tag{5.3}$$

where

E_d = design effect of the actions (e.g., sliding forces)
R_d = design resistance (dependent on soil strength)

The shear strength is expressed by the Mohr–Coulomb failure criterion:

$$\tau_f = c + \sigma \tan \phi \tag{5.4}$$

where

σ = total normal stress,
c = soil's cohesion based on the total stress,
ϕ = soil's internal friction angle based on the total stress.

The Mohr–Coulomb failure criterion can also be expressed using the effective stress:

$$\tau_f = c' + \sigma' \tan \phi' \tag{5.5}$$

where

σ' = effective normal stress,

c' = soil's cohesion based on the effective stress, also referred to as effective cohesion,

ϕ' = soil's internal friction angle based on the effective stress, also referred to as effective friction angle.

The slope stability analysis methods can be largely divided into two groups: the *total stress method* and the *effective stress method*. In the total stress method, c and ϕ that are based on the total stress are used. In the effective stress method, c' and ϕ' that are based on the effective stress are used.

In slope stability analyses, two "artificial" factors of safety are defined on the basis of c and ϕ, respectively:

$$\text{FS}_c = \frac{c}{c_m} \tag{5.6}$$

$$\text{FS}_\phi = \frac{\tan \phi}{\tan \phi_m} \tag{5.7}$$

where

FS_c = factor of safety on the basis of c,

FS_ϕ = factor of safety on the basis of ϕ,

c_m = mobilized cohesion that is actually developed along a slip surface, also denoted as c_d,

ϕ_m = mobilized internal friction angle that is actually developed along a slip surface, also denoted as ϕ_d.

Using a limit state design approach, partial factors of safety similar to those in equations (5.6) and (5.7) can be used to reduce the geotechnical parameters and increase the value of the disturbing forces involved in the verification of the corresponding limit state.

In a stable slope, the entire soil's strength may not be needed to maintain the force or moment equilibrium. The developed or mobilized shear strength that is needed for slope stability is represented by c_m and ϕ_m Therefore, $c_m \leq c$ and $\phi_m \leq \phi$. At equilibrium, the developed or mobilized shear stress can be expressed as:

$$\tau_m = c_m + \sigma \tan \phi_m \tag{5.8}$$

At the critical condition, the entire shear strength is needed for equilibrium. So, $c_m = c$, and $\phi_m = \phi$. Therefore, the minimum value for both FS_c and FS_ϕ is 1.0.

If: $\text{FS}_c = \text{FS}_\phi = \text{constant } a$

Then: $c = a \cdot c_m$, and $\tan \phi = a \cdot \tan \phi_m$

$$\text{FS} = \frac{\tau_f}{\tau_m} = \frac{c + \sigma \tan \phi}{c_m + \sigma \tan \phi_m} = \frac{a \cdot c_m + \sigma(a \cdot \tan \phi_m)}{c_m + \sigma \tan \phi_m} = a$$

$$\text{So}: \text{FS} = \text{FS}_c = \text{FS}_\phi \tag{5.9}$$

Chapter 5

This approach is very useful in the slope stability analyses that are described in this chapter. It should be noted that FS_c and FS_ϕ are not the real factor of safety of a slope; they are defined only to determine the factor of safety, FS, of a slope.

If the effective stress method is used, the above definitions follow the same approach:

$$FS_{c'} = \frac{c'}{c'_m} \tag{5.10}$$

$$FS_{\phi'} = \frac{\tan \phi'}{\tan \phi'_m} \tag{5.11}$$

$$\tau_m = c'_m + \sigma' \tan \phi'_m \tag{5.12}$$

$$\text{If} : FS_{c'} = FS_{\phi'}, \text{ then } FS = FS_{c'} = FS_{\phi'} \tag{5.13}$$

On the basis of the shape of the failure surface and the characteristics of a slope, the slope stability analyses follow the methods presented in Table 5.1, which are described in detail in this chapter. There are many other widely used methods of slope stability analyses that are not covered in this chapter.

As shown in Table 5.1, there are various methods available for different soil conditions and failure modes. However, the accuracy of slope stability analyses depends largely on accurately

Table 5.1   Summary of typical slope stability analysis methods (under static condition).

Mass methods (The sliding soil mass is analyzed as one entity; the mass methods are applicable only to homogeneous slopes.)	Infinite slope methods	Dry slopes
		Submerged slopes without seepage
		Submerged slopes with seepage
	Finite slope methods	Planar failure surfaces (Culmann's method)
		Curved failure surfaces	Undrained clay slope ($\phi = 0$)
			Taylor's chart for $c - \phi$ soil (both c and ϕ are not zero)
			Michalowski's chart for $c - \phi$ soil (considering pore water pressure)
Methods of slices (The sliding soil mass is divided into numerous slices and the stability of each slice is analyzed. The methods are applicable to homogeneous or heterogeneous slopes.)	Ordinary method of slices (Fellenius method of slices)	
	Bishop-simplified method of slices	No pore water pressure
		Consider pore water pressure (Bishop–Morgenstern method)
	Spencer method with consideration of pore water pressure	
	Morgenstern charts for rapid drawdown	
Finite element methods		

delineating the subsurface profile including thin but weak seams. Therefore, subsurface exploration is fundamentally important to the slope stability analyses.

5.3   Slope stability analyses – infinite slope methods

The infinite slope methods analyze the translational slope stability. As the surficial sliding layer is thin compared to the slope's height, the surficial layer is assumed to extend upslope and downslope infinitely. The analyses of infinite slope stability depend on the saturation and seepage conditions. The following three scenarios are presented.

5.3.1   Dry slopes

As shown in Figure 5.3, a section of the surficial layer is analyzed for its force equilibrium. The section's length is l, its thickness is H in the vertical (gravitational) direction, and the slope angle is β. The soil has cohesion c, internal friction angle ϕ, and unit weight γ. The weight of the section per unit length (perpendicular to the cross section) of the slope is:

$$W = \gamma(lH \cos \beta) \tag{5.14}$$

The reaction of the weight is R and it has two components – the shear resistance force T and the normal force N with respect to the slip surface:

$$T = W \sin \beta \tag{5.15}$$

$$N = W \cos \beta \tag{5.16}$$

The shear and normal stresses that are caused by the weight of the soil section are:

$$\tau = \frac{N}{l \cdot 1} = \frac{W \cos \beta}{l} = \gamma H \cos \beta \sin \beta \tag{5.17}$$

$$\sigma = \frac{N}{l \cdot 1} = \frac{W \cos \beta}{l} = \gamma H \cos^2 \beta \tag{5.18}$$

The factor of safety is:

$$FS = \frac{\tau_f}{\tau} = \frac{c + \sigma \tan \phi}{\tau} = \frac{c + \gamma H \cos^2 \beta \tan \phi}{\gamma H \cos \beta \sin \beta} = \frac{2c}{\gamma H \sin(2\beta)} + \frac{\tan \phi}{\tan \beta} \tag{5.19}$$

In the limit state design, the design effect of the actions (Ed) is given by the component of the weight of the section that is parallel to the slip surface (i.e., equation 5.15). Hence,

$$E_d = \gamma_G \gamma_k lH \cos \beta \sin \beta \tag{5.20}$$

where

γ_G = partial factor of safety (for a permanent, unfavorable force)
γ_k = characteristic value (cautious estimate) of the soil unit weight
l = section length
H = section thickness
β = slope angle

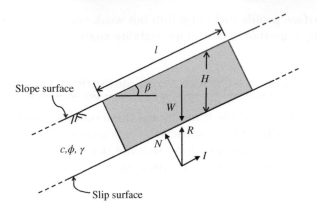

Fig. 5.3 Dry infinite slope without groundwater.

Similarly, the design resistance, R_d, can be obtained from the denominator in equation (5.19) such that:

$$R_d = \frac{l}{\gamma_{R_e}} \left[\frac{c'_k}{\gamma_c} + \frac{\gamma_k}{\gamma_\gamma} H \cos^2 \beta \frac{\tan \phi'_k}{\gamma_{\phi'}} \right] \tag{5.21}$$

where

γ_{R_e} = Partial factor of safety on the resistance
γ_c = Partial factor of safety on the cohesion
γ_γ = Partial factor of safety on the unit weight
$\gamma_{\phi'}$ = Partial factor of safety on the internal friction angle
γ_k = Cautious estimate of unit weight
c'_k = Cautious estimate of cohesion
ϕ'_k = Cautious estimate of internal friction angle

Therefore, the limit state can be verified as:

$$E_d \leq R_d \tag{5.22}$$

$$\gamma_G \gamma_k l H \cos \beta \sin \beta \leq \frac{l}{\gamma_{R_e}} \left[\frac{c'_k}{\gamma_c} + \frac{\gamma_k}{\gamma_\gamma} H \cos^2 \beta \frac{\tan \phi'_k}{\gamma_{\phi'}} \right] \tag{5.23}$$

5.3.2 Submerged slopes with no seepage

The slope is submerged below the groundwater table and there is no seepage down the slope. The analysis approach is the same as with the dry slope, except that the weight of the sliding soil is determined using the submerged unit weight, $\gamma' = \gamma_{\text{sat}} - \gamma_w$. The factor of safety is:

$$\text{FS} = \frac{\tau_f}{\tau} = \frac{c + \sigma \tan \phi}{\tau} = \frac{c + \gamma' H \cos^2 \beta \tan \phi}{\gamma' H \cos \beta \sin \beta} = \frac{2c}{\gamma' H \sin(2\beta)} + \frac{\tan \phi}{\tan \beta} \tag{5.24}$$

Alternatively, in the limit state design, the submerged unit weight needs to be accounted for; this limit state can be verified as

$$E_d \leq R_d \tag{5.25}$$

$$\gamma_G \gamma'_k l H \cos \beta \sin \beta \leq \frac{1}{\gamma_{R_e}} \left[\frac{c'_k}{\gamma_c} + \frac{\gamma'_k}{\gamma_\gamma} H \cos^2 \beta \frac{\tan \phi'_k}{\gamma_{\phi'}} \right] \qquad (5.26)$$

5.3.3 Submerged slopes with seepage parallel to the slope face

Figure 5.4 shows that the groundwater table coincides with the slope surface, and there is seepage parallel to the slope surface. The weight of the submerged soil is calculated using the saturated unit weight, γ_{sat}. The shear and normal stresses that are caused by the weight of the sliding section are:

$$\tau = \frac{T}{l \cdot 1} = \frac{W \sin \beta}{l} = \gamma_{sat} H \cos \beta \sin \beta \qquad (5.27)$$

$$\sigma = \frac{N}{l \cdot 1} = \frac{W \cos \beta}{l} = \gamma_{sat} H \cos^2 \beta \qquad (5.28)$$

As shown in Figure 5.4, the pore water pressure at any cross section (perpendicular to the flow direction) of the sliding soil section is:

$$u = \gamma_w (H \cos^2 \beta) \qquad (5.29)$$

Using the cohesion and friction angle on the basis of based on the effective stress, the factor of safety is:

$$\begin{aligned}
\mathrm{FS} &= \frac{\tau_f}{\tau} = \frac{c' + \sigma' \tan \phi'}{\tau} = \frac{c' + (\gamma_{sat} H \cos^2 \beta - \gamma_w H \cos^2 \beta) \tan \phi'}{\gamma_{sat} H \cos \beta \sin \beta} \\
&= \frac{2c'}{\gamma_{sat} H \sin(2\beta)} + \frac{\gamma'}{\gamma_{sat}} \frac{\tan \phi'}{\tan \beta}
\end{aligned} \qquad (5.30)$$

Or, following the same principles of the previous sections, using a limit state design approach, this inequality must be satisfied:

$$\gamma_G \gamma_{sat,k} l H \cos \beta \sin \beta \leq \frac{1}{\gamma_{R_e}} \left[\frac{c'_k}{\gamma_c} + \left(\frac{\gamma_{sat,k}}{\gamma_\gamma} H \cos^2 \beta - \gamma_w H \cos^2 \beta \right) \frac{\tan \phi'_k}{\gamma_{\phi'}} \right] \qquad (5.31)$$

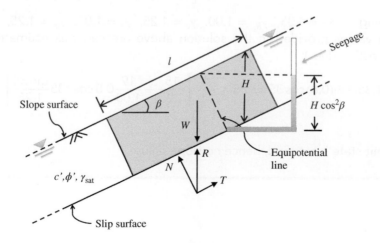

Fig. 5.4 Submerged slope with seepage parallel to the slope face.

where $\gamma_{sat,k}$ is a cautious estimate of the saturated unit weight of the soil and the other variables are the same as described in previous paragraphs. Note that the unit weight of the water (γ_w) is not factored by a partial factor of safety, but the worst case scenario should be used.

Sample Problem 5.1: Infinite slope method for dry slope.

A natural slope is 50-m high and the slope angle is 35 degrees. Subsurface investigation found that a 5-cm thick weak clay layer exists beneath the slope surface at a depth of 80 cm. The surficial soil layer's cohesion is 15 kN/m², the internal friction angle is 30 degrees, and the bulk unit weight is 19 kN/m³. Determine the factor of safety of the surficial soil layer on the weak seam against translational failure.

Solution:

The infinite slope method for dry slope is used. The factor of safety is:

$$FS = \frac{\tau_f}{\tau} = \frac{2c}{\gamma H \sin(2\beta)} + \frac{\tan \phi}{\tan \beta}$$

Given $c = 15 \, kN/m^2, \phi = 30°, \gamma = 19 \, kN/m^3, \beta = 35°, H$ (the vertical thickness of the surficial layer) = 0.8 m.

$$FS = \frac{2 \times 15}{19 \times 0.8 \times \sin(2 \times 35)} + \frac{\tan 30}{\tan 35} = 2.92 > 1.5, \quad \text{no slope failure}$$

Alternative solution using limit state design:

$$E_d \leq R_d$$

$$\gamma_G \gamma_k l H \cos \beta \sin \beta \leq \frac{l}{\gamma_{R_e}} \left[\frac{c'_k}{\gamma_c} + \frac{\gamma_k}{\gamma_\gamma} H \cos^2 \beta \frac{\tan \phi'_k}{\gamma_{\phi'}} \right]$$

Assuming $\gamma_G = 1.35, \; \gamma_{R_e} = 1.00, \; \gamma_c = 1.25, \; \gamma_\gamma = 1.00, \; \gamma_\phi = 1.25,$ and that the values provided in the solution above are cautious estimates (i.e., characteristic values):

$$1.35(19)(0.8) \cos 35 \sin 35 \leq \frac{1}{1.0} \left[\frac{15}{1.25} + \frac{19}{1.00} 0.8 \cos^2 35 \frac{\tan 30}{1.25} \right]$$

$$9.6 \leq 16.7$$

The limit state is satisfied, hence no slope failure

Sample Problem 5.2: Infinite slope method for saturated slope with seepage.

The same slope in sample problem 5.1 is submerged beneath the groundwater table. The seepage flows downward and is parallel to the slope surface. It is assumed the unit weight, cohesion, and internal friction angle are the same as in sample problem 5.1. Determine the FS.

Solution:

The infinite slope method with seepage parallel to the slope face is used. The FS is based on Equation (5.23).

$$FS = \frac{\tau_f}{\tau} = \frac{2c'}{\gamma_{sat}H\sin(2\beta)} + \frac{\gamma'}{\gamma_{sat}}\frac{\tan\phi'}{\tan\beta}$$

where: $c' = 15\ kN/m^2$, $\phi' = 30°$, $\gamma_{sat} = 19\ kN/m^3$, $\beta = 35°$, $\gamma' = 19 - 9.81 = 9.19\ kN/m^3$, $H = 0.8\ m$.

$$FS = \frac{2 \times 15}{19 \times 0.8 \times \sin(2 \times 35)} + \frac{9.19}{19} \times \frac{\tan 30}{\tan 35} = 2.5 > 1.5, \text{ no slope failure.}$$

Alternative solution using limit state design:

The same assumptions as for the alternative solution in sample problem 5.1 are made here. The inequality that needs to be satisfied is, however:

$$\gamma_G\gamma_{sat,k}lH\cos\beta\sin\beta \leq \frac{l}{\gamma_{R_e}}\left[\frac{c'_k}{\gamma_c} + \left(\frac{\gamma_{sat,k}}{\gamma_\gamma}H\cos^2\beta - \gamma_w H\cos^2\beta\right)\frac{\tan\phi'_k}{\gamma_{\phi'}}\right]$$

$$1.35(19)(0.8)\cos 35\sin 35 \leq \frac{1}{1.00}$$

$$\times\left[\frac{15}{1.25} + \left(\frac{19}{1.00}0.8\cos^2 35 - 9.81\cdot 0.8\cos^2 35\right)\frac{\tan 30}{1.25}\right]$$

$$9.6 \leq 14.27$$

The limit state is satisfied, hence no slope failure

5.4   Slope stability analyses – Culmann's method for planar failure surfaces

Culmann's method was proposed by Carl Culmann in 1875 (Culmann 1875). It is used to analyze a finite slope that slides on a planar failure surface, as shown in Figure 5.5.

In Figure 5.5, the plane AC is an assumed failure surface with inclination angle of θ. The soil is homogeneous and has cohesion c, internal friction angle ϕ, and bulk unit weight γ. The weight

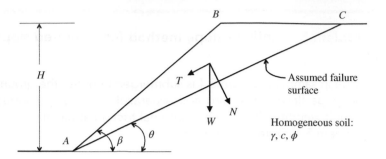

Fig. 5.5   Culmann's method with planar failure surface.

of the failure wedge ABC is:

$$W = \frac{1}{2}(\overline{BC})H(1)\gamma = \frac{1}{2}(H\cot\theta - H\cot\beta)H(1)\gamma = \frac{1}{2}\gamma H^2(\cot\theta - \cot\beta) \qquad (5.32)$$

The weight has two components: the tangential force parallel to the assumed failure plane AC and the normal force perpendicular to AC:

$$T = W\sin\theta \qquad (5.33)$$

$$N = W\cos\theta \qquad (5.34)$$

The length of AC is:

$$\overline{AC} = \frac{H}{\sin\theta} \qquad (5.35)$$

So, the shear and normal stresses on the assumed failure plane AC are:

$$\tau = \frac{T}{(\overline{AC})(1)} \qquad (5.36)$$

$$\sigma = \frac{N}{(\overline{AC})(1)} \qquad (5.37)$$

The shear stress τ is equal to the mobilized (or actually developed) shear stress that is expressed in Equation (5.8):

$$\tau_m = c_m + \sigma\tan\phi_m = \frac{T}{(\overline{AC})(1)} \qquad (5.38)$$

Substituting σ, T, and $\overline{AC}$ in Equation (5.38) with the expressions in (5.32) to (5.35) and (5.37), c_m can be derived as:

$$c_m = \frac{\frac{1}{2}\gamma H^2(\cot\theta - \cot\beta)\sin\theta}{\dfrac{H}{\sin\theta}} - \frac{\frac{1}{2}\gamma H^2(\cot\theta - \cot\beta)\cos\theta}{\dfrac{H}{\sin\theta}}\tan\phi_m$$

$$= \frac{1}{2}\gamma H(\cot\theta - \cot\beta)(\sin^2\theta - \sin\theta\cos\theta\tan\phi_m) \qquad (5.39)$$

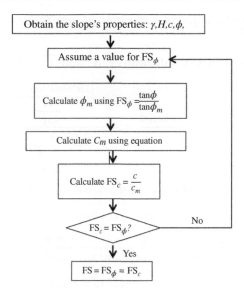

Fig. 5.6   Culmann's method to find the factor of safety for planar failure surface.

If the FS is known, then:

$$\left(\text{FS} = \frac{\tau_f}{\tau_m} \right) = \left(\text{FS}_c = \frac{c}{c_m} \right) = \left(\text{FS}_\phi = \frac{\tan\phi}{\tan\phi_m} \right) \tag{5.40}$$

For the critical failure plane for which the FS is the smallest, c_d should have the largest value, which can be found using:

$$\frac{\partial c_m}{\partial\theta} = 0 \tag{5.41}$$

Combining Equation (5.39) with (5.41) and solving for θ, the critical failure plane is at:

$$\theta_{cr} = \frac{\beta + \phi_m}{2} \tag{5.42}$$

Substituting θ in Equation (5.39) with Equation (5.42), c_m on the critical failure plane can be expressed as:

$$c_m = \frac{\gamma H}{4} \left[\frac{1 - \cos(\beta - \phi_m)}{\sin\beta \cos\phi_m} \right] \tag{5.43}$$

It is also possible to use partial factors of safety and a limit state design approach using Culmann's method. The principle behind this approach is that the component of the weight of the slice that is parallel to the slip surface is the disturbing action (force) that needs to be considered to estimate the design effect of the actions (E_d). Similarly, the design resistance R_d can be determined using equation (5.38) as the starting point.

Culmann's method has the following applications:

(a) *Find the factor of safety of a slope if the failure surface is a plane*
   The approach follows Equation (5.9) and is shown in Figure 5.6. The approach is illustrated in Sample Problem 5.3.

Chapter 5

(b) *Find the factor of safety of a slope on a specified planar failure surface*

As $FS = \dfrac{\tau_f}{\tau} = \dfrac{c + \sigma \tan \phi}{\tau}$

From Equations (5.36) and (5.37), substitute T, N, and $\overline{AC}$ using Equations (5.32)–(5.35):

$$FS = \frac{\tau_f}{\tau} = \frac{c + \sigma \tan \phi}{\tau} = \frac{c + \dfrac{\frac{1}{2}\gamma H^2(\cot\theta - \cot\beta)\cos\theta}{\dfrac{H}{\sin\theta}} \tan\phi}{\dfrac{\frac{1}{2}\gamma H^2(\cot\theta - \cot\beta)\sin\theta}{\dfrac{H}{\sin\theta}}}$$

$$= \frac{c + \frac{1}{2}\gamma H(\cot\theta - \cot\beta)\cos\theta \sin\theta \tan\phi}{\frac{1}{2}\gamma H(\cot\theta - \cot\beta)\sin^2\theta}$$

$$= \frac{2c}{\gamma H} \cdot \frac{1}{(\cot\theta - \cot\beta)\sin^2\theta} + \frac{\tan\phi}{\tan\theta} \qquad (5.44)$$

(c) *Given the factor of safety, determine the height of a slope*

Given: $FS = FS_c = FS_\phi$

Then: $c_m = \dfrac{c}{FS}$ and $\phi_m = \tan^{-1}\left(\dfrac{\tan\phi}{FS}\right)$

Substitute c_m and ϕ_m above in Equation (5.43); then H can be derived as follows:

$$H = \frac{4c_m}{\gamma}\left[\frac{\sin\beta \cos\phi_m}{1 - \cos\left(\beta - \phi_m\right)}\right] \qquad (5.45)$$

(d) *Determine the maximum height of a slope if the failure surface is a plane*

To obtain the maximum height of a slope, FS = 1.0. The approach is the same as in case (c) with FS = 1.0. The slope height with FS = 1.0 is referred to as the critical height:

$$H_{cr} = \frac{4c}{\gamma}\left[\frac{\sin\beta \cos\phi}{1 - \cos\left(\beta - \phi\right)}\right] \qquad (5.46)$$

Sample Problem 5.3: Find the factor of safety of a slope if the failure surface is a plane, using Culmann's method.

A natural slope is 50-m high and the slope angle is 35 degrees. Subsurface investigation found that the subsoil is homogeneous with the effective cohesion of 65 kN/m², effective internal friction angle of 30 degrees, and bulk unit weight of 19 kN/m³. Assume the potential failure surface is a plane. Determine the factor of safety.

Solution:

Given: $H = 50$ m, $\beta = 35°$, $\gamma = 19$ kN/m³, $c' = 65$ kN/m², $\phi' = 30°$.

Use Culmann's method and follow the approach in Figure 5.6. The effective stress method is used. The following equations are used. And, Table 5.2 is developed for the trial-and-error approach.

$$FS_{c'} = \frac{c'}{c_m'}, FS_{\phi'} = \frac{\tan \phi'}{\tan \phi_m'}, c_m' = \frac{\gamma H}{4}\left[\frac{1 - \cos\left(\beta - \phi_m'\right)}{\sin \beta \cos \phi_m'}\right]$$

Conclusion: the factor of safety of the slope is *2.30*; the slope is stable.

Table 5.2   Culmann's method, trial-and-error approach to find FS.

$FS_{\phi'}$(assumed)	ϕ_m'	$c_m'(kN/m^2)$	FS_c'
1.0	30°	1.82	35.72
2	16.1°	23.23	2.80
2.4	13.5°	29.56	2.20
2.3	14.1°	28.11	2.31
2.305	14.06°	28.19	2.306

Sample Problem 5.4:

For the same slope described in sample problem 5.3, what is the factor of safety on a planar failure surface that has an inclination of 30 degrees?

Solution:

Given: $H = 50$ m, $\beta = 35°$, $\gamma = 19$ kN/m³, $c' = 65$ kN/m², $\phi' = 30°$, and $\theta = 30°$. Use Culmann's method and Equation (5.36).

$$FS = \frac{2c}{\gamma H}\cdot\frac{1}{(\cot\theta - \cot\beta)\sin^2\theta} + \frac{\tan\phi}{\tan\theta}$$

$$= \frac{2\times 65}{19\times 50}\times\frac{1}{(\cot 30° - \cot 35°)\times \sin^2 30°} + \frac{\tan 30°}{\tan 30°}$$

$$= \mathbf{2.80}$$

It can be seen that the FS (2.80) on a plane of 30° is larger than the minimum FS of 2.30 that was obtained in Sample Problem 5.3. Therefore, the plane of 30° inclination is not the critical failure plane.

Chapter 5

5.5   Slope stability analyses – curved failure surfaces

To simplify the analysis of slope stability, a curved slope failure surface can be assumed to be an arc of a circle, also referred to as a circular failure surface. On the basis of their locations, the circular failure surfaces may include *slope circles* that cut through the slope surface, *toe circles* that go through the toe of a slope, or *deep-seated circles* that extend into the foundation soil beneath the slope (Figure 5.7). Among deep-seated circles is a *mid-point circle* that is often used in stability analyses (such as in the Taylor's chart described below). As shown in Figure 5.7(c), the mid-point circle is so defined that the rotational center O is at the mid-point between the toe and the crest of the slope.

The analyses of circular failure surfaces comprise two major categories depending on the soil's conditions: (1) undrained clay that has $\phi = 0$, and (2) a soil that has both nonzero c and ϕ values, referred to as $c - \phi$ soil.

5.5.1   Undrained clay slope ($\phi = 0$)

Analytical method

For undrained clay, it is assumed $\phi = 0$; the cohesion is referred to as "undrained cohesion", c_u. As shown in Figure 5.8, for an *assumed* failure circle with center O and radius R, the weight of the sliding portion that acts on the centroid is W. The rotational arm is l. The total driving moment is:

$$M_D = W \cdot l \tag{5.47}$$

The total maximum resisting moment is caused by the cohesion along the slip circle (as friction is zero):

$$M_R = [c_u \cdot (R\theta) \cdot (1)] \cdot R \tag{5.48}$$

So the factor of safety is:

$$FS = \frac{M_R}{M_D} = \frac{c_u R^2 \theta}{W \cdot l} \tag{5.49}$$

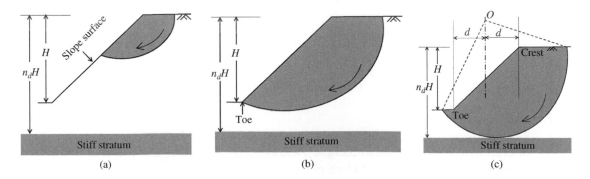

Fig. 5.7   Types of circular failure surfaces. (a) Slope circle, (b) toe circle, (c) deep-seated circle.

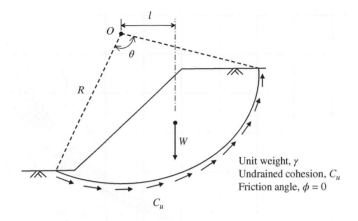

Fig. 5.8   Illustration of circular failure surface of undrained clay slope.

To obtain the real FS of an undrained clay slope, many trial failure surfaces should be analyzed by varying the locations of the center and the radii of the circles, so that the minimum FS can be obtained.

Alternatively, the limit state approach can be used, and the usual inequality needs to be satisfied:

$$E_d \leq R_d \tag{5.50}$$

In this case, however, the design effect of the actions (E_d) refers to the overturning moment, and the design resistance (R_d) is the moment provided by the shear strength along the slip surface. Therefore,

$$\gamma_G Wl \leq c_u R^2 \theta / \gamma_{cu} \tag{5.51}$$

where γ_G and γ_{cu} are partial factors of safety on the disturbing (permanent, unfavorable) moment and the undrained shear strength, respectively.

Taylor's chart for undrained clay

Another widely used method to determine the factor of safety of undrained clay slopes is Taylor's chart (Figure 5.9). Taylor (1937) identified three groups of failure circles: slope circles, toe circles, and deep-seated, mid-point circles, as shown in Figure 5.7. The type of failure circles depends on the slope angle and the depth of a stiff stratum beneath the slope, which is represented by $n_d H_c$,; H_c is the clay slope height. For a slope with an angle of at least 53°, the failure circle is a toe circle, regardless of the depth of stiff stratum. For a slope with an angle of less than 53°, when $n_d \geq 4$, the failure circle becomes deep-seated, mid-point circle; when $n_d < 4$, the failure circle can be any of the three types of circles, depending on the slope angle and the stiff stratum's depth.

In Taylor's chart, a dimensionless stability number is defined:

$$N_s = \frac{\gamma H_c}{s_{u(mob)}} \tag{5.52}$$

Chapter 5

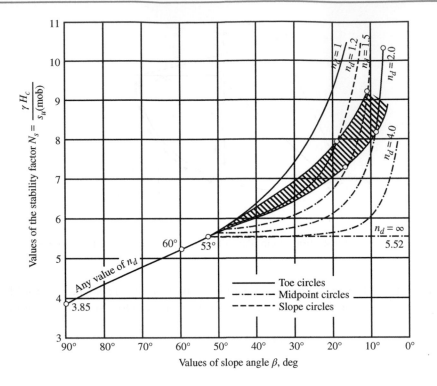

Fig. 5.9 Taylor's chart for undrained clay with $\phi = 0$. (From Terzaghi et al. 1996, based on Taylor 1937, reprinted with permission of John Wiley & Sons.)

where

γ = unit weight of the clay slope,
H_c = height of the clay slope,
$s_{u(mob)}$ = mobilized undrained shear strength.

Because it is assumed $\phi = 0$ for undrained clay, the undrained shear strength is:

$$s_u = c_u \tag{5.53}$$

and the factor of safety is:

$$FS = \frac{c_u}{s_{u(mob)}} \tag{5.54}$$

Given the slope angle β and the depth of a stiff stratum (if any), N_s can be obtained from Figure 5.9. Knowing the soil's unit weight, γ, and the slope height, H_c, $s_{u(mob)}$ can be calculated. Given c_u, the factor of safety can be calculated using Equation (5.44). Taylor's chart is easy to use and it gives the minimum FS. The disadvantage of the method is that it does not give the location of the critical slip surface.

5.5.2   c – φ soil (both c and φ are not zero)

Taylor (1937) provided a chart to obtain the FS for a slope with nonzero c and ϕ (or c' and ϕ'). The chart is shown in Figure 5.10. The chart can be used in the total stress method and the effective stress method. In the figure, a stability number is defined:

$$N_s = \frac{\gamma H_c}{c'_m} \tag{5.55}$$

where: c'_m is the mobilized cohesion. And, ϕ'_m in Figure 5.10 is the mobilized internal friction angle. It is noted that $c'_m \leq c'$ and $\phi'_m \leq \phi'$. In Figure 5.10, for intermediate values of ϕ'_m that are not shown, linear interpolation can be used; for values of $\phi'_m > 25°$, linear extrapolation can be used. The figure does not provide stability numbers for slopes with low β (less than 18°) and high ϕ' (higher than 25°).

The methodology expressed in Equation (5.9) is used. The iterative approach is shown in Figure 5.11.

When considering a limit state design approach that involves partial factors of safety, the use of simple charts such as that in Figure 5.10 is more complex and it is not described here. However, a detailed discussion including calculation examples is provided by Bond and Harris (2008).

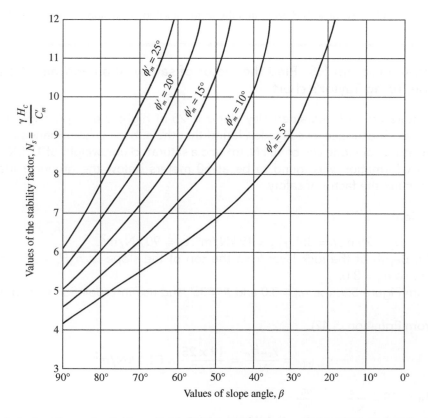

Fig. 5.10   Taylor's chart for $c - \phi$ soil. (From Terzaghi et al. 1996, based on Taylor 1937, reprinted with permission of John Wiley & Sons.)

Chapter 5

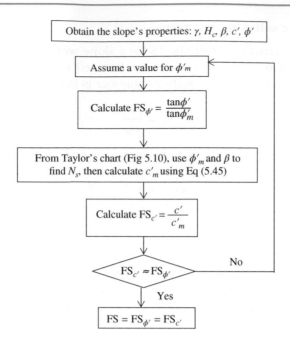

Fig. 5.11   Use Taylor's chart to find the factor of safety for a slope with $c - \phi$ soil.

Sample Problem 5.5: Find the factor of safety of an undrained clay slope, using Taylor's chart.

A saturated and undrained clay slope is 25-m high; the slope angle is 35 degrees. Subsurface investigation found that the subsoil is homogeneous clay with undrained cohesion of 90 kN/m² and a saturated unit weight of 19 kN/m³. A rock formation was found to be at 50 meters below the ground surface. Determine the factor of safety.

Solution:

Given: $H_c = 25$ m, $\beta = 35°$, $\gamma_{sat} = 19$ kN/m³, $c_u = 90$ kN/m².

The depth of the rock layer from the top of the slope is: $n_d H_c = 25 + 50 = 75$ m, so $n_d = 3.0$.

From Figure 5.9, use $n_d = 3.0$ and $\beta = 35°$ and use interpolation to find $N_s \approx 5.65$.

From Equation (5.42),

$$s_{u(mob)} = \frac{\gamma_{sat} H_c}{N_s} = \frac{19 \times 25}{5.65} = 84.1 \text{ kN/m}^2$$

Then: $FS = \dfrac{c_u}{s_{u(mob)}} = \dfrac{90}{84.1} = \mathbf{1.07}$

Conclusion: The FS is barely larger than 1.0. Therefore, the slope is considered unstable.

Sample Problem 5.6: Find the factor of safety of a slope with $c'-\phi'$ soil, using the Taylor's chart.

A natural slope is 25-m high and the slope angle is 35 degrees. Subsurface investigation found that the subsoil is homogeneous with effective cohesion of 90 kN/m^2, effective internal friction angle of 30 degrees, and unit weight of 19 kN/m^3. Determine the factor of safety.

Solution:

Given: $H_c = 25$ m, $\beta = 50°$, $\gamma = 19$ kN/m^3, $c' = 90$ kN/m^2, $\phi' = 30°$.
   Following the approach in Figure 5.11, Table 5.3 is developed for the iterative calculation.
   Conclusion: **FS = 2.08**, the slope is stable.

Table 5.3   Taylor's chart for $c'-\phi'$, use trial-and-error approach to find FS.

ϕ'_m (assumed)	$FS_{\phi'} = \dfrac{\tan\phi'}{\tan\phi'_m}$	$N_s = \dfrac{\gamma H_c}{c'_m}$ (from Figure 5.10)	$c'_m = \dfrac{\gamma H_c}{N_s}$ (kN/m^2)	$FS_{c'} = \dfrac{c'}{c_m'}$
10°	3.27	8.4	56.5	1.59
15°	2.15	10.7	44.4	2.02
17°	1.89	11.6 (use interpolation)	40.9	2.20
16°	2.01	11.2 (use interpolation)	42.4	2.12
15.5°	2.08	11.0 (use interpolation)	43.2	2.08

5.6   Slope stability analyses – methods of slices

In the methods of slices, the soil mass above an assumed failure surface is divided into vertical slices, and the force and moment equilibriums for each slice are calculated. Then all the slices are combined to derive the factor of safety of the slope for the assumed failure surface. To obtain the true factor of safety of the slope, numerous trial surfaces are analyzed that provide the minimum factor of safety.

   Fellenius (1927, 1936) first developed the method of slices, now commonly known as the *ordinary method of slices* or *Fellenius method of slices*. Since then, a number of methods of slices have been developed to consider noncircular failure surfaces, interslice forces, seepage and pore water pressure, and seismic forces. A thorough review of the methods of slices was provided by Abramson et al. (2002). In this chapter, the ordinary method of slices and the Bishop method of slices are introduced.

5.6.1   Ordinary method of slices (Fellenius method of slices)

The ordinary method of slices assumes the failure surface is circular. Both the total stress method and the effective stress method can be used in the ordinary method of slices. As shown in

Chapter 5

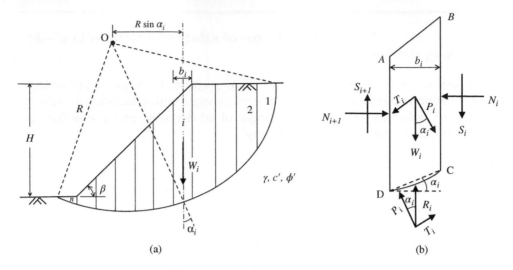

Fig. 5.12   Ordinary method of slices. (a) Slices and failure circle, (b) forces on the ith slice.

Figure 5.12(a), the soil mass above an assumed circular failure surface is divided into n slices. The failure circle's center is O and radius is R. The soil has cohesion c, internal friction angle ϕ, and unit weight γ. The slope height is H and angle is β.

The ith slice is separated out and the forces on the slice are analyzed (Figure 5.12b). The slice is subjected to gravity W_i, a reaction force R_i, a normal force from the upslope slice N_i, a shear force from the upslope slice S_i, a normal force from the downslope slice N_{i+1}, and a shear force from the downslope slice S_{i+1}. The ordinary method of slices assumes that the interslice forces on both sides of each slice cancel each other; that is, the resultant of the upslsope interslice forces is equal to the resultant of the downslope interslice forces:

$$\vec{N_i} + \vec{S_i} = \overrightarrow{N_{i+1}} + \overrightarrow{S_{i+1}} \tag{5.56}$$

In the above equation, the arrows indicate the forces are vectors that comprise direction and magnitude.

The total driving moment that causes the slice to slide (rotate) per unit length of the slope is:

$$M_d = W_i(R \sin \alpha_i) \tag{5.57}$$

The total *maximum* resisting moment per unit length of the slope is because of the cohesion and friction at the slip surface:

$$M_r = \left[(c' + \sigma' \tan \phi') \, (\overline{CD})(1) \right] R \tag{5.58}$$

where
$\overline{CD}$ = the length of the straight line CD, which is to approximate the length of arc CD.

$$\overline{CD} = \frac{b_i}{\cos \alpha_i} \qquad (5.59)$$

where

b_i = width of the ith slice; the widths of different slices can be different,

α_i = the angle of the ith slice as depicted in Figure 5.12.

σ' in Equation (5.58) is caused by the normal force P_i that is perpendicular to the line CD:

$$\sigma' = \frac{P_i}{(\overline{CD})(1)} = \frac{W_i \cos \alpha_i}{\dfrac{b_i}{\cos \alpha_i}} = \frac{W_i \cos^2 \alpha_i}{b_i} \qquad (5.60)$$

Substitute Equations (5.59) and (5.60) into Equation (5.58):

$$
\begin{aligned}
M_r &= \left[\left(c' + \frac{W_i \cos^2 \alpha_i \tan \phi'}{b_i} \right) \frac{b_i}{\cos \alpha_i} \right] R \\
&= \left(\frac{c' b_i}{\cos \alpha_i} + W_i \cos \alpha_i \tan \phi' \right) R
\end{aligned} \qquad (5.61)
$$

So, the total driving moment of the n slices in the slope is:

$$\sum_{i=1}^{n} M_d = R \sum_{i=1}^{n} (W_i \sin \alpha_i) \qquad (5.62)$$

and the total resisting moment of the n slices in the slope is:

$$\sum_{i=1}^{n} M_r = R \sum_{i=1}^{n} \left(\frac{c' b_i}{\cos \alpha_i} + W_i \cos \alpha_i \tan \phi' \right) \qquad (5.63)$$

The factor of safety of the slope on the assumed circular slip surface is:

$$FS = \frac{\displaystyle\sum_{i=1}^{n} M_r}{\displaystyle\sum_{i=1}^{n} M_d} = \frac{\displaystyle\sum_{i=1}^{n} \left(\frac{c' b_i}{\cos \alpha_i} + W_i \cos \alpha_i \tan \phi' \right)}{\displaystyle\sum_{i=1}^{n} (W_i \sin \alpha_i)} \qquad (5.64)$$

The ordinary method of slices assumes that the interslice forces on both sides of each slice cancel one another. Therefore, the factor of safety derived from this method is overconservative and is lower than other methods of slices.

Furthermore, in terms of limit state design, standards such as BS EN 1997-1:2004 require that both the vertical and momentum equilibrium of the sliding mass are checked. In addition, if horizontal equilibrium is not verified, the standard also suggests that interslice forces are assumed to be horizontal. This requirement imposes limitations on the use of the ordinary method of slices because the interslice forces are neglected. Also, on the derivation of the factor of safety the vertical and horizontal equilibriums are not both satisfied.

Chapter 5

Sample Problem 5.7: Use the ordinary method of slices to determine the FS of a slope with assumed circular failure surface.

A natural slope is shown in Figure 5.13. It contains two soil strata. The slope configuration and the soil characteristics are shown in the figure. A potential toe circle with radius of 70.0 m passes the coordinate of (65 m, 30 m). The toe is at the origin (0,0). Use the ordinary method of slices to determine the factor of safety for this potential failure circle.

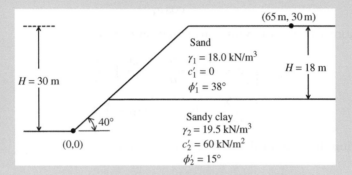

Fig. 5.13   Slope configuration for sample problem 5.7.

Solution:

The ordinary method of slices is used. The factor of safety follows Equation (5.64).

$$FS = \frac{\sum\limits_{i=1}^{n} M_r}{\sum\limits_{i=1}^{n} M_d} = \frac{\sum\limits_{i=1}^{n}\left(\dfrac{c'b_i}{\cos\alpha_i} + W_i\cos\alpha_i\tan\phi'\right)}{\sum\limits_{i=1}^{n}(W_i\sin\alpha_i)}$$

   Figure 5.14 shows the assumed toe failure circle. The slope is divided into 14 slices. For easy calculation, the bottom of each slice should be entirely in one soil stratum. The widths of the slices can vary. AutoCAD is used to obtain the area, angle α_i, width, and centroid of each slice. Table 5.4 lists the parameters of the slices that are used in the calculation of the FS. It is noted that this sample solution is to illustrate the method. In practice, a sufficient number of slices should be used to obtain an accurate FS. For example, Spencer (1967) found that increasing the number of slices above 32 results in minimal improvements in accuracy.

$$FS = \frac{\sum\limits_{i=1}^{n} M_r}{\sum\limits_{i=1}^{n} M_d} = \frac{\sum\limits_{i=1}^{n}\left(\dfrac{c'b_i}{\cos\alpha_i} + W_i\cos\alpha_i\tan\phi'\right)}{\sum\limits_{i=1}^{n}(W_i\sin\alpha_i)} = \frac{7805.35}{6422.30} = 1.22$$

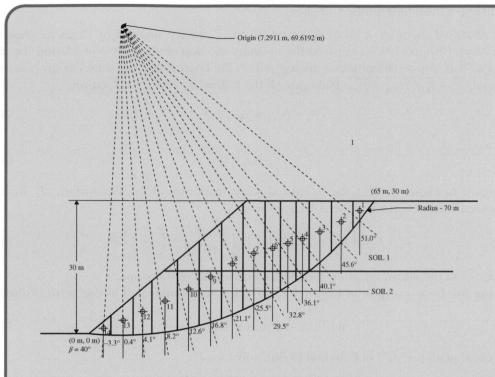

Fig. 5.14   Ordinary method of slices solution.

Table 5.4   Calculations using the ordinary method of slices.

(1) Slice number	(2) Slice area in layer 1 (m²)	(3) Slice area in layer 2 (m²)	(4) Total area (m²)	(5) Slice weight (kN/m)	(6) α_i (deg)	(7) b_i (m)	(8) $W_i \sin\alpha_i$ (kN/m)	(9) $\dfrac{c'b_i}{\cos\alpha_i} + W_i\cos\alpha_i\tan\phi'$ (kN/m)
1	16.76	0.00	16.76	301.59	51.00	5.00	234.38	148.29
2	45.56	0.00	45.56	820.11	45.60	5.00	585.95	448.30
3	68.95	0.00	68.95	1241.13	40.10	5.00	799.44	741.72
4	50.13	0.00	50.13	902.42	36.10	5.00	531.70	195.37
5	46.88	20.86	67.75	1250.73	32.80	2.96	677.53	459.26
6	55.99	18.68	74.68	1372.22	29.50	3.53	675.71	532.08
7	55.54	61.75	117.30	2203.94	25.50	5.00	948.82	833.02
8	65.56	49.12	114.68	2137.95	21.10	5.00	769.66	834.45
9	44.59	57.71	102.30	1927.95	16.80	5.00	557.24	794.55
10	23.61	64.25	87.86	1677.89	12.60	5.00	366.02	738.76
11	4.10	75.99	80.09	1555.66	8.20	5.00	221.88	712.58
12	0.00	44.71	44.71	871.94	4.10	5.00	62.34	533.04
13	0.00	33.29	33.29	649.23	0.40	5.00	4.53	473.96
14	0.00	11.50	11.50	224.19	−3.30	5.00	−12.91	359.97
						Σ	6422.30	7805.35

5.6.2　Bishop's modified method of slices

Bishop's modified method of slices is based on the ordinary method of slices as shown in Figure 5.12(a). Bishop (1955) improved the ordinary method of slices by considering the inter-slice forces. This improvement makes its use within the context of limit state design possible.

In general, $\overrightarrow{N_i} + \overrightarrow{S_i} \neq \overrightarrow{N_{i+1}} + \overrightarrow{S_{i+1}}$. Bishop took the following into consideration:

$$N_i - N_{i+1} = \Delta N_i \neq 0 \tag{5.65}$$

and,

$$S_i - S_{i+1} = \Delta S_i \neq 0 \tag{5.66}$$

With regard to Figure 5.15(a), the mobilized shear resistance at the slip surface, T_i, includes the mobilized friction force and the cohesive force:

$$T_i = P_i \tan \phi'_m + c'_m \overline{CD}(1) = P_i \left(\frac{\tan \phi'}{FS} \right) + \left(\frac{c'}{FS} \right) \overline{CD} \tag{5.67}$$

The value "1" in the above equation represents unit length of the slope.

Following the force polygon in Figure 5.15(b), the force equilibrium in the vertical direction gives:

$$W_i + \Delta S_i = P_i \cos \alpha_i + T_i \sin \alpha_i \tag{5.68}$$

Substitute Equation (5.67) in Equation (5.68), solve for P_i:

$$P_i = \frac{W_i + \Delta S_i - \dfrac{c'(\overline{CD}) \sin \alpha_i}{FS}}{\cos \alpha_i + \dfrac{\tan \phi' \sin \alpha_i}{FS}} \tag{5.69}$$

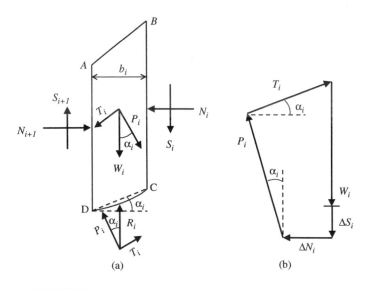

(a)　　　　(b)

Fig. 5.15　Bishop's modified method of slices. (a) Forces on the ith slice, (b) force polygon for equilibrium.

When considering the sliding and resisting moments of the entire failure mass above the slip surface, the moments caused by the interslice forces cancel. The moment equilibrium of the entire failure mass gives:

$$\sum_{i=1}^{n}(W_i R \sin \alpha_i) = \sum_{i=1}^{n}(T_i R) \tag{5.70}$$

Substitute Equation (5.69) in Equation (5.67), then substitute Equation (5.67) into Equation (5.70), solve for FS and find:

$$FS = \frac{\sum_{i=1}^{n}\left[(c'b_i + W_i \tan \phi' + \Delta S_i)\frac{1}{m_i}\right]}{\sum_{i=1}^{n}(W_i \sin \alpha_i)} \tag{5.71}$$

where

$$m_i = \cos \alpha_i + \frac{\tan \phi' \sin \alpha_i}{FS} \tag{5.72}$$

Equations (5.71) and (5.72) together are known as *Bishop's modified method of slices*. As the interslice shear force difference, ΔS_i, is difficult to evaluate, the application of the method is limited. Bishop (1955) simplified Equation (5.71) by assuming:

$$\Delta S_i = 0 \text{ for each slice} \tag{5.73}$$

So, Equation (5.71) becomes:

$$FS = \frac{\sum_{i=1}^{n}(c'b_i + W_i \tan \phi')\frac{1}{m_i}}{\sum_{i=1}^{n}(W_i \sin \alpha_i)} \tag{5.74}$$

Equation (5.74) is referred to as *Bishop's simplified method of slices*. It is noted that the FS exists on both sides of the equation. Therefore, the trial-and-error method is used: for an assumed failure surface, assume an FS and calculate m_i, then use Equation (5.74) to calculate the FS. If the calculated FS is different from the assumed FS, the average of the two FS is used as a new assumed FS. This process is repeated until the calculated FS is equal to the assumed FS. To obtain the true FS of a slope, numerous failure surfaces should be evaluated by varying the center and the radius so that the minimum factor of safety can be identified.

Chapter 5

Sample Problem 5.8: Use Bishop's simplified method of slices to determine the FS of a slope with an assumed circular failure surface.

The problem statement is the same as in the sample problem 5.7 and is shown in Figure 5.13. Use Bishop's simplified method of slices to determine the factor of safety for this potential failure circle, and compare the FS with the one obtained using the ordinary method of slices.

Solution:

Use Bishop's simplified method of slices. The factor of safety follows Equations (5.72) and (5.74).

$$FS = \frac{\sum_{i=1}^{n}\left[(c'b_i + W_i \tan \phi')\dfrac{1}{m_i}\right]}{\sum_{i=1}^{n}(W_i \sin \alpha_i)}$$

$$m_i = \cos \alpha_i + \frac{\tan \phi' \sin \alpha_i}{FS}$$

Table 5.5   Calculations using Bishop's simplified method of slices.

(1)	(2)	(3)	(4)	(5)	(6)	(7)	(8)	(9)	(10)
Slice #	Slice area in layer 1 (m²)	Slice area in layer 2 (m²)	Total area (m²)	Slice weight (kN/m)	α_i (deg)	b_i (m)	$W_i \sin\alpha_i$ (kN/m)	m_i	$(c'b_i + W_i\tan\phi')\dfrac{1}{m_i}$ (kN/m)
1	16.76	0.00	16.76	301.59	51.0	5.00	234.38	1.08	219.03
2	45.56	0.00	45.56	820.11	45.6	5.00	585.95	1.11	577.19
3	68.95	0.00	68.95	1241.13	40.1	5.00	799.44	1.13	854.38
4	50.13	0.00	50.13	902.42	36.1	5.00	531.70	0.92	561.67
5	46.88	20.86	67.75	1250.73	32.8	2.96	677.53	0.95	531.34
6	55.99	18.68	74.68	1372.22	29.5	3.53	675.71	0.97	592.15
7	55.54	61.75	117.30	2203.94	25.5	5.00	948.82	0.99	898.08
8	65.56	49.12	114.68	2137.95	21.1	5.00	769.66	1.00	870.65
9	44.59	57.71	102.30	1927.95	16.8	5.00	557.24	1.01	809.33
10	23.61	64.25	87.86	1677.89	12.6	5.00	366.02	1.02	741.25
11	4.10	75.99	80.09	1555.66	8.2	5.00	221.88	1.02	709.52
12	0.00	44.71	44.71	871.94	4.1	5.00	62.34	1.01	530.97
13	0.00	33.29	33.29	649.23	0.4	5.00	4.53	1.00	473.73
14	0.00	11.50	11.50	224.19	−3.3	5.00	−12.91	0.99	360.86
						Σ	6422.30		8730.14

The assumed toe failure circle and slices are the same as shown in Figure 5.14. AutoCAD is used to obtain the area, angle α_i, width, and centroid of each slice. Using trial-and-error method and an assumed FS of 1.36, the calculations in Table 5.5 are prepared. It is noted that columns (1) to (8) are the same as in Table 5.4 (ordinary method of slices).

$$FS = \frac{\sum_{i=1}^{n}\left[(c'b_i + W_i \tan \phi')\dfrac{1}{m_i}\right]}{\sum_{i=1}^{n}(W_i \sin \alpha_i)} = \frac{8730.14}{6422.30} = 1.36 \text{ (equal to the assumed FS)}$$

The FS (1.36) obtained using Bishop's simplified method of slices is larger than the FS (1.22) using the ordinary method of slices.

Despite that this method satisfies the equilibrium requirements and the assumptions regarding interslice forces as required by BS EN 1997-1:2004, it is not an easy one to verify ultimate limit states because the factor of safety (F in equation 5.74) is a function of an unfactored disturbing moment and a factored shear resistance. Furthermore, as discussed in previous chapters, limit state design considers different values for partial factors of safety for favorable and unfavorable forces. Within that context, knowing that the center and radius of the slip circle will influence whether the weight of a corresponding slice is favourable or unfavourable is very difficult to know before completing the analysis. Details of such method are therefore not discussed here, but interested readers can refer to Smith (2007) and Barnes (2010) as both have calculation examples for this case.

5.7   Slope stability analyses – consideration of pore water pressure

5.7.1   Bishop–Morgenstern method

Bishop–Morgenstern method (Bishop and Morgenstern 1960) is based on the effective stress method and considers the effect of pore water pressure on slope stability. Figure 5.16 shows a slope with an assumed failure surface. The pore water pressure at any location along a failure surface is expressed using a pore pressure ratio, r_u, defined by Bishop and Morgenstern (1960):

$$r_u = \frac{u}{\gamma b} \qquad (5.75)$$

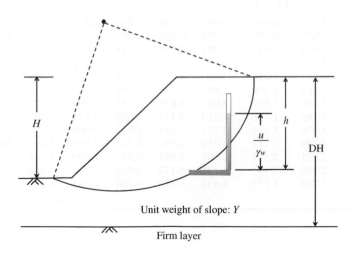

Fig. 5.16   Illustration of pore water pressure in a slope.

Table 5.6   Values of m and n in Bishop–Morgenstern method.

ϕ'(deg)	Slope 2(H):1(V)		Slope 3:1		Slope 4:1		Slope 5:1	
	m	n	m	n	m	n	m	n
(a) $\frac{c'}{\gamma H} = 0$ and for all values of D								
10	0.353	0.441	0.529	0.588	0.705	0.749	0.882	0.917
12.5	0.443	0.554	0.665	0.739	0.887	0.943	1.109	1.153
15	0.536	0.670	0.804	0.893	1.072	1.139	1.340	1.393
17.5	0.631	0.789	0.946	1.051	1.261	1.340	1.577	1.639
20	0.728	0.910	1.092	1.213	1.456	1.547	1.820	1.892
22.5	0.828	1.035	1.243	1.381	1.657	1.761	2.071	2.153
25	0.933	1.166	1.399	1.554	1.865	1.982	2.332	2.424
27.5	1.041	1.301	1.562	1.736	2.082	2.213	2.603	2.706
30	1.155	1.444	1.732	1.924	2.309	2.454	2.887	3.001
32.5	1.274	1.593	1.911	2.123	2.548	2.708	3.185	3.311
35	1.400	1.750	2.101	2.334	2.801	2.977	3.501	3.639
37.5	1.535	1.919	2.302	2.558	3.069	3.261	3.837	3.989
40	1.678	2.098	2.517	2.797	3.356	3.566	4.196	4.362
(b) $\frac{c'}{\gamma H} = 0.025$ and $D = 1.00$								
10	0.678	0.534	0.906	0.683	1.130	0.846	1.365	1.031
12.5	0.790	0.655	1.066	0.849	1.337	1.061	1.620	1.282
15	0.901	0.776	1.224	1.014	1.544	1.273	1.868	1.543
17.5	1.012	0.898	1.380	1.179	1.751	1.485	1.212	1.789
20	1.124	1.022	1.542	1.347	1.962	1.698	2.380	2.050
22.5	1.239	1.150	1.705	1.518	2.177	1.916	2.646	2.317
25	1.356	1.282	1.875	1.696	2.400	2.141	2.921	2.596
27.5	1.478	1.421	2.050	1.882	2.631	2.375	3.207	2.886
30	1.606	1.567	2.235	2.078	2.873	2.622	3.508	3.191
32.5	1.739	1.721	2.431	2.285	3.127	2.883	3.823	3.511
35	1.880	1.885	2.635	2.505	3.396	3.160	4.156	3.849
37.5	2.030	2.060	2.855	2.741	3.681	3.458	4.510	4.209
40	2.190	2.247	3.090	2.993	3.984	3.778	4.885	4.592
(c) $\frac{c'}{\gamma H} = 0.025$ and $D = 1.25$								
10	0.737	0.614	0.901	0.726	1.085	0.867	1.285	1.014
12.5	0.878	0.759	1.076	0.908	1.299	1.089	1.543	1.278
15	1.019	0.907	1.253	1.093	1.515	1.312	1.803	1.545
17.5	1.162	1.059	1.433	1.282	1.736	1.541	2.065	1.814
20	1.309	1.216	1.618	1.478	1.961	1.775	2.334	2.090
22.5	1.461	1.379	1.808	1.680	2.194	2.017	2.610	2.373
25	1.619	1.547	2.007	1.891	2.437	2.269	2.897	2.669
27.5	1.783	1.728	2.213	2.111	2.689	2.531	3.196	2.976
30	1.956	1.915	2.431	2.342	2.953	2.806	3.511	3.299
32.5	2.139	2.112	2.659	2.585	3.231	3.095	3.841	3.638
35	2.331	2.321	2.901	2.841	3.524	3.400	4.191	3.998
37.5	2.536	2.541	3.158	3.112	3.835	3.723	4.563	4.379
40	2.753	2.775	3.431	3.399	4.164	4.064	4.958	4.784

(continued overleaf)

Table 5.6   (*continued*)

ϕ'(deg)	Slope 2(H):1(V)		Slope 3:1		Slope 4:1		Slope 5:1	
	m	*n*	*m*	*n*	*m*	*n*	*m*	*n*

(d) $\dfrac{c'}{\gamma H} = 0.05$ and $D = 1.00$

ϕ'(deg)	*m*	*n*	*m*	*n*	*m*	*n*	*m*	*n*
10	0.913	0.563	1.181	0.717	1.469	0.910	1.733	1.069
12.5	1.030	0.690	1.343	0.878	1.688	1.136	1.995	1.316
15	1.145	0.816	1.506	1.043	1.904	1.353	2.256	1.567
17.5	1.262	0.942	1.671	1.212	2.117	1.565	2.517	1.825
20	1.380	1.071	1.840	1.387	2.333	1.776	2.783	2.091
22.5	1.500	1.202	2.104	1.568	2.551	1.989	3.055	2.365
25	1.624	1.338	2.193	1.757	2.778	2.211	3.336	2.651
27.5	1.753	1.480	2.380	1.952	3.013	2.444	3.628	2.948
30	1.888	1.630	2.574	2.157	3.261	2.693	3.934	3.259
32.5	2.029	1.789	2.777	2.370	3.523	2.961	4.256	3.585
35	2.178	1.958	2.990	2.592	3.803	3.253	4.597	3.927
37.5	2.336	2.138	3.215	2.826	4.013	3.574	4.959	4.288
40	2.505	2.332	3.451	3.071	4.425	3.926	5.344	4.668

(e) $\dfrac{c'}{\gamma H} = 0.05$ and $D = 1.25$

ϕ'(deg)	*m*	*n*	*m*	*n*	*m*	*n*	*m*	*n*
10	0.919	0.633	1.119	0.766	1.344	0.886	1.594	1.042
12.5	1.065	0.792	1.294	0.941	1.563	1.112	1.850	1.300
15	1.211	0.950	1.471	1.119	1.782	1.338	2.109	1.562
17.5	1.359	1.108	1.650	1.303	2.004	1.567	2.373	1.831
20	1.509	1.266	1.834	1.493	2.230	1.799	2.643	2.107
22.5	1.663	1.428	2.024	1.690	2.463	2.038	2.921	2.392
25	1.822	1.595	2.222	1.897	2.705	2.287	3.211	2.690
27.5	1.988	1.769	2.428	2.113	2.957	2.546	3.513	2.999
30	2.161	1.950	2.645	2.342	3.221	2.819	3.829	3.324
32.5	2.343	2.141	2.837	2.583	3.500	3.107	4.161	3.665
35	2.535	2.344	3.114	2.839	3.795	3.413	4.511	4.025
37.5	2.738	2.560	3.370	3.111	4.109	3.740	4.881	4.405
40	2.953	2.791	3.642	3.400	4.442	4.090	5.273	4.806

(f) $\dfrac{c'}{\gamma H} = 0.05$ and $D = 1.50$

ϕ'(deg)	*m*	*n*	*m*	*n*	*m*	*n*	*m*	*n*
10	1.022	0.751	1.170	0.828	1.343	0.974	1.547	1.108
12.5	1.202	0.936	1.376	1.043	1.589	1.227	1.829	1.399
15	1.383	1.122	1.583	1.260	1.835	1.480	2.112	1.690
17.5	1.565	1.309	1.795	1.480	2.084	1.734	2.398	1.983
20	1.752	1.501	2.011	1.705	2.337	1.993	2.690	2.280
22.5	1.943	1.698	2.234	1.937	2.597	2.258	2.990	2.585
25	2.143	1.903	2.467	2.179	2.876	2.534	3.302	2.902
27.5	2.350	2.117	2.709	2.431	3.148	2.820	3.626	3.231
30	2.568	2.342	2.964	2.696	3.443	3.120	3.967	3.577
32.5	2.798	2.580	3.232	2.975	3.753	3.436	4.326	3.940
35	3.041	2.832	3.515	3.269	4.082	3.771	4.707	4.325
37.5	3.299	3.102	3.817	3.583	4.431	4.128	5.112	4.735
40	3.574	3.389	4.136	3.915	4.803	4.507	5.543	5.171

Chapter 5

Where

u = pore water pressure at any location along a failure surface,
(γb) = total overburden stress at the location where u is measured.

Bishop and Morgenstern (1960) proposed a simplified expression for the calculation of the factor of safety:

$$FS = m - n \cdot r_u \qquad (5.76)$$

where: m and n are the stability coefficients depending on the slope height (H), slope inclination, unit weight of the slope (γ), the soil's effective cohesion (c') and effective friction angle (ϕ'), and the location of the critical failure circle. The location of the critical failure circle depends on the elevation of a firm layer beneath the slope, which is expressed by DH, where D is a unitless coefficient. On the basis of Bishop's simplified method of slices and the evaluation of numerous assumed failure surfaces, Bishop and Morgenstern (1960) provided the m and n values in a table format (Table 5.6), which is indexed by a unitless parameter $c'/\gamma H$ of three values (0, 0.025, and 0.05) and the depth coefficient (D) of a firm layer.

If there is a firm layer beneath the slope, the slip surface is usually tangent to the firm layer. Therefore, the elevation of the firm layer within a shallow depth determines the location of the failure surface. If there is no firm layer beneath a slope, or a firm layer is far below a slope, the location of the critical failure surface can be found by varying D and finding the minimum factor of safety, as shown in Sample Problem 5.9.

Chandler and Peiris (1989) further extended the Bishop–Morgenstern slope stability tables by providing slope inclinations from 0.5:1 to 5:1 (H:V), $c'/\gamma H$ values from 0 to 0.15, and ϕ' values from 20° to 40°. Tables 5.7(a)–(c) list the supplementary m and n values for $c'/\gamma H = 0$, 0.025, and 0.05, and slope inclinations of 0.5:1 and 1:1; Tables 5.7(d)–(o) list the m and n values for $c'/\gamma H = 0.075$, 0.100, 0.125, and 0.150 (Table 5.7).

Determination of pore pressure ratio, r_u

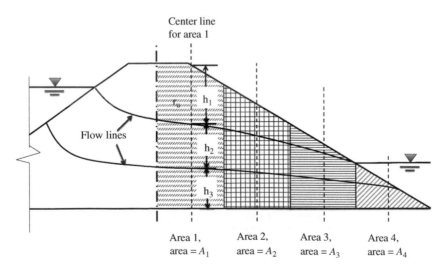

Fig. 5.17 Example of a nonuniform pore pressure distribution and calculation of average r_u.

Table 5.7   Values of m and n in Bishop–Morgenstern method (Chandler and Peiris 1989).

(a) $\dfrac{c'}{\gamma H} = 0$ and for all values of D

ϕ'	Slope 0.5(H):1(V)		Slope 1:1	
	m	n	m	n
20°	0.182	0.910	0.364	0.728
25°	0.233	1.166	0.466	0.933
30°	0.289	1.443	0.577	1.155
35°	0.350	1.751	0.700	1.400
40°	0.420	2.098	0.839	1.678

(b) $\dfrac{c'}{\gamma H} = 0.025$

ϕ'	$D = 1.00$				$D = 1.25$			
	Slope 0.5(H):1(V)		Slope 1:1		Slope 0.5:1		Slope 1:1	
	m	n	m	n	m	n	m	n
20°	0.523	0.733	0.707	0.764	0.996	0.943	1.076	1.036
25°	0.586	0.740	0.839	0.974	1.224	1.198	1.328	1.321
30°	0.667	0.875	0.983	1.202	1.472	1.483	1.602	1.631
35°	0.755	1.010	1.142	1.454	1.737	1.773	1.905	1.974
40°	0.857	1.183	1.318	1.731	2.045	2.118	2.242	2.362

(c) $\dfrac{c'}{\gamma H} = 0.05$

ϕ'	$D = 1.00$				$D = 1.25$				$D = 1.50$			
	Slope 0.5(H):1(V)		Slope 1:1		Slope 0.5:1		Slope 1:1		Slope 0.5:1		Slope 1:1	
	m	n	m	n	m	n	m	n	m	n	m	n
20°	0.688	0.783	0.912	0.818	1.172	0.988	1.253	1.084	1.491	1.289	1.561	1.343
25°	0.797	1.000	1.069	1.042	1.405	1.242	1.509	1.363	1.826	1.637	1.910	1.709
30°	0.908	1.217	1.222	1.247	1.656	1.518	1.783	1.669	2.187	2.015	2.287	2.104
35°	1.032	1.417	1.379	1.469	1.935	1.830	2.087	2.007	2.587	2.436	2.704	2.541
40°	1.148	1.617	1.599	1.755	2.245	2.174	2.429	2.390	3.040	2.915	3.175	3.036

(d) $\dfrac{c'}{\gamma H} = 0.075; D = 1.00$

ϕ'	Slope 0.5(H):1(V)		Slope 1:1		Slope 2:1		Slope 3:1		Slope 4:1		Slope 5:1	
	m	n	m	n	m	n	m	n	m	n	m	n
20°	0.845	0.800	1.088	0.837	1.610	1.100	2.141	1.443	2.664	1.801	3.173	2.130
25°	0.950	1.013	1.245	1.053	1.872	1.386	2.502	1.815	3.126	2.259	3.742	2.715
30°	1.064	1.238	1.416	1.296	2.142	1.686	2.884	2.201	3.623	2.758	4.357	3.331
35°	1.190	1.485	1.605	1.564	2.443	2.030	3.306	2.659	4.177	3.331	5.024	4.001
40°	1.332	1.762	1.798	1.824	2.772	2.386	3.775	3.145	4.785	3.945	5.776	4.759

(e) $\dfrac{c'}{\gamma H} = 0.075; D = 1.25$

ϕ'												
20°	1.336	1.023	1.387	1.087	1.688	1.285	2.071	1.543	2.492	1.815	2.954	2.173
25°	1.575	1.284	1.656	1.386	2.004	1.641	2.469	1.957	2.972	2.315	3.523	2.730
30°	1.830	1.560	1.943	1.701	2.352	2.015	2.888	2.385	3.499	2.857	4.149	3.357
35°	2.109	1.865	2.245	2.025	2.728	2.385	3.357	2.870	4.079	3.457	4.831	4.043
40°	2.424	2.210	2.583	2.403	3.154	2.841	3.889	3.428	4.729	4.128	5.603	4.830

(continued overleaf)

Chapter 5

Table 5.7 (*continued*)

ϕ'	Slope 0.5(H):1(V)		Slope 1:1		Slope 2:1		Slope 3:1		Slope 4:1		Slope 5:1	
	m	*n*	*m*	*n*	*m*	*n*	*m*	*n*	*m*	*n*	*m*	*n*
(f) $\dfrac{c'}{\gamma H} = 0.075; D = 1.50$												
20°	1.637	1.305	1.706	1.349	1.918	1.514	2.199	1.728	2.548	1.985	2.931	2.272
25°	1.977	1.663	2.052	1.708	2.308	1.914	2.660	2.200	3.083	2.530	3.552	2.915
30°	2.340	2.041	2.426	2.100	2.735	2.355	3.158	2.714	3.659	3.128	4.218	3.585
35°	2.741	2.459	2.841	2.537	3.211	2.854	3.708	3.285	4.302	3.786	4.961	4.343
40°	3.193	2.931	3.310	3.031	3.742	3.397	4.332	3.926	5.026	4.527	5.788	5.185
(g) $\dfrac{c'}{\gamma H} = 0.100; D = 1.00$												
20°	0.993	0.797	1.263	0.871	1.841	1.143	2.421	1.472	2.982	1.815	3.549	2.157
25°	1.106	1.025	1.422	1.078	2.102	1.430	2.785	1.845	3.358	2.303	4.131	2.743
30°	1.222	1.259	1.592	1.306	2.378	1.714	3.183	2.258	3.973	2.830	4.751	3.372
35°	1.347	1.508	1.781	1.576	2.692	2.086	3.612	2.715	4.516	3.359	5.426	4.059
40°	1.489	1.788	1.995	1.879	3.025	2.445	4.103	3.230	5.144	4.001	6.187	4.831
(h) $\dfrac{c'}{\gamma H} = 0.100; D = 1.25$												
20°	1.489	1.036	1.529	1.095	1.874	1.301	2.283	1.558	2.751	1.843	3.253	2.158
25°	1.735	1.313	1.799	1.394	2.197	1.642	2.681	1.972	3.233	2.330	3.833	2.758
30°	1.997	1.602	2.091	1.718	2.540	2.000	3.112	2.415	3.753	2.858	4.451	3.372
35°	2.280	1.908	2.414	2.076	2.922	2.415	3.588	2.914	4.333	3.458	5.141	4.072
40°	2.597	2.253	2.763	2.453	3.345	2.855	4.119	3.457	4.987	4.142	5.921	4.872
(i) $\dfrac{c'}{\gamma H} = 0.100; D = 1.50$												
20°	1.778	1.314	1.863	1.371	2.079	1.528	2.387	1.742	2.768	2.014	3.158	2.285
25°	2.119	1.674	2.211	1.732	2.477	1.942	2.852	2.215	3.297	2.542	3.796	2.927
30°	2.489	2.063	2.586	2.122	2.908	2.385	3.349	2.728	3.881	3.143	4.468	3.614
35°	2.892	2.484	3.000	2.553	3.385	2.884	3.900	3.300	4.520	3.800	5.211	4.372
40°	3.347	2.957	3.469	3.046	3.924	3.441	4.524	3.941	5.247	4.542	6.040	5.200
(j) $\dfrac{c'}{\gamma H} = 0.125; D = 1.00$												
20°	1.121	0.808	1.425	0.881	2.042	1.148	2.689	1.541	3.263	1.784	3.868	2.124
25°	1.254	1.051	1.596	1.112	2.323	1.447	3.062	1.908	3.737	2.271	4.446	2.721
30°	1.376	1.267	1.769	1.337	2.618	1.777	3.457	2.298	4.253	2.810	5.073	3.368
35°	1.505	1.530	1.956	1.586	2.929	2.115	3.880	2.705	4.823	3.407	5.767	4.048
40°	1.612	1.743	2.171	1.891	3.272	2.483	4.356	3.183	5.457	4.060	6.551	4.893
(k) $\dfrac{c'}{\gamma H} = 0.125; D = 1.25$												
20°	1.642	1.057	1.671	1.102	2.054	1.324	2.492	1.579	2.983	1.861	3.496	2.167
25°	1.888	1.326	1.941	1.402	2.377	1.671	2.894	1.993	3.481	2.379	4.078	2.753
30°	2.156	1.626	2.234	1.727	2.727	2.042	3.324	2.431	4.009	2.916	4.712	3.405
35°	2.447	1.948	2.557	2.085	3.110	2.452	3.801	2.928	4.586	3.500	5.414	4.128
40°	2.767	2.295	2.922	2.490	3.542	2.913	4.338	3.494	5.237	4.161	6.207	4.945
(l) $\dfrac{c'}{\gamma H} = 0.125; D = 1.50$												
20°	1.920	1.322	2.015	1.385	2.234	1.545	2.565	1.749	2.963	2.004	3.400	2.287
25°	2.261	1.683	2.368	1.754	2.638	1.972	3.028	2.229	3.500	2.550	4.019	2.913
30°	2.631	2.073	2.745	2.145	3.072	2.425	3.529	2.749	4.083	3.149	4.692	3.598
35°	3.039	2.504	3.160	2.577	3.549	2.923	4.084	3.324	4.727	3.813	5.436	4.362
40°	3.497	2.982	3.628	3.065	4.089	3.485	4.712	3.980	5.456	4.566	6.278	5.226

(*continued overleaf*)

Table 5.7   *(continued)*

ϕ'	Slope 0.5(H):1(V)		Slope 1:1		Slope 2:1		Slope 3:1		Slope 4:1		Slope 5:1	
	m	*n*	*m*	*n*	*m*	*n*	*m*	*n*	*m*	*n*	*m*	*n*
(m) $\dfrac{c'}{\gamma H} = 0.150; D = 1.00$												
20°	1.248	0.813	1.585	0.886	2.261	1.170	2.895	1.448	3.579	1.806	4.230	2.159
25°	1.386	1.034	1.761	1.126	2.536	1.462	3.259	1.814	4.052	2.280	4.817	2.765
30°	1.525	1.260	1.944	1.370	2.836	1.791	3.657	2.245	4.567	2.811	5.451	3.416
35°	1.660	1.539	2.134	1.619	3.161	2.153	4.098	2.721	5.137	3.408	6.143	4.117
40°	1.805	1.832	2.346	1.901	3.512	2.535	4.597	3.258	5.782	4.083	6.913	4.888
(n) $\dfrac{c'}{\gamma H} = 0.150; D = 1.25$												
20°	1.796	1.079	1.813	1.107	2.229	1.334	2.701	1.600	3.225	1.837	3.780	2.182
25°	2.042	1.344	2.083	1.409	2.560	1.692	3.107	2.015	3.724	2.384	4.363	2.769
30°	2.309	1.639	2.377	1.734	2.909	2.065	3.542	2.464	4.262	2.941	4.995	3.406
35°	2.605	1.971	2.700	2.094	3.295	2.457	4.018	2.946	4.846	3.534	5.697	4.129
40°	2.934	2.335	3.066	2.449	3.728	2.938	4.556	3.509	5.498	4.195	6.490	4.947
(o) $\dfrac{c'}{\gamma H} = 0.150; D = 1.50$												
20°	2.061	1.335	2.164	1.391	2.394	1.550	2.748	1.756	3.174	2.020	3.641	2.308
25°	2.402	1.691	2.520	1.768	2.798	1.978	3.212	2.237	3.711	2.561	4.259	2.924
30°	2.772	2.082	2.902	2.168	3.236	2.441	3.718	2.758	4.293	3.156	4.931	3.604
35°	3.181	2.514	3.319	2.600	3.715	2.940	4.269	3.333	4.938	3.819	5.675	4.364
40°	3.643	3.000	2.788	3.088	4.255	3.503	4.896	3.983	5.667	4.569	6.517	5.228

The value of r_u is in general not constant over the entire cross section of a slope. In most slope stability problems, an average value of r_u can be readily calculated and used in the stability analyses with little loss in accuracy (Bishop and Morgenstern 1960). For steady seepage conditions in earthen dams or levees and the long-term stability of a natural slope, the pore water pressure distribution can be measured using piezometers, estimated using a flow net, or determined using analytical or numerical analyses. Bishop and Morgenstern (1960) presented a method to determine the weighted average of r_u for a slope with seepage. The approach is illustrated in Figure 5.17 and described as follows.

(a) Take a cross section as shown in Figure 5.17 and divide the slope into several areas with equal base. For earthen embankments, start from the middle of the crest; for natural slopes, start at the top of the slope at a distance of $H/4$ from the crest, H is the slope height.
(b) Divide each area into several "pore pressure zones" using flow lines as boundaries. Determine the pore water pressure at the middle point of the bottom of each zone. Figure 5.17 shows three zones in Area 1.
(c) Determine the height (h_i) of each zone along the centerline of the zone; determine the area of each area ($A_{(n)}$).
(d) Determine the total overburden stress (γh) for each zone; then calculate r_u for each zone using Equation (5.75): $r_u = u/\gamma h$.

Chapter 5

(e) For each area, determine the average r_u:

$$r_{u(n)} = \frac{\sum (b_i r_{ui})}{\sum b_i} \tag{5.77}$$

Where

b_i = height of each pore pressure zone, as shown in Figure 5.17,
r_{ui} = pore pressure in a zone in an area

(f) Determine the overall average r_u for the entire slope:

$$r_u = \frac{\sum (A_{(n)} r_{u(n)})}{\sum A_{(n)}} \tag{5.78}$$

For $c'/\gamma H$ values falling between the values given in Tables 5.6 and 5.7, the FS should first be calculated using the two closest values of $c'/\gamma H$ in the tables; then linear interpolation can be used on the calculated FS values to determine the FS for the given $c'/\gamma H$ value.

Sample Problem 5.9: Use Bishop–Morgenstern method to determine the FS with the consideration of pore pressure.

A natural slope is 45 meters high, and the slope inclination is 3(H):1(V). The slope's properties are: $c' = 30 \, kN/m^2$, $\phi' = 25°$, and $\gamma = 19.0 \, kN/m^3$. No firm layer was encountered beneath the slope during the subsurface investigation. Steady seepage is parallel to the slope surface. The average pore water pressure ratio for the slope was determined to be 0.35. Determine the minimum factor of safety for the slope, considering the effect of pore water pressure.

Solution:

Use Bishop–Morgenstern method and consider the effect of pore pressure.
$\dfrac{c'}{\gamma H} = \dfrac{30}{19.0 \times 45} = 0.035$, it is between 0.025 and 0.05.
Given the slope inclination of 3 : 1, $\phi' = 25°$, and average pore pressure ratio $r_u = 0.35$, the following trial-and-error approach is used to find the minimum FS:

• From Table 5.6(b): $c'/\gamma H = 0.025$ and assume a failure surface passes $D = 1.00$:

$$m = 1.875, \quad n = 1.696. \quad FS = m - n r_u = 1.875 - 1.696 \times 0.35 = 1.281$$

- From Table 5.6(c): $c'/\gamma H = 0.025$ and assume a failure surface passes $D = 1.25$:

$$m = 2.007, n = 1.891. \quad FS = m - nr_u = 2.007 - 1.891 \times 0.35 = 1.345$$

- From Table 5.6(d): $c/\gamma H = 0.05$ and assume a failure surface passes $D = 1.00$

$$m = 2.193, n = 1.757. \quad FS = m - nr_u = 2.193 - 1.757 \times 0.35 = 1.578$$

- From Table 5.6(e): $c'/\gamma H = 0.05$ and assume a failure surface passes $D = 1.25$

$$m = 2.222, n = 1.897. \quad FS = m - nr_u = 2.222 - 1.897 \times 0.35 = 1.558$$

Using interpolation, find the FS for $c'/\gamma H = 0.035$ for $D = 1.00$:

$$FS = 1.281 + \frac{0.035 - 0.025}{0.05 - 0.025} \times (1.578 - 1.281) = 1.40$$

Use interpolation, find the FS for $c'/\gamma H = 0.035$ for $D = 1.25$:

$$FS = 1.345 + \frac{0.035 - 0.025}{0.05 - 0.025} \times (1.558 - 1.345) = 1.43$$

Therefore, the minimum FS $= 1.40$ and the bottom of the critical failure surface is at the same level as the toe ($D = 1.0$).

5.7.2   Spencer charts

Spencer (1967) provided charts to determine the minimum factor of safety, known as "Spencer charts" (Figure 5.18). The charts are based on the *effective stress method*. In developing the charts, Spencer assumed the slope and the underlying soil are homogeneous and have similar properties, and the position of the critical slip circle is determined by the height, angle, and properties of the slope, not by the presence of a hard stratum, that is, the depth of a hard stratum is sufficiently great and does not affect the slip circle. The charts were obtained using the method of slices that assumed the interslices are parallel and satisfied the equilibrium of both forces and moments. Note that all these assumptions satisfy all the requirements of limit state design as stated in BS EN 1997-1:2004). The Spencer charts used three pore pressure ratios: $r_u = 0$ (no pore pressure), $r_u = 0.25$, and $r_u = 0.5$. For each value of r_u, the relationship between the mobilized effective internal friction angle (ϕ'_m) and the stability number ($c'/FS \cdot \gamma H$) are obtained for four different slopes of 1.5:1, 2:1, 3:1, and 4:1 (H:V). If r_u falls between 0 and 0.25 or between 0.25 and 0.5, linear interpolation can be used with sufficient accuracy for practical purposes (Spencer 1967).

The trial-and-error approach is used in obtaining the minimum FS. The approach is shown in Figure 5.19.

Chapter 5

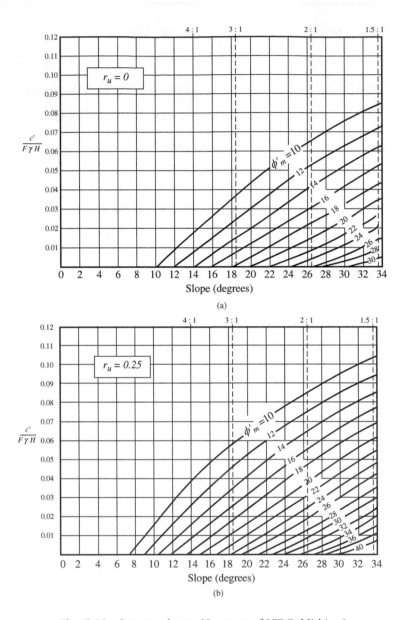

Fig. 5.18 Spencer charts. (Courtesy of ICE Publishing.)

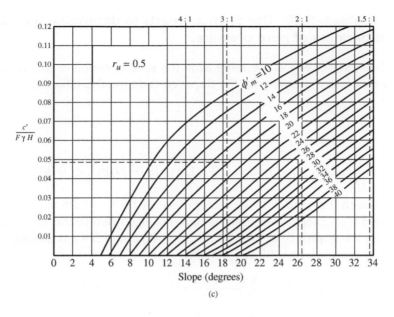

Fig. 5.18   (*continued*)

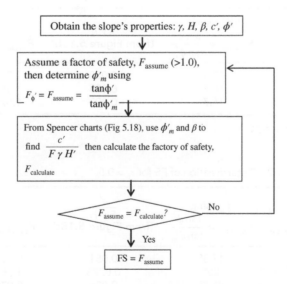

Fig. 5.19   Approach to determine FS using Spencer chart.

Sample Problem 5.10: **Use Spencer charts to determine the FS with the consideration of pore pressure.**

The problem statement is the same as in the sample problem 5.9. Determine the minimum factor of safety for the slope, considering the effect of pore water pressure.

Solution:

Given:

$H = 45$ m,   slope   inclination $3 : 1, \gamma = 19.0$ kN/m³, $c' = 30$ kN/m², $\phi' = 25°, r_u = 0.35$.

r_u is between 0.25 and 0.5. The factor of safety for $r_u = 0.25$ and $r_u = 0.5$ are determined first. Then linear interpolation is used to determine the FS for $r_u = 0.35$. Using the approach in Figure 5.19, Tables 5.8 and 5.9 are developed.

Use linear interpolation; FS at $r_u = 0.35$:

$$FS = 1.53 - \frac{0.35 - 0.25}{0.5 - 0.25} \times (1.53 - 1.13) = 1.37$$

Note: the FS obtained using Bishop–Morgenstern method for the same problem is 1.40.

Table 5.8   Determination of FS for $r_u = 0.25$.

F_{assume}	ϕ'_m using $F_{assume} = \dfrac{\tan \phi'}{\tan \phi'_m}$	$\dfrac{c'}{F \gamma H}$ (from Figure 5.18b)	$F_{calculate}$
1.5	17.3°	0.020	1.75
1.6	16.2°	0.025	1.40
1.55	16.7°	0.023	1.52

Take FS = 1.53

Table 5.9   Determination of FS for $r_u = 0.5$.

F_{assume}	ϕ'_m using $F_{assume} = \dfrac{\tan \phi'}{\tan \phi'_m}$	$\dfrac{c'}{F \gamma H}$ (from Figure 5.18c)	$F_{calculate}$
1.5	17.3°	0.051	0.69
1.0	25°	0.0225	1.56
1.2	21.2°	0.036	0.97
1.1	22.9	0.030	1.17

Take FS = 1.13

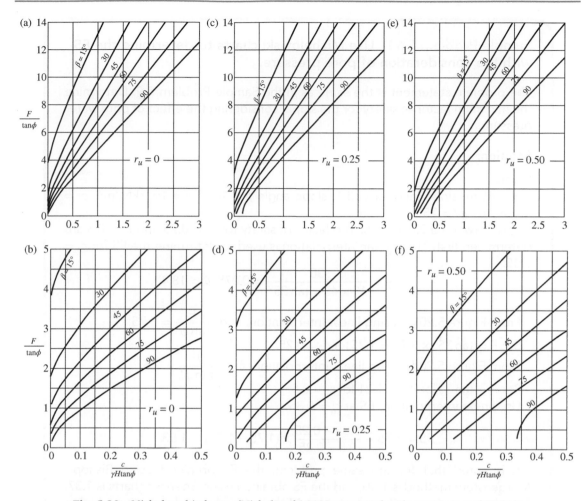

Fig. 5.20   Michalowski charts. (Michalowski 2002, reprinted with permission of ASCE.)

5.7.3   Michalowski charts

Michalowski (2002) provided charts (Figure 5.20) to directly obtain the minimum factor of safety. The charts eliminate the necessity for iterations. The charts are based on the kinematic approach of limit analyses using log-spiral slip surfaces. A modified stability number, $c/\gamma H \tan \phi$, is used. Therefore, the charts cannot be used for slopes with zero internal friction angle. The charts are for three pore pressure ratios: $r_u = 0$ (no pore pressure), $r_u = 0.25$, and $r_u = 0.5$. If r_u falls between 0 and 0.25 and between 0.25 and 0.5, the factor of safety can first be determined using the nearest upper and lower limits of $r_{u,}$, then linear interpolation can be used to obtain the factor of safety for the given r_u.

Chapter 5

Sample Problem 5.11: Use Michalowski charts to determine the FS with the consideration of pore pressure.

The problem statement is the same as in the Sample Problem 5.9. Determine the minimum factor of safety for the slope, considering the effect of pore water pressure.

Solution:

Given:

$H = 45$ m, slope inclination 3:1, slope angle $\beta = 18.4°, \gamma = 19.0$ kN/m³, $c\prime = 30$ kN/m², $\phi\prime = 25°, r_u = 0.35$.

r_u is between 0.25 and 0.5. The factor of safety for $r_u = 0.25$ and $r_u = 0.5$ are determined first. Then linear interpolation is used to determine the FS for $r_u = 0.35$.

$$\frac{c\prime}{\gamma H \tan \phi\prime} = \frac{30}{19 \times 45 \times \tan 25°} = 0.075, \beta = 18.4°$$

From Figure 5.20(d), using linear interpolation, find: $\dfrac{F}{\tan \phi\prime} = 3.75$.

So: $F = \tan 25° \times 3.75 = 1.75$ for $r_u = 0.25$.

From Figure 5.20(f), using linear interpolation, find: $\dfrac{F}{\tan \phi\prime} = 2.6$.

So: $F = \tan 25° \times 2.6 = 1.21$ for $r_u = 0.5$.

Use linear interpolation, FS at $r_u = 0.35$:

$$FS = 1.75 - \frac{0.35 - 0.25}{0.5 - 0.25} \times (1.75 - 1.21) = 1.53$$

It is noted that for the same problem, the FS obtained using Bishop–Morgenstern method is 1.40, and the FS obtained using Spencer charts is 1.37. This does not necessarily indicate that Michalowski charts always give a higher FS as there are approximation errors in reading the charts.

5.8   Morgenstern charts for rapid drawdown

When a reservoir's water level drops quickly, the pore water in the earth embankment cannot drop as fast as the water level in the reservoir. This process is referred to as rapid drawdown. As the pore water pressure in the embankment cannot dissipate quickly, the slope stability may be adversely affected. Morgenstern (1963) provided charts for the rapid drawdown scenario. The development of the charts was based on effective stress. Morgenstern (1963) assumed the soil is homogeneous and the initial water level is at the crest. As shown in Figure 5.21, the slope height is H, the drawdown depth is L, and L/H is defined as the drawdown ratio. The Morgenstern

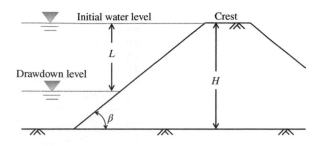

Fig. 5.21 Illustration of rapid drawdown of reservoir.

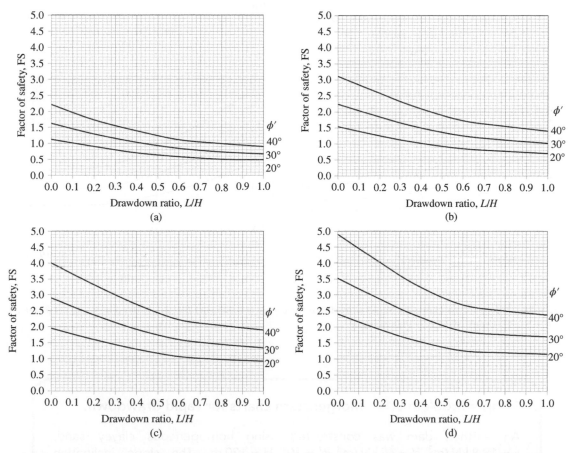

Fig. 5.22 Morgenstern charts for rapid drawdown for $c'/\gamma H = 0.0125$. (a) $\beta = 2 : 1$, (b) $\beta = 3 : 1$, (c) $\beta = 4 : 1$, (d) $\beta = 5 : 1$. (Courtesy of ICE Publishing.)

charts for rapid drawdown are shown in Figures 5.22–5.24 for three $(c'/\gamma H)$ values: 0.0125, 0.025, and 0.05. For each $(c'/\gamma H)$ value, the factor of safety depends on the drawdown ratio, effective friction angle ϕ', and the slope angle (inclination) β. Linear interpolation can be used to obtain the FS for intermediate $(c'/\gamma H)$ values.

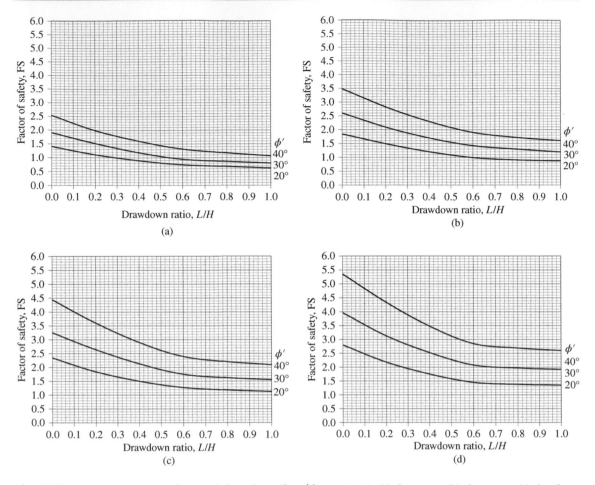

Fig. 5.23 Morgenstern charts for rapid drawdown for $c'/\gamma H = 0.025$. (a) $\beta = 2 : 1$, (b) $\beta = 3 : 1$, (c) $\beta = 4 : 1$, (d) $\beta = 5 : 1$. (Courtesy of ICE Publishing.)

Sample Problem 5.12: Morgenstern charts for rapid drawdown.

An earthen dam was constructed using homogeneous clayey sand, $\gamma = 19.8 \text{ kN/m}^3$, $c' = 65 \text{ kN/m}^3$, $\phi' = 30°$, H = 120 m. The slope inclination is 2:1 (H:V). Assume the initial water level is at the crest. A rapid partial drawdown was needed. The water level was quickly reduced to 60 m below the crest. Determine the factor of safety following the rapid drawdown.

Solution:

Morgenstern charts for rapid drawdown are used.

$$\frac{c'}{\gamma H} = \frac{65}{19.8 \times 120} = 0.027 \approx 0.025$$

Drawdown ratio: $\dfrac{L}{H} = \dfrac{60}{120} = 0.5$

Use Figure 5.23(a), find FS = 1.0.

Conclusion: the rapid drawdown could make the earthen dam unstable.

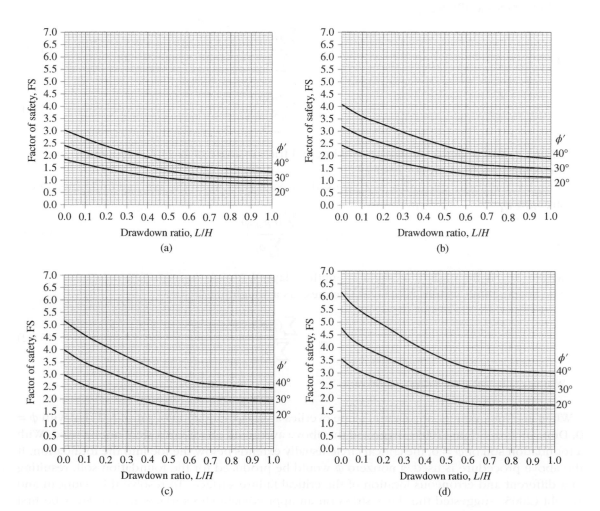

Fig. 5.24 Morgenstern charts for rapid drawdown for $c'/\gamma H = 0.05$. (a) $\beta = 2 : 1$, (b) $\beta = 3 : 1$, (c) $\beta = 4 : 1$, (d) $\beta = 5 : 1$. (Courtesy of ICE Publishing.)

5.9   Averaging unit weights and shear strengths in stratified slopes

The charts for slope stability analyses apply only to homogeneous slopes. To apply the charts to nonhomogeneous (or stratified) slopes, it is necessary to approximate the real conditions with an equivalent homogeneous slope. Duncan and Wright (2005) provided a method to average unit weights and shear strengths in stratified slopes. The approach is illustrated in Figure 5.25 and outlined in the following steps.

1. Identify the shear strength, unit weight, and thickness of each soil layer.
2. Identify or approximate the critical slip surface.
3. Measure the central angle of arc for each layer. Figure 5.25 shows an example of three layers, and the central angles of arc are δ_1, δ_2, and δ_3.
4. The central angles are used as weighting factors to calculate the weighted average strength parameters, c_{av} and ϕ_{av}:

$$c_{av} = \frac{\sum(\delta_i c_i)}{\sum \delta_i} \tag{5.79}$$

$$\phi_{av} = \frac{\sum(\delta_i \phi_i)}{\sum \delta_i} \tag{5.80}$$

where: c_i and ϕ_i are the cohesion and internal friction angle of layer i, respectively, and δ_i is the central angle of arc for layer i.
If undrained shear strength $s_u = c_u$ is used for each layer, then the average undrained shear strength is:

$$s_{u(av)} = \frac{\sum(\delta_i s_{u(i)})}{\sum \delta_i} \tag{5.81}$$

where: $s_{u(i)}$ is the undrained shear strength of layer i.
5. The average unit weight uses layer thickness as a weighting factor:

$$\gamma_{av} = \frac{\sum(\gamma_i h_i)}{\sum h_i} \tag{5.82}$$

where: γ_i is the unit weight of layer i.

When an embankment or a natural slope overlies a weak foundation of saturated clay with $\phi = 0$, Duncan and Wright (2005) suggested the above averaging procedures shall not be used. With a foundation soil of $\phi = 0$, the critical circle usually extends below the toe into the foundation. If the above procedure is used, a nonzero ϕ would be produced for the foundation soil, resulting in a different and erroneous location of the critical failure surface. To resolve this, Duncan and Wright (2005) suggested the shear stress on an approximate slip surface in each layer be first calculated using:

$$\tau_i = c_i + \sigma_{i(av)} \tan \phi_i \tag{5.83}$$

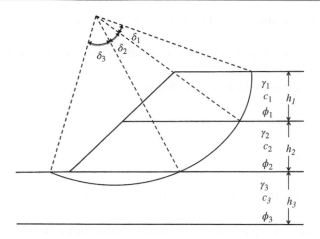

Fig. 5.25   Nonhomogeneous slope (Duncan and Wright 2005).

where: $\sigma_{i(av)}$ is the average normal stress on the slip surface in layer i. Then, τ_i is treated as the undrained shear strength $s_{u(i)}$ for the layer. Finally, Equation (5.79) can be used to obtain the equivalent average undrained shear strength for the entire slope and foundation.

5.10   Slope stability analyses – finite element methods

Finite element method (FEM) is a numerical technique for finding approximate solutions to differential equations. Depending on the nature of a problem, there are various differential equations that represent various characteristics of a system. For example, the governing differential equation for one-dimensional elastic deformation can be expressed using:

$$EA\frac{d^2u}{dx^2} + F = 0 \tag{5.84}$$

Where

E = Young's modulus,
A = cross-sectional area of the elastic material on which the force is applied,
F = force,
u = deformation in the x direction.

The FEM first geometrically divides a complex system into small element and solves the differential equations that govern the characteristics of the element relative to each other. Figure 5.26 shows an example of the finite element analyses of slope stability (Griffiths and Lane 1999), using the Mohr–Coulomb failure criterion. Detailed discussion of FEM is outside the scope of this chapter, but the FEM has several advantages over the other analysis methods on slope stability, such as (Griffiths and Lane 1999):

1. No assumption needs to be made in advance about the shape or location of the failure surface. Failure occurs through the soil zone where the soil's shear strength is unable to resist the applied shear stress.

Chapter 5

2. There is no concept of slices in FEM; therefore, there is no need for the assumption of interslice forces.
3. If realistic soil compressibility data are available, the FEM solutions can give the slope deformations at working stress level.
4. FEM can monitor progressive failure of a slope.

5.11   Slope stabilization measures

Slope stabilization and repair fall largely into two categories:

- Preventive treatments that are applied to currently stable but potentially unstable slopes.
- Remedial or corrective treatments that are applied to currently unstable, moving, or already failed slopes.

The choice and implementation of slope stabilization measures often go beyond technical considerations. Many factors should be considered in assessing slope stabilization measures, namely,

- Consequences of the failure: is the slope failure acceptable or must it be fixed?
- Time constraint: is a quick repair needed?
- Cost of the slope repair.
- Subsoil conditions.
- The current or potential failure mode: surficial, shallow and rotational, or deep-seated failure.
- Present and required future topography of the slope.
- Physical constraints, such as property line, existing building, right of way.
- Availability of materials, equipment, and expertise.
- Accessibility of materials and equipment to the site.

In this section, several common slope stabilization measures are presented.

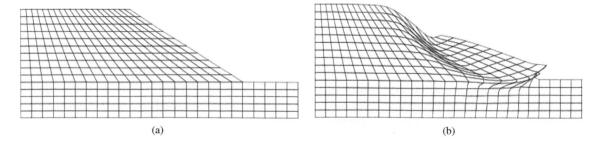

| (a) | (b) |

Fig. 5.26 Finite element method in slope stability analyses (obtained using the FEM software by Prof. D.V. Griffiths, Colorado School of Mines). (a) Finite element mesh before slope failure, (b) finite element mesh after failure.

(a)   (b)

Fig. 5.27   Slope surface drainage. (a) Interceptor trench drain, (b) interconnected surface drains.

5.11.1   Surface drainage

Drainage is the most important slope stabilization method. Uncontrolled slope surface runoff can cause significant surficial erosion and eventually slope failure. High groundwater elevation in the slope reduces the soil's shear strength and induces seepage forces. Drainage measures are used to prevent a newly built or an existing slope from failure or to treat a slope at an early stage of failure. They include surface runoff controls and subsurface drainage.

Trenches or a network of interconnected trenches can be built on a slope surface to intercept, collect, and drain runoff in a controlled manner. The purpose of the drain is to intercept surface runoff on the slope to reduce the sheet flow on the slope face and the amount of infiltration into the slope. Figure 5.27(a) shows a typical concrete drainage swale, also known as a "V" drain or "V" ditch. The swales can be interconnected, as shown in Figure 5.27(b), to guide the surface water away from a slope. Slope drains can vary in width from 0.3 m to over 3 meters, depending on the application and size of the slope.

Surface erosion can cause slope instability, contamination of downstream water-receiving bodies, and increased solid loading in drainage channels. To minimize surface runoff and soil erosion, sloping areas should be planted with perennial ground covers or turf. Straw tubes (Figure 5.28) are also commonly used to reduce surface runoff and soil erosion. Surface erosion control using geosynthetics that are incorporated with various vegetation and landscaping methods is becoming increasingly popular because of its fast installation, reduced cost, and sustainable functionality.

5.11.2   Internal drainage

* Horizontal drains
  Horizontal drains are usually perforated pipes that are installed in the slope with slight inclination (Figure 5.29). They are used to lower the groundwater table in the slope. Hollow-stem augers are used to drill holes near the toe of the slope, and then perforated or slotted PVC pipes are inserted into the holes. Because of the difficulty of installing filters, there is no filter between the soil and the pipe.

Chapter 5

- Drainage galleries

  Drainage galleries are hollow tunnels that are excavated deep inside a slope. Smaller drainage pipes can be installed within the tunnel and conduct drainage into the tunnel, as shown in Figure 5.30.

- Gravity drains

  Gravity drains, also known as drain wells or gravity wells, are vertical holes that are drilled at or near the top of the slope and penetrate various soil strata; the holes are then backfilled with gravels or rocks. They are used when soil strata with varying permeability exist in a slope, making the horizontal drains alone less effective. Gravity wells can be connected to horizontal drains at the bottom of a slope to facilitate the drainage, as shown in Figure 5.31.

- Well points and deep wells

  Well points, also known as suction wells, are small-diameter pipes that are driven into the soil. The lower portion of the pipe is perforated and is connected to a riser pipe that is not perforated. A vacuum pump is connected to the top of a well point and the vacuum draws the water up. Deep wells are larger-diameter holes drilled in the soil that contain submersible pumps located at the bottom of the holes to pump the groundwater to the ground surface.

5.11.3   Unloading

The height of a slope can be shortened or the slope inclination can be reduced (Figure 5.32), so that the load on the slope can be reduced because of gravity, thus reducing the shear stress or

Fig. 5.28   Surface erosion control using straw tubes.

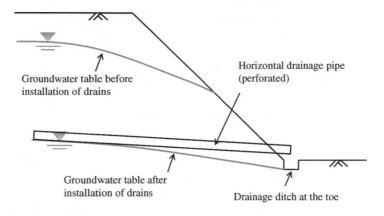

Groundwater table before
installation of drains

Horizontal drainage pipe
(perforated)

Groundwater table after
installation of drains

Drainage ditch at the toe

Fig. 5.29   Illustration of horizontal drains.

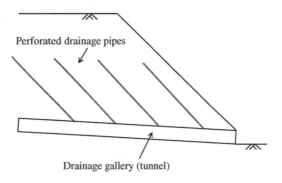

Perforated drainage pipes

Drainage gallery (tunnel)

Fig. 5.30   Illustration of drainage gallery.

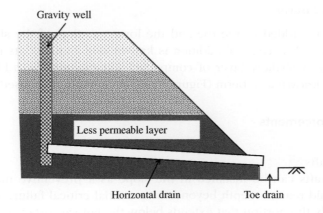

Gravity well

Less permeable layer

Horizontal drain

Toe drain

Fig. 5.31   Illustration of gravity drains.

Chapter 5

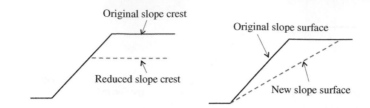

Fig. 5.32   Unloading of slope.

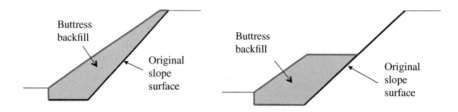

Fig. 5.33   Buttressing.

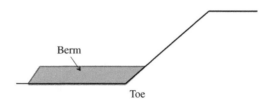

Fig. 5.34   Berm.

rotational moment that can cause a slope failure. For a new embankment, lightweight materials can be used as the backfill. Unloading is a common method for slope stabilization.

5.11.4   Buttress and berm

Compacted soil can be added to the toe and the lower portion of the slope or to the entire slope surface (Figure 5.33). The soil addition is known as a buttress. Its purpose is to reduce the sliding moments. A shallow layer of compacted soil can also be added to the toe of an embankment; this is known as a berm (Figure 5.34). Berms are often used to stabilize levees.

5.11.5   Slope reinforcements

- Piles or drilled shafts
  Piles or drilled shafts can be installed in the slope as a preventive measure. The piles or drilled shafts should reach a depth beyond the potential critical failure surface, as shown in Figure 5.35, as only the portion that extends below the failure surface can provide resistance to the rotation. The location, number, and length of piles or drilled shafts are designed on the basis of the slope stabilization needs, budget, and the slope and subsoil conditions.

- Soil nailing

  Soil nailing is the reinforcement of existing walls and slopes by installing closely spaced steel bars (i.e., soil nails) into the soil. Figure 5.36 illustrates a typical cross section of a soil nail-reinforced slope. Soil nailing has been successfully used in temporary or permanent soil retaining applications, including roadway cuts and widening, repairs or reconstructions of existing retaining structures, and excavations in urban environments. The detailed design of soil nail walls is presented in Chapter 7.

- Geosynthetic reinforced soil (GRS) slopes

  When constructing a new slope, geosynthetics (such as geogrids and geotextiles) can be installed inside the slope as reinforcement layers to resist slope sliding. Figure 5.37 shows the typical layout of a reinforced soil slope, and Figure 5.38 shows an example of a reinforced slope during construction. The detailed design of geosynthetic reinforced slopes is presented in Chapter 8.

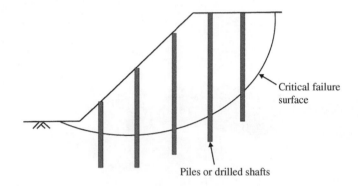

Fig. 5.35   Piles or drilled shafts reinforcing a slope.

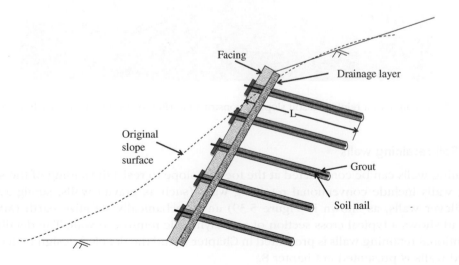

Fig. 5.36   Typical cross section and basic elements of a soil nail wall (grout is optional).

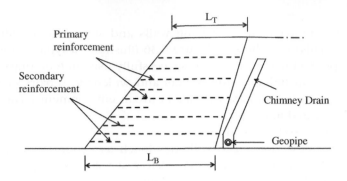

Fig. 5.37   Typical layout of a reinforced soil slope.

Fig. 5.38   Example of a reinforced soil slope in construction. (Photo courtesy of Tensar International.)

5.11.6   Soil retaining walls

Soil retaining walls can be constructed at the toe of a slope to resist the sliding of the slope. Soil retaining walls include conventional retaining walls (such as gravity walls, semigravity walls, and cantilever walls, as shown in Figure 5.39) and mechanically stabilize earth (MSE) walls. Figure 5.40 shows a typical cross section of a geosynthetic reinforced wall. The detailed design of conventional retaining walls is presented in Chapter 7, and the detailed design of geosynthetic reinforced walls is presented in Chapter 8.

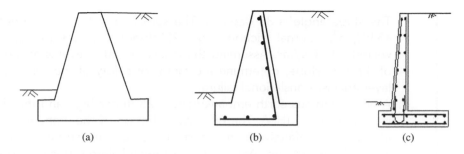

Fig. 5.39   Conventional retaining walls. (a) Gravity wall, (b) semigravity wall, (c) cantilever wall.

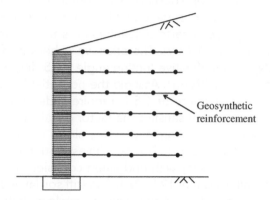

Geosynthetic
reinforcement

Fig. 5.40   Geosynthetic reinforced wall.

Homework Problems

1. A natural slope is 45-m high and the slope angle is 38 degrees. The surficial soil is loose silty sand with cohesion 45 kN/m^2, internal friction angle 25 degrees, and bulk unit weight 18.0 kN/m^3. The thickness of the topsoil is 1.0 m in the vertical direction. The slope is dry. Determine the factor of safety of the surficial soil layer against translational failure.

2. A saturated natural slope is 45-m high and the slope angle is 38 degrees. The surficial soil is loose silty sand with effective cohesion 45 kN/m^2, effective internal friction angle 25 degrees, and saturated unit weight 19.0 kN/m^3. The thickness of the topsoil is 1.0 m in the vertical direction. The downward seepage is parallel to the slope surface. Determine the factor of safety of the surficial soil layer against translational failure.

3. A reservoir is 45-m deep and the side slope of the reservoir develops a loose surficial layer that is 1.0 m thick (in the vertical direction).

The slope angle is 38 degrees. The surficial soil layer has cohesion 45 kN/m², internal friction angle 25 degrees, and saturated unit weight 19.0 kN/m³. Assuming the reservoir water level is at the top of the side slope, determine the factor of safety of the surficial soil layer against translational failure.

4. A homogeneous earth embankment is 25-meter high and the slope inclination is 2:1 (H:V). The effective cohesion of the embankment is 45 kN/m², its effective friction angle is 25 degrees, and the bulk unit weight is 19.2 kN/m³. Assume the potential failure surface is a plane. Determine the minimum factor of safety of the slope.

5. For the same slope described in Problem 4, what is the factor of safety on a planar failure surface that has an inclination angle of 20 degrees?

6. A homogeneous earth embankment is to be constructed. The slope inclination is 2:1 (H:V). The effective cohesion of the embankment is 45 kN/m², its effective friction angle is 25 degrees, and the bulk unit weight is 19.2 kN/m³. Assume the potential failure surface is a plane. If a factor of safety of 1.5 is required, determine the height of the embankment that satisfies the factor of safety.

7. A saturated and undrained clayey slope is 10-m high and the slope angle is 40 degrees. Subsurface investigation found that the subsoil is homogeneous clay with undrained cohesion of 110 kN/m² and saturated unit weight of 19.5 kN/m³. A stiff soil layer exists 5 meters below the toe of the slope. A potential toe circle with the radius of 20 meters passes the coordinate of (25 m, 10 m). The toe is at the origin (0, 0).

(1) Determine the factor of safety along the assumed slip circle using the analytical mass method.

(2) Determine the minimum factor of safety of the slope using Taylor's chart.

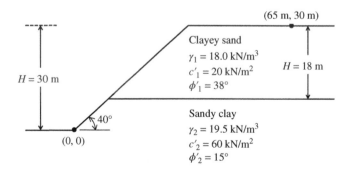

Fig. 5.41   Slope profile of problem 8.

8. A natural slope is shown in Figure 5.41. It contains two soil strata. The slope configuration and the soil characteristics are shown in the figure. A potential toe circle with radius of 70.0 m passes the coordinate of (65 m, 30 m). The toe is at the origin (0, 0).
   (1) Determine the average cohesion, friction angle, and unit weight of the slope and foundation soils.
   (2) Determine the factor of safety of the slope along the assumed failure circle.
   (3) Determine the minimum factor of safety of the slope, using Taylor's chart (Figure 5.41).
9. A homogeneous silty sand embankment is built on soft clay. The soil profile is shown in Figure 5.42. A potential critical slip surface is also shown.
   (1) Use the ordinary method of slices to determine the FS on the assumed failure plane.
   (2) Use Bishop's simplified method of slices to determine the FS on the assumed failure plane and compare the FS with the one obtained in (1).
   (3) Calculate the average cohesion, friction angle, and unit weight of the slope and foundation soil.
   (4) Use Taylor's chart to determine the minimum FS of the slope using the parameters obtained in (3) (Figure 5.42).

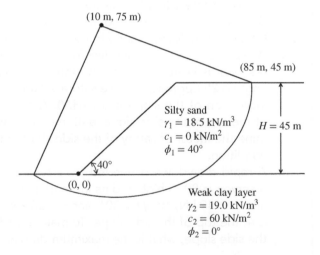

(10 m, 75 m)

(85 m, 45 m)

Silty sand
$\gamma_1 = 18.5$ kN/m^3
$c_1 = 0$ kN/m^2
$\phi_1 = 40°$

$H = 45$ m

40°

(0, 0)

Weak clay layer
$\gamma_2 = 19.0$ kN/m^3
$c_2 = 60$ kN/m^2
$\phi_2 = 0°$

Fig. 5.42   Slope configuration of problem 9.

Chapter 5

10. A homogeneous natural slope is 20 m high. The same type of soil extends to great depth. The slope angle is 35 degrees. The effective cohesion is 60 kN/m^2 and the effective internal friction angle is 25 degrees. The bulk density is 1800 kg/m^3. Use Bishop's simplified method of slices to determine the minimum factor of safety, using:
    (1) One slope circle.
    (2) One toe circle.
    (3) One deep-seated, mid-point circle.

11. A natural slope is 35 meters high and the slope inclination is 2(H):1(V). The slope's properties are: $c\prime = 50$ kN/m^2, $\phi\prime = 25°$, and $\gamma = 19.0$ kN/m^3. No firm layer was encountered beneath the slope during the subsurface investigation. Downward steady seepage that is parallel to the slope surface occurs in the slope. The average pore water pressure ratio for the slope was determined to be 0.4.
    (1) Use Bishop–Morgenstern method to determine the minimum factor of safety (FS) for the slope.
    (2) Use Spencer charts to determine the FS.
    (3) Use Michalowski charts to determine the FS.

12. A sandy soil embankment is 10 meters high and the slope inclination is 2(H):1(V). The slope's properties are: $c\prime = 0$ kN/m^2, $\phi\prime = 30°$, and $\gamma = 19.5$ kN/m^3. A firm layer exists 5 meters beneath the toe of the slope. The average pore water pressure ratio for the slope is 0.25.
    (1) Use Bishop–Morgenstern method to determine the minimum factor of safety (FS) for the slope.
    (2) Use Spencer charts to determine the FS.
    (3) Use Michalowski charts to determine the FS.

13. A rapid drawdown is needed for a reservoir. The side slope for the reservoir is 84-m high, its cohesion is 62 kN/m^2, the internal friction angle is 25 degrees, and the saturated unit weight is 18.5 kN/m^3. The inclination of the slope is 3:1 (H:V). Assume the initial water level is at the top of the side slope; a 42-meter drawdown is needed. Determine the factor of safety of the side slope under the rapid drawdown condition.

14. A rapid drawdown is needed for a reservoir. The side slope for the reservoir is 80-m high and its inclination is 2:1. The slope's properties are: $c\prime = 40$ kN/m^2, $\phi\prime = 30°$, and $\gamma = 20$ kN/m^3. The initial water level is at the top of the side slope. To maintain a factor of safety of 1.5 for the side slope, what is the maximum drawdown depth (L)?

References

Abramson, L.W., Lee, T.S., Sharma, S., and Boyce, G.M. (2002). *Slope Stability and Stabilization Methods*. John Wiley & Sons, Inc. New York, USA.

Barnes, G. (2010). Soil Mechanics. Principles and Practice. Third Edition. Palgrave MacMillan. UK.

Bishop, A.W. (1955). "The use of the slip circle in the stability analysis of slopes." *Geotechnique, Vol.* 5, pp 7–17.

Bishop, A.W., and Morgenstern, N.R. (1960). "Stability coefficients for earth slopes." *Geotechnique*, Vol. 10, No. 4, pp 129-150.

Bond, A., and Harris, A. (2008). Decoding Eurocode 7, Taylor & Francis, London.

Chandler, R.J., and Peiris, T.A. (1989). "Further extensions to the Bishop & Morgenstern slope stability charts." *Ground Engineering*, Vol. 22, No. 4, pp 33-38.

Culmann, C. (1875). *Die graphische Statik* (*The Graphic Statics*, in Germany). Meyer & Zeller, Zürich.

Duncan, J.M., and Wright, S.G. (2005). *Soil Strength and Slope Stability*, John Wiley & Sons, Inc, Hoboken, NJ, USA.

Fellenius, W. (1927). Erdstatische Berechnungen mit Reibung und Kohäsion (Adhäsion) und unter Annahme kreiszylindrischer Gleitflächen (Static calculations with friction and cohesion (adhesion), and assuming a circular cylindrical sliding surfaces, in German), Ernst & Sohn, Berlin, Germany.

Fellenius, W. (1936). Calculations of the stability of earth dams. Transaction of the 2nd Congress on Large Dams, 4, Washington, D.C.

Griffiths, D.V., and Lane, P.A. (1999). "Slope stability analysis by finite elements." *Geotechnique*, Vol. 49, No. 3, pp 387-403.

Lee, K.L., and Duncan, J.M. (1975). *Landslide of April 25, 1974 on the Mantaro River, Peru*, report of inspection submitted to the Committee on Natural Disasters, National Research Council, National Academy of Sciences, Washington, D.C.

Michalowski, R. L. (2002). "Stability Charts for Uniform Slopes." *ASCE Journal of Geotechnical and Geoenvironmental Engineering*, Vol. 128, No. 4, pp 351-355.

Morgenstern, N.R. (1963). "Stability charts for earth slopes during rapid drawdown." *Geotechnique*, Vol. 13, No. 2, pp 121-133.

Smith, I. (2007). Smith's Elements of Soil Mechanics. Eighth Edition. Blackwell Publishing. Oxford, UK.

Spencer, E. (1967). "A method of analysis of the stability of embankments assuming parallel inter-slice forces." *Geotechnique*, Vol. 17, No. 1, pp 11-16.

Taylor, D.W. (1937. "Stability of Earth Slopes." *Journal of the Boston Society of Civil Engineers, Vol.* 24, pp 197-246.

Terzaghi, K., Peck, R.B., and Mesri, G. (1996). *Soil Mechanics in Engineering Practice*, third edition. John Wiley & Son, Hoboken, NJ.

Chapter 6

Filtration, Drainage, Dewatering, and Erosion Control

6.1 Basics of saturated flow in porous media

Saturated flow of water in porous media, such as soils and geotextiles, is described by Darcy's law:

$$v = k \cdot i \tag{6.1}$$

or

$$q = k \cdot i \cdot A \tag{6.2}$$

where

q = flow rate $[L^3/T]$,
A = cross-sectional area of the porous medium that water flows through,
v = flow velocity, also known as Darcy's velocity,
k = coefficient of permeability, also known as hydraulic conductivity,
i = hydraulic gradient,

and

$$q = v \cdot A \tag{6.3}$$

$$i = \frac{\Delta h}{L} \tag{6.4}$$

Δh = total hydraulic head difference (or loss) across a flow path of L.

Darcy's law can apply only to saturated and laminar flow. If the flow is unsaturated or turbulent, Darcy's law cannot be used.

The seepage of water under saturated condition can be quantified using a flow net. Figure 6.1 shows an example of a flow net under a sheet pile wall, and Figure 6.2 shows a flow net through an earthen dam resting on an impermeable soil. A flow net comprises two families of intercepting lines: flow lines and equipotential lines. The flow lines follow the actual flow paths of the seepage, and each equipotential line has the same total hydraulic head (or potential) on the line. The seepage (flow rate) in homogeneous and isotropic soil is determined using:

$$q = k \cdot \Delta h \cdot \frac{N_f}{N_d} \tag{6.5}$$

Geotechnical Engineering Design, First Edition. Ming Xiao.
© 2015 John Wiley & Sons, Ltd. Published 2015 by John Wiley & Sons, Ltd.
Companion Website: www.wiley.com/go/Xiao

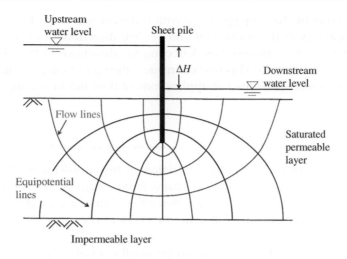

Fig. 6.1   Flow net for seepage around a sheet pile wall.

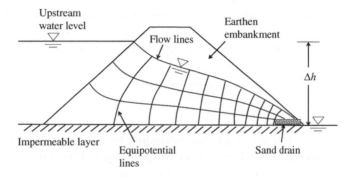

Fig. 6.2   Flow net for seepage in an earthen dam.

where:

q = seepage or flow rate per unit length in the longitudinal direction,

Δh = total hydraulic head difference (or loss) from the entrance of the flow net to the exit of the flow net (see Figures 6.1 and 6.2),

N_f = number of flow channels; one flow channel is constructed by two adjacent flow lines; for example, $N_f = 4$ in Figure 6.1 and $N_f = 3$ in Figure 6.2.

N_d = number of equipotential drops; a drop from one equipotential line to the immediately downstream equipotential line is counted as one equipotential drop; for example, $N_d = 8$ in Figure 6.1 and $N_d = 11$ in Figure 6.2

Equation (6.5) *only* applies to a flow net in which the flow lines and the equipotential lines are orthogonal and each cell in the flow net is "square." Being "square" means the corners of each cell have right angles, and the center lines of each cell are equal in length, as shown in Figures 6.1 and 6.2. Flow nets with "square cells" exist only in homogeneous and isotropic soils. Constructing a flow net in homogeneous and anisotropic soils entails drawing the horizontal and vertical dimensions of the seepage zone with different scales on the basis of the relative values of the permeability in the horizontal and vertical directions. Constructing a flow net in heterogeneous and isotropic soils involves deflecting the directions of flow lines when seepage enters into a different soil stratum. This book considers the readers are familiar with the concept of flow net, and this chapter presents the applications of flow net in solving seepage problems.

6.2   Filtration methods and design

Filters are porous media that prevent the migration of fine particles while allowing water to pass freely. Filters are often used to protect drains from clogging in various civil engineering applications. Figure 6.3 illustrates a few examples where filters are used.

A satisfactory filter should meet the following two major criteria.

- **Soil retention**: the voids of the filter should be small enough to prevent the passage of the protected base soil; otherwise, the particles from the base soil can clog the filter and drainage system. If the pore spaces in a filter are small enough to hold D_{85} of the base soil to be retained, then it is generally deemed that the finer soil particles of the base soil can also be retained. The pore spaces in a filter can be represented using an effective diameter of the pores. Terzaghi and Peck (1967) recommended $(1/5)D_{15}$ be used as the effective diameter and provided the following widely used soil retention criterion:

$$\frac{D_{15(F)}}{D_{85(S)}} \leq 4 \text{ to } 5 \tag{6.6}$$

where:

$D_{15(F)}$ = particle diameter of the filter corresponding to 15% finer by mass,
$D_{85(S)}$ = particle diameter of the base soil corresponding to 85% finer by mass.

- **Drainage**: the permeability of the filter should be sufficiently large, that is, the voids of the filter should be sufficiently large to permit the seepage without building up hydraulic pressure. Terzaghi and Peck (1967) provided the following widely used drainage criterion:

$$\frac{D_{15(F)}}{D_{15(S)}} \geq 4 \text{ to } 5 \tag{6.7}$$

where:

$D_{15(S)}$ = particle diameter of the base soil corresponding to 15% finer by mass.

The formulae (6.6) and (6.7) provide a range of the D_{15} of the granular filter:

$$(4 \text{ to } 5)D_{15(S)} \leq D_{15(F)} \leq (4 \text{ to } 5)D_{85(S)} \tag{6.8}$$

On the basis of the grain size distribution (GSD) of the base soil, the range of $D_{15(F)}$ can be determined. The upper and the lower bounds of the GSD of the acceptable filter should pass through the upper and lower limits of $D_{15(F)}$ and should follow the shape of the GSD of the natural (base) soil. The acceptable grain size distribution of the filter should fall within the upper and lower bounds, as shown in the shaded zone in Figure 6.4.

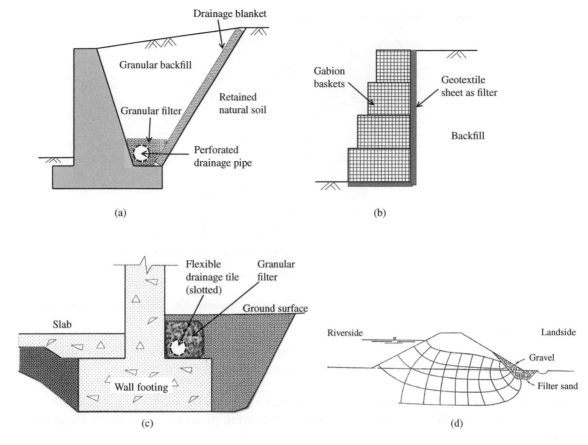

Fig. 6.3 Examples of filters in civil engineering. (a) Granular filter protecting a perforated drainage pipe behind a retaining wall, (b) geotextile filter behind a gabion wall, (c) granular filter protecting a drainage pipe around the exterior of a shallow foundation, (d) granular filter protecting the toe drain in a levee.

There are many other soil retention criteria that were proposed by researchers on the basis of laboratory testing; these criteria were summarized by Reddi (2003) and Das (2008). The International Commission on Large Dams (ICOLD 1994) provided further required characteristics of ideal granular filters as follows:

- Filter materials should not segregate during processing, handling, placing, spreading, or compaction. This requires the gradation of granular filters to be sufficiently uniform. Finer and coarser soil particles tend to segregate (i.e., separate), resulting in layers of extremely fine and coarse materials, thus affecting the function of the filters.
- Filter materials should not change in gradation; they should not degrade or break down during processing, handing, placing, spreading, and/or compaction, and they should not degrade with time.
- Filter materials should not have apparent or real cohesion or the ability to cement as a result of chemical, physical, or biological reactions.

Chapter 6

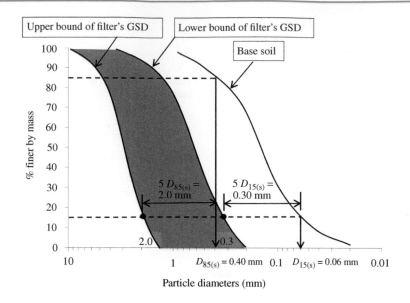

Fig. 6.4 Determination of grain size distribution of filter using formula (6.8). (Modified after Das 2010.)

- Filter materials should be internally stable, that is, the fine particles of a filter should not migrate within the filter and consequently clog the downstream filter section.

The widely referenced *Soils and Foundations Design Manual* by the US Naval Facilities Engineering Command (NAVFAC) (1986) provided the following general requirements for granular filters:

1. To avoid head loss in filters (drainage requirement):

$$\frac{D_{15(F)}}{D_{15(S)}} > 4 \tag{6.9}$$

and the permeability of the filter must be large enough to suffice any particular drainage system.

2. To avoid movement of particles from the base soil (soil retention requirement):

$$\frac{D_{15(F)}}{D_{85(S)}} < 5 \tag{6.10}$$

$$\frac{D_{50(F)}}{D_{50(S)}} < 25 \tag{6.11}$$

$$\frac{D_{15(F)}}{D_{15(S)}} < 20 \tag{6.12}$$

For very uniform base material ($c_u < 1.5$) : $D_{15(F)}/D_{85(S)}$ may be increased to 6.
For broadly graded base material ($c_u > 4$) : $D_{15(F)}/D_{15(S)}$ may be increased to 40.

3. To avoid movement of the grains of a filter into drain pipe perforations or joints (as illustrated in Figure 6.3a):

$$\frac{D_{85(F)}}{\text{Slot width}} > 1.2 \text{ to } 1.4 \tag{6.13}$$

$$\frac{D_{85(F)}}{\text{Hole diameter}} > 1.0 \text{ to } 1.2 \tag{6.14}$$

4. To avoid segregation, a filter should contain no particles larger than 7.6 cm (3 inch).
5. To avoid internal movement of fines within a filter, the filter should have no more than 5% passing No. 200 sieve.

In terms of limit state design such as advocated in the structural eurocodes, it is suggested that filter criteria is used to limit the danger of material transport by internal erosion. To that extent, criteria such as that in equations (6.9)–(6.15) can be used. BS EN 1997-1:2004 also recommends using noncohesive soils as filter materials; it also states that in certain cases more than one filter layer may be necessary to ensure that the particle size distribution changes in a stepwise fashion.

Another alternative suggested by BS EN 1997-1:2004 is as follows. If the filter criteria above are not satisfied, it is required to demonstrate that the design value (i_d) (i.e., a cautious estimate affected by a partial factor of safety) of the hydraulic gradient is well below the critical hydraulic gradient (i_{cr}), considering direction of flow, grain size distribution, and shape of grains, as well as the stratification of the soil. In other words, the following equation should be satisfied:

$$i_d \ll i_{cr} = \frac{\gamma'}{\gamma_w} = \frac{G_s - 1}{1 + e} \tag{6.15}$$

where:

γ' = submerged unit weight of the soil,

$$\gamma' = \gamma_{sat} - \gamma_w \tag{6.16}$$

γ_{sat} = saturated unit weight,
γ_w = unit weight of water,
G_s = specific gravity of soil solid,
e = void ratio of the soil.

Although BS EN 1997-1:2004 does not provide a recommended factor of safety for the calculation, Bond and Harris (2008) state that on the basis of previous experience it should be at least equal to 4.0.

The above design methodology applies to granular filters. The design requirements for geotextile filters are presented in Chapter 8.

6.3 Dewatering and drainage

Drainage is the removal of percolating water or groundwater from soils and rocks by natural or artificial means. Dewatering systems have the following main purposes (Cedergren 1977):

(a) Intercepting seepage that otherwise would enter excavations and interfere with the construction.
(b) Improving the stability of slopes, thus preventing sloughing or slope failures.

Chapter 6

(c) Preventing the bottoms of excavations from heaving because of excessive hydrostatic pressure.

(d) Improving the compaction characteristics of soils in the bottom of excavations for basements, freeways, and so on.

(e) Drying up borrow pits so that excavated materials can be properly compacted in embankments.

(f) Reducing earth pressures on temporary supports and sheeting.

This section presents five common construction dewatering methods.

6.3.1   Open pumping

As shown in Figure 6.5, where a sloped open excavation is acceptable and does not cause slope instability, the inflowing groundwater is allowed to flow into the ditches and/or sumps at selected locations at the foot of the sloped cut, and the collected water is removed by pumps. This method is called open pumping or sump pumping. Open pumping may not be used where the inflow may cause sand boils, piping, or slope failure. The inflow is driven by gravity only. So, this method is effective mainly in coarse-grained soils with high permeability.

6.3.2   Well points

As shown in Figure 6.6, a well point system includes a well point, a riser pipe, a header pipe (manifold), and a pump. A well point is a closed-end pipe with perforations along its lower section and conical steel end with holes. It is installed to the desired depth below groundwater table. During the well point installation, jetting fluid is pumped through the holes at the end of the well point and cuts the soil in order to advance the well point. During the dewatering, the groundwater enters the perforations of the well point and is brought to the ground surface through the riser pipe, where it is collected by the horizontal header pipe (manifold pipe). Pumps on the ground surface are connected to the header pipes to remove the water. Usually, a series of well points are installed and connected to the same header pipe. The spacing of well points is from 1 to 3 meters and depends on the soil type.

The perforated section of a well point is protected with screens that keep the subsoil from clogging the perforations. Where finer soils are dewatered, a hole is first drilled, then the well point is lowered in the hole, and filter sand is backfilled around the perforations in the hole. The selection of filter follows the methods described in Section 6.2.

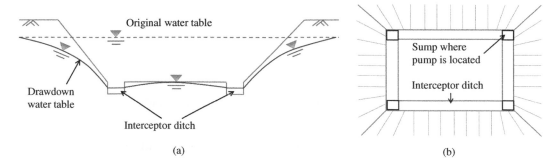

Fig. 6.5   Illustration of open pumping. (a) Section view, (b) plan view.

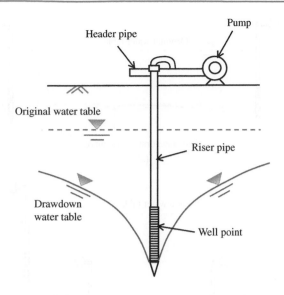

Fig. 6.6 Components of a well point system.

Table 6.1 Approximate slopes of drawdown curves because of a single row of well points (after McCarthy 2007).

Type of soil that is dewatered	Slope of drawdown curve $(\%) = (h/d) \times 100\%$
Coarse sand	1–3
Medium sand	2–5
Fine sand	5–20
Silt–clay	20–35

The depth of dewatering using one level of well point system is limited. Lowering the groundwater table to a greater depth may require multistage of installation of well point systems, as shown in Figure 6.7. The well points of the first stage are installed on a perimeter line outside the actual excavation area. Excavation proceeds within the perimeter formed by the well points to the depth where the groundwater is encountered. Then well points of the second stage are installed within the excavation to further lower the groundwater table. Further stages of well points system may be required until the groundwater table reaches the designed excavation depth.

The dewatering depth and the horizontal extent that are affected by the dewatering depend on the depth of the well points and the permeability of the soil. McCarthy (2012) provided the approximate slopes of drawdown curves in various types of soils, as listed in Table 6.1 and illustrated in Figure 6.7.

6.3.3 Deep wells

If dewatering of the groundwater to a great depth (e.g., more than 10 meters) is needed, deep wells and deep well pumps can be used. Deep wells are relatively large diameter holes (e.g., 60 cm) that are drilled in the subsoil, where a perforated protective casing is installed in the bore

Chapter 6

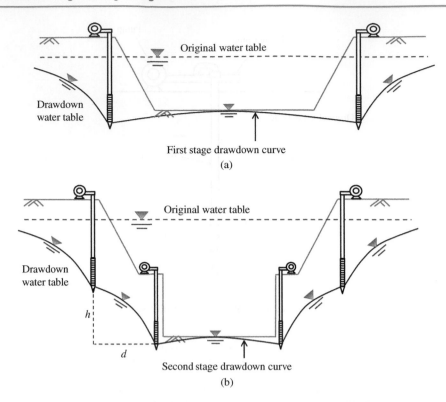

Fig. 6.7　Multistage dewatering using well point system (not to scale). (a) First stage drawdown using well point system, (b) second stage drawdown using well point system.

hole and a submersible deep well pump is placed at the bottom of the bore hole. Coarse filter material is placed between the casing and the wall of the drilled hole to protect the perforated pipe from clogging. Deep wells are usually located around an excavation's outside edges.

6.3.4　Vacuum dewatering

Soils that are too fine to be drained by wells and well points may be effectively drained by the vacuum method or electroosmosis.

　When a well point system is used to dewater silt, gravity flow is restricted because of the soil's low permeability. By applying a vacuum to the system, the flow in silt to the well points can be accelerated. Vacuum dewatering uses the same setup of a well point system. In addition, a vacuum pump is connected to the header and is used to maintain a vacuum in the riser pipes and the well points, so that water can flow to the well points through the silt more efficiently. Closer well point spacing in the vacuum dewatering is usually required than in the conventional well point system. Multistage vacuum-assisted well points may also be used if greater dewatering depth is needed.

6.3.5　Electroosmosis

Clay particles typically have negative electrical charge on the surface, and it attracts positive ions (cations) in the pore fluid. When an electric field is established in the saturated clayey soil, the

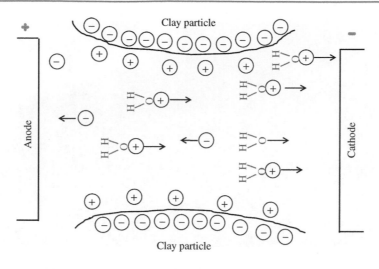

Fig. 6.8   Mechanism of electroosmosis.

excess cations close to the clay surface move toward the negatively charged electrode (cathode). The movement of these positively charged species and dipolar water molecules (H_2O) that are closely associated with these species results in a net strain on the pore fluid surrounding the soil particles. This strain causes a shear force on the pore fluid because of the viscosity of the fluid. As there is usually an excess amount of cations close to the soil surface, the net force and momentum toward the cathode results in a pore fluid flux in the same direction. This pore fluid flux, as a result of the electrical potential gradient, is named electroosmosis. Figure 6.8 illustrates the mechanism of electroosmosis.

In the field application, arrays of positive and negative electrodes are alternatively installed in the subsoil, and direct current (DC) power supply connects to the electrodes, as shown in Figure 6.9. The cathodes are installed in predrilled holes, and a coarse soil (gravel or coarse sand) is backfilled in the hole surrounding the cathode. Pumps are used to withdraw the water collected in the boreholes that contain cathodes. Figure 6.10 showcases a field application using

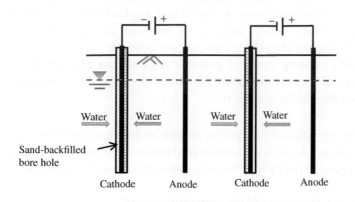

Fig. 6.9   Illustration of the electroosmosis setup.

Chapter 6

Fig. 6.10   A field project using electroosmosis. (Photo courtesy of Electrokinetic Limited, UK.)

electroosmosis to dewater and stabilize a slope in the United Kingdom. A DC potential of $60-80$ V was used. The treatment took six weeks and resulted in the following improvements:

- Dewatering from the cathodes.
- A reduction in plasticity and shrinkage characteristics.
- An increase in groundwater temperature from 10°C to 20°C.
- A modest DC power consumption of only $11.5 \, \text{kWhrs/m}^3$ of soil treated.
- Improvements in soil's shear strength.
- A 263% improvement in the bond strength of the anodes acting as nails.
- A cessation of slope movement.

Following the electroosmosis treatment the anodes have been retained as permanent soil nails, and the horizontal cathodes have been retained as permanent drainage.

The effectiveness of electroosmosis depends on the subsoil type, characteristics of the pore fluid (such as pH, temperature, ionic strength) and applied electric gradient. The cost of electroosmosis is usually high, mainly because of the high energy cost. It is used when other dewatering methods cannot effectively dewater the subsoil.

Most standards and existing codes of practice suggest that any dewatering method must be based on the results of a geotechnical and/or hydrogeological investigation. In general terms, the following issues should be considered:

- Soil movement at the sides of excavation as well as heave at their base needs to be avoided.
- Dewatering should not lead to settlement or damage of structures on its proximity.
- A margin of pumping capacity and back-up capacity should be available in case of any breakdown.
- Dewatering schemes should avoid the contamination of existing aquifers and water catchment areas.

6.4 Surface erosion and control

6.4.1 Surface erosion on embankments and slopes

Soil erosion is the dislodging and transport of soil particles caused by water, wind, or other physical disturbances. Surface erosion of soil can damage the productivity of cropland, and the eroded soil sediments can impair streams, lakes, and reservoirs. Soil erosion of roadside embankments forms rills and gullies on the embankments and can lead to increased surface runoff, which in turn can cause more soil erosion and eventually slope failure.

To predict the soil loss because of surface erosion, the US Department of Agriculture (USDA) developed a *universal soil loss equation* (USLE) (Wischmeier et al. 1971; Wischmeier 1976; Wischmeier and Smith 1978), which has been widely used to predict the average annual rate of sheet and rill erosion in agricultural fields. The equation considers rainfall pattern, soil type, topography, crop system, and management practices, and is expressed as:

$$A = R \times K \times (LS) \times C \times P \tag{6.17}$$

where:

A = average annual soil loss; it is conventionally expressed in tons/ac/yr,
R = rainfall and runoff factor; it depends on the rainfall intensity and duration,
K = soil erodibility factor; it represents a soil's ability to resist erosion and is determined by the soil texture, soil structure, organic matter content, and soil permeability,
L = slope length,
S = steepness factor,
C = cover and management factor; it is the ratio of soil loss in an area with specified cover and management to the corresponding soil loss in a clean-tilled and continuously fallow condition. For bare ground, $C = 1.0$,
P = support practice factor; it is the ratio of soil loss with a support practice such as contouring, strip-cropping, or implementing terraces compared to up-and-down-the-slope cultivation. For construction sites such as roadside embankment, P is not used in the equation

In 1996, the USDA published the *Revised Universal Soil Loss Equation* (RUSLE) (Renard et al. 1996) to supersede the original USLE. The RUSLE retains the form and factors of the original USLE, but the technology for the factor evaluations was altered and new data were introduced. For example, the database for the factor R was expanded; the factor K was revised to be time-varying to reflect freeze-thaw conditions and consolidations; the topographic factors for slope length and steepness, LS, were revised to reflect the ratio of rill to interrill erosion; the cover-management factor C was altered from the seasonal soil loss ratios to a continuous function that is the product of four subfactors; and the factor P was expanded to consider conditions for rangelands, contouring, strip-cropping, and terracing.

6.4.2 Surface erosion control measures

Various surface erosion control measures are utilized in the field practice. Sustainable, economical, and innovative methods continue to emerge. This section of the book describes a few currently popular methods of surface erosion control.

(a)   (b)

Fig. 6.11 Rock riprap protection of eroded banks of the Sacramento River (California, USA). (a) River bank erosion (critical condition), (b) river bank after rock riprap repair.

Riprap

According to the US Federal Highway Administration (1989), riprap is "a flexible channel or bank lining or facing consisting of a well graded mixture of rock, broken concrete, or other material, usually dumped or hand-placed, which provides protection from erosion." Riprap is often used to protect and stabilize embankments, side slopes of rivers, channels, lakes, dams, and slope drains and storm drains. Riprap revetments may include rock riprap, wire-enclosed rock, grouted rock, precast concrete block revetments, and paved lining.

Rock riprap is the most common surface erosion protection method for river and channel banks; an example is shown in Figure 6.11. The individual stones are typically angular in shape and well graded so that they can interlock. This interlocking property combines with the weight of the stone to form a solid mass that can resist erosion. Rock riprap can be unstable on steep slopes, especially when rounded rock is used. For slopes steeper than 2:1, other materials such as geosynthetics matting should be considered. The design of rock riprap may include the following considerations:

- Rock size
- Rock gradation
- Riprap layer thickness
- Filter design
- Material quality
- Edge treatment
- Construction considerations

Many government agencies provide riprap design guidelines. The following design recommendations for rock riprap were provided by Smolen et al. (1988).

- Gradation: Use a well-graded mixture of rock sizes instead of one uniform size.
- Quality of stone: Use riprap material that is durable so that the freeze and thaw cycles do not decompose it in a short time; most igneous stones, such as granite, have suitable durability.

- Riprap depth: Make the riprap layer at least two times as thick as the maximum stone diameter used in the riprap.
- Filter material: Apply a filter material, usually a synthetic cloth or a layer of gravel, before applying the riprap. This prevents the underlying soil from moving through the riprap.
- Riprap limits: Place riprap in such a way that it extends to the maximum flow depth, or to a point where vegetation will be satisfactory to control erosion.
- Curved flow channels: Ensure that riprap extends to five times the bottom width upstream and downstream of the beginning and ending of the curve and the entire curved section.
- Riprap size: The size of the riprap material depends on the shear stress of the flows the riprap will be subjected to; it ranges from an average size of 5 cm to 60 cm in diameter.

Compost

Applying compost in highly erodible areas can decrease erosion and allow quicker establishment of vegetation. Compost is the product of an aerobic process during which microorganisms decompose organic matter into a stable amendment. Composting turns wastes into useful materials that possess substantial economic and environmental benefits; composting also diverts the wastes from going to landfills. Various types of composts have been used in surface erosion control on embankments and natural slopes, including (1) green material compost made from yard trimmings, clippings, and agricultural byproducts, (2) manure compost such as from dairy and poultry manures, (3) co-compost material such as biosolids and green material mixed together, and (4) wood chips and forestry residual composts. Food scraps and municipal solid waste composts have also been used for erosion control. Depending on the sources and manufacturing of compost, their characteristics may vary significantly, in terms of pH, soluble salts, moisture content, organic matter content, maturity, stability, particle size, pathogen, physical contaminants, and so on. Compost that is suitable for engineering applications may be specified by different agencies. For example, the USDA and the United States Compost Council (USCC) have set forth the "Test Methods for the Examination of Composting and Compost" to ensure the quality of the final dry compost (USDA and USCC 2003).

Vegetation

Vegetative cover is another common erosion control method. It protects the soil surface from raindrop impact, shields the soil surface from the scouring effect of overland flow, and decreases the erosive capacity of the flowing water by reducing its velocity. The roots of vegetation can hold the soil, increase its erosion resistance and the rate of infiltration, and decrease runoff. Vegetative cover is relatively inexpensive to achieve and maintain, and it provides an aesthetically pleasing environment. It is often used with other erosion control methods, such as compost blankets, geosynthetic covers, mulches. The factors that should be considered in the design and establishment of vegetative covers are:

- Soil characteristics, namely, acidity, moisture retention, drainage, texture, organic matter, fertility.
- Site condition, namely, slopes, area of vegetative cover.
- Climate, such as temperatures, wind, precipitations.

Chapter 6

- Species selection, which depends on the regional climate, planting season, water requirement, soil preparation, weed control, postconstruction land use, and the expected level of maintenance such as irrigation and cost.
- Establishment methods.
- Maintenance procedures.

Erosion control using geosynthetics

Geosynthetics became increasingly popular in erosion and sediment control and slope stabilization. A great variety of methods that use geosynthetics are available, and new methods are emerging, namely, degradable rolled erosion control products (RECPs), nondegradable RECPs, and hard armors. Degradable products can be used to enhance the establishment of vegetation on rehabilitated lakeshores and riverbanks and alongside recently constructed roadways. These products are used where vegetation alone can provide sufficient site protection once the erosion control product has degraded. Nondegradable products provide long-term reinforcement of vegetation. They are used in more challenging erosion control applications where immediate, high-performance erosion protection is required. The materials extend the erosion resistance of soil, rock, and other materials by permanently reinforcing the vegetative root structure. Figure 6.12 shows a nondegradable erosion control measure. The geosynthetic erosion control measures often serve multiple functions such as surface runoff collection and drainage, filtration, separation, reinforcement, and establishment and maintenance of vegetative covers.

The concept and applications of geosynthetics are introduced in Chapter 8.

Fig. 6.12   Geocells are placed on the slope surface and then filled with gravel, as a permanent nondegradable erosion control and slope protection layer. (Photo courtesy of *Geosynthetics* magazine.)

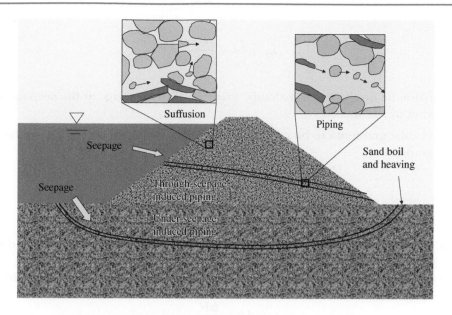

Fig. 6.13   Subsurface erosion in earthen embankment.

6.5   Subsurface erosion and seepage control methods

6.5.1   Subsurface erosion

Subsurface erosion has been one of the most prevalent causes of catastrophic failures of levees and earthen dams. Failures caused by surface erosion and scour are specific limit states that need to be considered in most existing design codes. Such examples include the 1972 failure of the Buffalo Creek dam in West Virginia, USA (Wahler 1973), and the 1990 collapsing of an earthen dam in South Carolina, USA (Leonards and Deschamps 1998). Subsurface erosion has various forms as shown in Figure 6.13. They are (1) piping – soil grains inside the soil matrix are mobilized and washed out of the matrix by concentrated seepage, resulting in a tubular channel, or pipe, that progressively forms from downstream to upstream; the pipe can develop into a large tunnel that can cause significant loss of soil and structural integrity; (2) suffusion – the mobilization and transportation of fine grains within a coarser soil matrix; suffusion may occur in the presence of discontinuity or segregation of soil grains; and (3) dispersion, which is a chemically induced erosion.

Piping may be the most significant and mostly studied subsurface erosion. As it generally initiates at the landside toe of an earthen embankment, the hydraulic gradient of the seepage that exits the subsoil at the landside toe is evaluated for piping potential. The factor of safety against piping is commonly expressed as:

$$\text{FS} = \frac{i_{\text{cr}}}{i_{\text{exit}}} \tag{6.18}$$

where:

i_{cr} = critical hydraulic gradient, the maximum hydraulic gradient that the soil can sustain before piping occurs; it depends on the unit weight of the soil solids and the compaction of the soil,

Chapter 6

$$i_{cr} = \frac{\gamma'}{\gamma_w} = \frac{G_s - 1}{1 + e} \qquad (6.19)$$

i_{exit} = exit hydraulic gradient, the hydraulic gradient of the seepage at the seepage's exit section in the subsoil.

The exit hydraulic gradient can be approximated using a flow net. According to the definition of hydraulic gradient:

$$i = \frac{\Delta b}{L} \qquad (6.20)$$

where,

Δb = total hydraulic head difference across a flow path of length L.

As shown in the flow net in Figure 6.14, Δb is the last equipotential drop at the downstream section, and L is the length of the flow lines within the last equipotential drop. Δb can be determined using the total head difference (Δb) and the number of equipotential drops (N_d) in the entire seepage zone:

$$\Delta b = \frac{\Delta H}{N_d} \qquad (6.21)$$

The highest exit hydraulic gradient occurs with the shortest flow path L, which is L_1 in Figure 6.14 that is nearest to the sheet pile wall.

6.5.2   Underseepage control methods in levees and earthen dams

Underseepage is the seepage through the foundation soil under a levee, an earthen dam, or a water basin. Without control, underseepage in pervious foundation soils beneath levees or earthen dams may result in (1) excessive hydrostatic pressures beneath an impervious top stratum on the landside, (2) sand boils, and (3) piping beneath the earthen structure. Principal

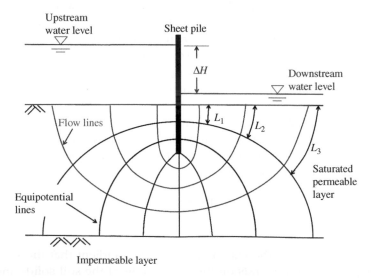

Fig. 6.14   Determining exit hydraulic gradient using flow net.

underseepage control measures are cutoff walls, riverside impervious blankets, landside seepage berms, pervious toe trenches, and pressure relief wells.

Cutoff walls

As shown in Figure 6.15, a cutoff wall is an impermeable barrier that is installed in an earthen embankment and penetrates the underlying permeable foundation soil. It is considered the most positive means of eliminating seepage and subsurface erosion problems and is widely used. Cutoff walls include slurry walls and sheet pile walls. A slurry wall is installed by first excavating a trench that is then backfilled with slurry. There are various types of slurry materials, such as cement-bentonite slurry, soil-cement-bentonite slurry, and soil-bentonite slurry. The slurry solidifies into an impermeable wall along the longitudinal direction of the levee or earthen dam. A sheet pile wall is made of interlocking steel sheet piles that are driven through the embankment and into the foundation soil using a vibratory hammer. A cutoff wall should penetrate most (such as 95%) of the permeable soil stratum. If the pervious foundation soil has significant depth, installing cutoff walls may not be economical.

Riverside impervious blankets

A riverside impervious blanket is usually a thin layer of compacted clay on the bottom of the river channel or reservoir, as shown in Figure 6.16. The blanket prevents water from seeping into the pervious foundation soil, thus reducing the seepage. The effectiveness of the blanket depends on its thickness, length, distance to the levee riverside toe, and permeability and can be evaluated using flow net or mathematical methods. The riverside impervious blankets can be natural or constructed. Protecting the blankets from damages is important against seepage and erosion.

Landside seepage berms

An impermeable clay layer (berm) can be constructed at the landside toe of a levee or earthen dam. Seepage berms may reinforce an existing impervious or semipervious top stratum, or if none exists, be placed directly on pervious deposits. A seepage berm can reduce the underseepage in the pervious foundation soil and the uplifting pressure on the top impervious stratum,

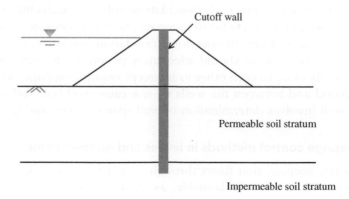

Fig. 6.15   Cutoff wall in levees.

Chapter 6

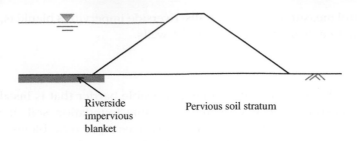

Fig. 6.16   Riverside impervious blanket.

if it exists, at the landside toe. Consequently, a seepage berm can prevent heaving and rupture of the top impervious stratum and sand boil. A berm can also provide some protection against the sloughing of the landside slope. Berms are very simple to construct and require little maintenance. Subsurface profiles must be carefully studied in selecting berm widths. For example, if a levee is founded on a thin top impervious stratum that becomes thicker landward, as shown in Figure 6.17(a), the seepage berm should extend landward to overlap the thick deposit (Figure 6.17(b)). Otherwise, a concentration of seepage and high exit hydraulic gradient may occur between the berm toe and the landward edge of the thick clay deposit.

Toe trenches

As shown in Figure 6.18, a drainage trench installed at the landside toe of an embankment is used to control shallow underseepage and protect the area in the vicinity of the levee toe from heaving and sand boils. Toe drains can be gravels or perforated pipes, and they should be protected by filter materials against clogging. Sand or geotextile filters can be used. Deeper seepage may bypass and exit beyond the toe trench. So, toe trenches may be used together with relief well systems: the wells collect the deeper seepage and the toe trench collects the shallow seepage.

Pressure relief wells

As shown in Figure 6.19, pressure relief wells are coarse-material-filled wells that are installed along the landside toe of an embankment. They are used to reduce uplift pressure that may otherwise cause sand boils and piping of the foundation soils. The wells intercept deeper underseepage and provide controlled outlets for the seepage, so that it does not emerge uncontrolled at the landside. They are used when the pervious foundation soil is too thick to be penetrated by cutoff walls. Pressure relief wells should adequately penetrate the pervious foundation soil and be placed sufficiently close to each other to intercept enough seepage, so that the hydraulic pressures acting beyond and between the wells do not cause soil boil and piping. The design of a pressure relief well involves determination of well spacing, size, and penetration.

6.5.3   Through-seepage control methods in levees and earthen dams

Through-seepage is the seepage that flows through an earthen embankment and exits at the downslope of the embankment at the landside, as illustrated in Figure 6.20. If uncontrolled, through-seepage can soften the soil in the vicinity of the landside toe, cause sloughing of the slope, and lead to piping and slope failure.

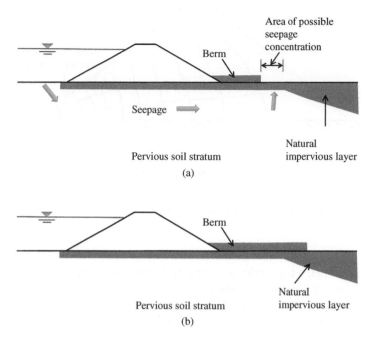

Fig. 6.17 Example of incorrect and correct berm width on the basis of existing foundation condition. (After USACE 2000). (a) Incorrect installation of berm, (b) correct installation of berm.

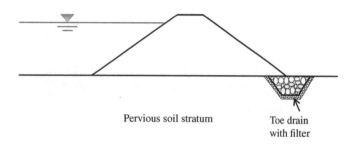

Fig. 6.18 Pervious toe trench. (After USACE 2000.)

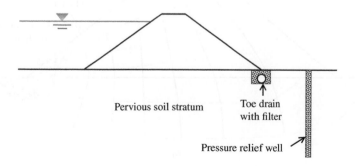

Fig. 6.19 Pressure relief wells. (After USACE 2000.)

Chapter 6

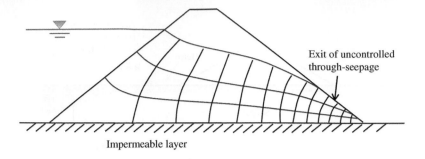

Fig. 6.20   Uncontrolled through-seepage in an earthen embankment.

Three commonly used through-seepage control measures are presented as follows.

Toe drains

As shown in Figure 6.21(a), a toe drain is compacted gravel or rock that is placed at the landside toe of an embankment. It provides ready exit for seepage through the embankment and can lower phreatic surface line (the top boundary of the saturated seepage zone), so that no seepage can emerge at the landside toe beyond the drain. A toe drain can also be combined with partially

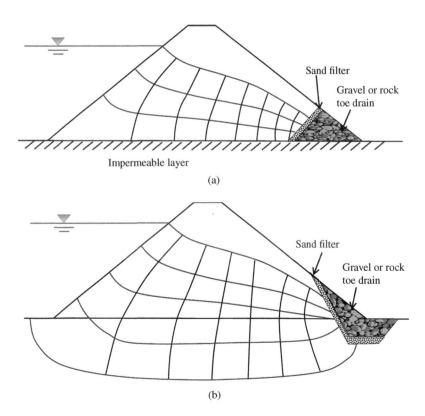

Fig. 6.21   Toe drain in embankment. (After USACE 2000). (a) Pervious toe drain in levee on impervious foundation, (b) pervious toe drain combined with toe trench.

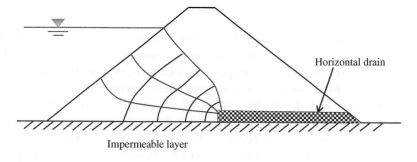

Fig. 6.22   Horizontal drainage layer in embankment. (After USACE 2000.)

penetrating toe trenches to simultaneously control shallow underseepage and through-seepage, as shown in Figure 6.21(b).

Horizontal drainage layers

Horizontal drainage layers, as shown in Figure 6.22, essentially serve the same purpose as the toe drain. They can also protect the base of the embankment against high uplift pressures, where shallow underseepage occurs.

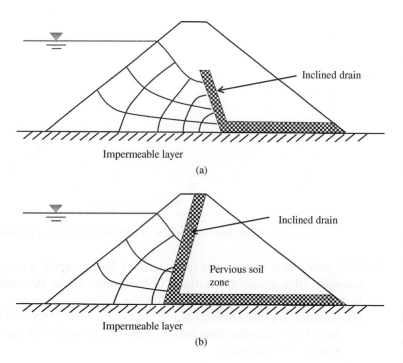

Fig. 6.23   Inclined drainage layers in embankment. (After USACE 2000). (a) In homogeneous embankment. (b) In zoned embankment.

Chapter 6

Inclined drainage layers

Inclined drainage layers, sometimes known as chimney drains, are used extensively in earthen dams. They are rarely used in levee constructions because of their added cost, but they may be justified for short levee reaches in important locations where landside slopes must be steep and the levee has high water table for prolonged periods. They intend to completely intercept through-seepage and should be tied into horizontal drainage layers in order to drain the intercepted seepage. The use of this type drain allows the landside of an embankment to be built using any stable materials regardless of permeability, as shown in Figure 6.23. If used in zoned embankments, the inclined drain should also function as a filter to prevent the soil migration from the riverside section into the pervious landside section.

Homework Problems

1. A granular filter is to be designed around a perforated pipe around a shallow foundation. The natural soil's grain size distribution is shown in Figure 6.24. Specify the grain size distribution of the granular filter.

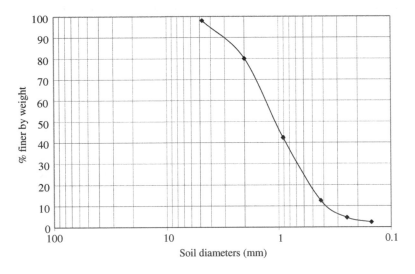

Fig. 6.24   Grain size distribution of natural base soil for Problem 1.

2. Well points are to be used to dewater a large trench excavation that extends 10 m below the surface of the water table. Assume that the bottom width of the trench excavation is 20 m.
   (1) Develop a simple sketch of the necessary two-stage well point system. Use separate diagrams to show the progress of the necessary stages of installation.

(2) Assume that half of the necessary lowering of the water table elevation is achieved using the outer row of well points. If the soil being dewatered is a fine-to-medium sand, approximately at what distance beyond the well point location is the original groundwater table that is not affected by well point dewatering?

3. What are the conditions that are suitable for vacuum dewatering?

4. What is electroosmosis? What is the soil condition that is suitable for electroosmosis?

5. In designing riprap, what should be determined?

6. What are the typical types of compost? In selecting compost for surface erosion control, what are the characteristics of compost that need to be evaluated?

7. What are the factors that should be considered in the design and establishment of vegetative covers?

8. Present a case study where compost is used as a surface erosion control measure.

9. Present a case study where degradable erosion control products are used.

10. Present a case study where nondegradable geosynthetic erosion control products are used.

11. Briefly describe the forms of subsurface erosion.

12. A concrete dam section is shown in Figure 6.25. The foundation soil is homogeneous and isotropic; its permeability is 4×10^{-5} mm/s. Also given are $G_s = 2.65$ and $e = 0.5$.
    (1) Draw flow net.
    (2) Calculate seepage beneath the dam.
    (3) Find exit hydraulic gradient, i_{exit}.
    (4) Determine the factor of safety against piping at the toe of the dam.
    (5) Calculate the hydraulic lift force beneath the dam.

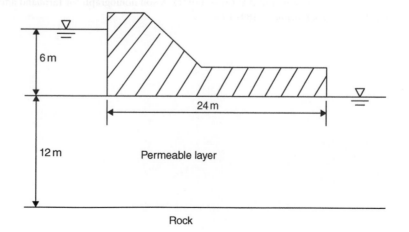

6 m

24 m

12 m   Permeable layer

Rock

Fig. 6.25   A dam section for Problem 12.

Chapter 6

References

Bond, A., and Harris, A. (2008). Decoding Eurocode 7, Taylor & Francis, London.

Cedergren, H.R. (1977). Seepage, Drainage, and Flow Nets, 2nd edition, John Wiley & Sons, Inc, New York, NY, USA.

Das, B.M. (2010). Principles of Geotechnical Engineering, 7th edition, Cengage Learning, Stamford, CT, USA.

Das, B.M. (2008). Advanced Soil Mechanics, 3rd edition, Taylor & Francis, New York, NY, USA.

International Commission on Large Dams (ICOLD) (1994). Embankment Dams – Granular Filters and Drains, Bulletin 95, published by the ICOLD, Paris. 256 pp.

Leonards, G.A., and Deschamps, R.J., 1998, "Failure of Cyanide Overflow Pond Dam," *Journal of Performance of Constructed Facilities,* Vol. 12(1), 3–11.

McCarthy, D.F. (2007). Essentials of Soil Mechanics and Foundations: Basic Geotechnics, 7th edition. Pearson Education, Inc. Upper Saddle River, New Jersey, USA.

Reddi, L.N. 2003. Seepage in Soils, John Wiley & Sons, Inc., Hoboken, NJ, USA.

Renard, K.G., Foster, G.R., Weesies, G.A., McCool, D.K., and Yoder, D.C. (1996). Predicting Soil Erosion by Water: A Guide to Conservation Planning with the Revised Universal Soil Loss Equation (RUSLE). United States Department of Agriculture, Agricultural Research Service (USDA-ARS), Agriculture Handbook Number 703. Washington, DC, USA.

Smolen, M.D., Miller, D.W., Wyatt, L.C., Lichthardt, J., and Lanier, A.L. (1988). Erosion and Sediment Control Planning and Design Manual. North Carolina Sedimentation Control Commission; North Carolina Department of Environment, Health and Natural Resources, and Division of Land Resources Land Quality Section, Raleigh, NC.

Terzaghi, K., and Peck, R.B. (1967). Soil Mechanics in Engineering Practice, 2nd Edition, John Wiley & Sons, New York, NY, USA.

US Army Corps of Engineers (USACE) (2000). Design and Construction of Levees, EM 1110-2-1913, 30 April, 2000. Department of the Army, US Army Corps of Engineers, Washington, DC.

US Federal Highway Administration (FHWA) (1989). Design of Riprap Revetment, HEC 11, FHWA Publication Number: IP-89-016, FHWA, United State Department of Transportation, Washington, DC, USA.

US Naval Facilities Engineering Command (NAVFAC) (1986). Soils and Foundations Design Manuals, DM7.01, Soil Mechanics. U.S. Naval Facilities Engineering Command, Alexandria, VA, USA.

United States Department of Agriculture (USDA) and US Composting Council (USCC). (2003). Test Methods for the Examination of Composting and Compost (TMECC) 02.02-B.

Wahler, W.A., 1973, "Analysis of Coal Refuse Dam Failure, Middle Fork Buffalo Creek, Saunders West Virginia," National Technical Service Rep. PB-215, Washington, DC, pp. 142-143.

Wischmeier, W.H. (1976). Use and misuse of the universal soil loss equation. *Journal of Soil and Water Conservation* 31:5–9.

Wischmeier, W.H. and D.D. Smith. (1978). Predicting Rainfall Erosion Losses: A Guide to Conservation Planning. USDA Agricultural Handbook, No. 537.

Wischmeier, W.H., C.B. Joshson, and B.V. Cross. (1971). A soil nomograph for farmland and construction sites. *Journal of Soil and Water Conservation* 26:189–193.

Chapter 6

Chapter 7

Soil Retaining Structures

7.1 Introduction to soil retaining structures

Soil retaining structures are used to retain slopes, steep or vertical cuts, fills, foundation excavations, bridge abutments, and so on. Figures 7.1–7.4 show four common types of soil retaining structures: retaining wall, sheet pile wall, bracing for excavation, and soil nail wall that are used to retain soils for various purposes.

In this chapter, the designs of three types of soil retaining structures are covered: conventional retaining walls, sheet pile walls, and soil nail walls. The design of mechanically stabilized earth (MSE) walls is covered in Chapter 8.

Prior to the design of any soil retaining structure, geotechnical investigation should be thoroughly carried out, so that the subsurface soil strata and their characteristics, the backfill soil types and properties, and the interaction parameters (e.g., adhesion and external friction angle) between the soil and the retaining structure can be adequately determined. Moreover, the system loads should be correctly considered and determined. The system loads are earth pressures, surcharge loads, water loads, namely, hydrostatic pressure, seepage effects, and wave action, and other loads such as wind loads, earthquake forces, ice forces. The design of earth retaining structures should be based on the possible failure modes of the structures.

In this chapter, the methods of determining lateral earth pressures are reviewed first.

7.2 Lateral earth pressures

The lateral earth or soil pressure is the lateral pressure exerted on the soil retaining structure by the soil behind the retaining structure. In the retaining wall design, the soil behind the wall is also known as the backfill. There are three types of lateral earth pressures based on the direction of the retaining wall movement. They are:

1. *At-rest earth pressure*, when the retaining wall stands vertically and does not move either away from or into the backfill.
2. *Active earth pressure*, when the retaining wall moves away from the backfill, that is, the backfill actively pushes the wall.
3. *Passive earth pressure*, when the retaining wall moves into the backfill, that is, the backfill is passively pushed. The rotation of the wall into the backfill can be caused by external lateral forces on the wall.

Geotechnical Engineering Design, First Edition. Ming Xiao.
© 2015 John Wiley & Sons, Ltd. Published 2015 by John Wiley & Sons, Ltd.
Companion Website: www.wiley.com/go/Xiao

Fig. 7.1   Retaining wall on the campus of California State University, Fresno, USA.

Fig. 7.2   Sheet pile wall for vertical cut. (Photo courtesy of Earthwork Engineering, Inc. Hollis, New Hampshire.)

The three types of lateral soil pressures are illustrated in Figure 7.5. The determination of lateral soil pressures is based on the limit equilibrium principle: the backfill is at the onset of failure. So the active and the passive earth pressures are defined as the pressures when a failure shear surface is developed in the backfill (Figures 7.5(b) and (c)), and the failure occurs only when there is a sufficient wall movement. Figure 7.6 illustrates this concept.

Fig. 7.3   Braced sheet pile wall. (Photo courtesy of Earthwork Engineering, Inc. Hollis, New Hampshire.)

Fig. 7.4   Soil nail wall used to repair a landslide in Montana, USA. (Photo courtesy of Soil Nail Launcher, Inc.)

It is important to note that lateral soil pressure should be calculated using effective stress.

7.2.1   At-rest earth pressure

$$\sigma_{\text{at-rest}} = K_0 \sigma' \tag{7.1}$$

where:

$\sigma_{\text{at-rest}}$ = the at-rest earth pressure,
K_0 = the at-rest earth pressure coefficient,
σ' = the vertical effective stress in the backfill along the back of the wall.

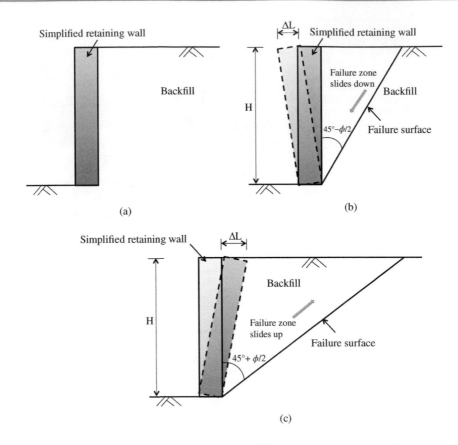

Fig. 7.5 Lateral earth pressures. (a) At-rest earth pressure, (b) active earth pressure, (c) passive earth pressure.

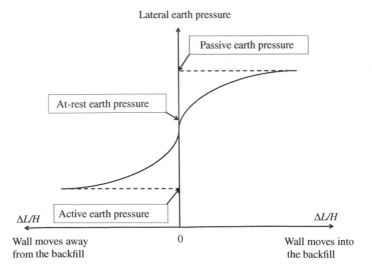

Fig. 7.6 Definitions of lateral earth pressures are based on the maximum wall movement until failure.

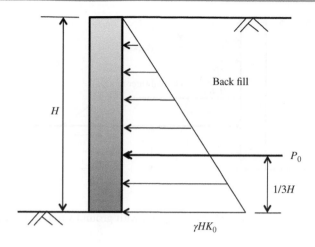

Fig. 7.7   At-rest pressure distribution diagram.

The at-rest earth pressure coefficient, K_0, depends on the backfill conditions, such as the soil type, density, strength parameters, consolidation, and the inclination of the backfill.

For coarse-grained or normally consolidated soils behind a vertical wall (Caltrans Bridge Design Specification 2004):

$$K_0 = (1 - \sin \phi')(1 - \sin \beta) \tag{7.2}$$

where:

ϕ' = effective friction angle of the backfill,
β = inclination angle of the backfill.

For overconsolidated soils whose types can range from clay to gravel (Mayne and Kulhawy 1982):

$$K_0 = (1 - \sin) \cdot (\text{OCR})^{\sin \phi'} \tag{7.3}$$

where:

OCR = overconsolidation ratio.

For the same type of backfill, K_0 is constant. So the at-rest earth pressure varies linearly with the depth of the backfill, as shown in Figure 7.7.

The total resultant lateral earth force per unit length of the wall, P_0, is equal to the area of the earth pressure distribution diagram; and the point of application is $H/3$ from the bottom of the wall.

$$P_0 = \frac{1}{2}\gamma H^2 K_0 \tag{7.4}$$

If the backfill is partially submerged under the groundwater table, the at-rest pressure varies with the variation of the effective stress. An example of the at-rest earth pressure distribution is shown in Figure 7.8. The total resultant lateral force, P_0, are the lateral earth force (shown by areas ①, ②, and ③) and the hydrostatic force (shown by area ④).

$$P_0 = \text{area}① + \text{area}② + \text{area}③ + \text{area}④$$
$$= \frac{1}{2}\gamma H_1^2 K_0 + \gamma H_1 H_2 K_0 + \frac{1}{2}\gamma' H_2^2 K_0 + \frac{1}{2}\gamma_w H_2^2 \tag{7.5}$$

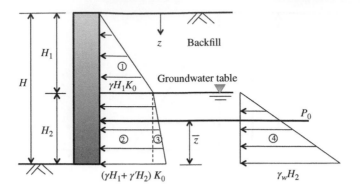

Fig. 7.8   At-rest pressure distribution diagram with groundwater table in the backfill.

To determine the point of application of P_0, moment equilibrium about the toe of the wall is used:

$$P_0\,\bar{z} = \text{area}① \times \textbf{arm} + \text{area}② \times \textbf{arm} + \text{area}③ \times \textbf{arm} + \text{area}④ \times \textbf{arm}$$

$$= \frac{1}{2}\gamma H_1^2 K_0 \left(\frac{H_1}{3} + H_2 \right) + \gamma H_1 H_2 K_0 \left(\frac{H_2}{2} \right) + \frac{1}{2}\gamma' H_2^2 K_0 \left(\frac{H_2}{3} \right) + \frac{1}{2}\gamma_w H_2^2 \left(\frac{H_2}{3} \right) \qquad (7.6)$$

Then the point of application, $\bar{z}$, from the bottom of the wall can be calculated.

7.2.2   Rankine's theory

In 1857, William Rankine, a Scottish civil engineer, physicist, and mathematician, proposed the Rankine theory to determine the active and passive earth pressures on a retaining wall. The theory was based on the following three conditions:

- the back of the wall is vertical,
- the backfill is horizontal, and
- there is no friction between the back of the wall and the backfill.

Rankine active earth pressure

The stress condition of a two-dimensional soil element can be seen in Figure 7.9. When a failure surface is developed in the backfill, the Mohr's circle that represents the stress condition is tangent to the Mohr–Coulomb failure criterion line. At failure, the minor principal stress, σ'_a, is defined as the active earth pressure. Using the knowledge of trigonometry, the Rankine active earth pressure is:

$$\sigma'_a = \sigma'_0 \tan^2 \left(45° - \frac{\phi'}{2} \right) - 2c' \tan \left(45° - \frac{\phi'}{2} \right) \qquad (7.7)$$

where:

σ'_0 = effective vertical stress behind the retaining wall,
ϕ' = effective internal friction angle of the backfill,
c' = effective cohesion of the backfill.

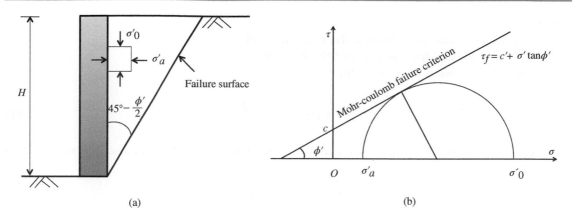

Fig. 7.9   Rankine active earth pressure. (a) 2-D soil element behind a retaining wall, (b) Mohr's circle and failure criterion.

It is defined:

$$K_a = \tan^2\left(45° - \frac{\phi'}{2}\right) \tag{7.8}$$

K_a is the Rankine's active earth pressure coefficient. So, Equation (7.7) can also be expressed as:

$$\sigma'_a = \sigma'_0 K_a - 2c'\sqrt{K_a} \tag{7.9}$$

Rankine passive earth pressure

The stress condition of a two-dimensional soil element for passive pressure case can be seen in Figure 7.10. When a failure surface is developed in the backfill, the Mohr's circle that represents the stress condition is tangent to the Mohr–Coulomb failure envelope. At failure, the major principal stress, σ'_p, is defined as the passive earth pressure. Using the knowledge of trigonometry, the Rankine passive earth pressure is:

$$\sigma'_p = \sigma'_0 \tan^2\left(45° + \frac{\phi'}{2}\right) + 2c' \tan\left(45° + \frac{\phi'}{2}\right) \tag{7.10}$$

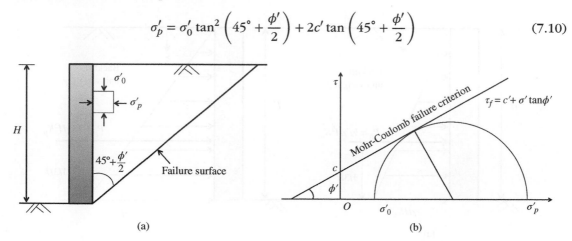

Fig. 7.10   Rankine passive earth pressure. (a) 2-D soil element behind retaining wall, (b) Mohr's circle and failure criterion.

Chapter 7

where:

σ_0' = effective vertical stress behind the retaining wall,
ϕ' = effective internal friction angle of the backfill,
c' = effective cohesion of the backfill.

It is defined:

$$K_p = \tan^2 \left(45° - \frac{\phi'}{2} \right)$$ (7.11)

K_p is Rankine's passive earth pressure coefficient. So, Equation (7.10) can also be expressed as:

$$\sigma_p' = \sigma_0' K_p + 2c' \sqrt{K_p}$$ (7.12)

Pressure distribution diagram with granular backfill (c' = 0)

(a) Active earth pressure

When ($c' = 0$) and there is no groundwater in the backfill, the Rankine active earth pressure can be expressed as:

$$\sigma_a' = \gamma z K_a$$ (7.13)

The pressure distribution is shown in Figure 7.11(a). The total earth force per unit length of the wall is equal to the area of the pressure distribution diagram, and the point of application is $H/3$ from the bottom of the wall.

(b) Passive earth pressure

When $c' = 0$ and there is no groundwater in the backfill, the passive earth pressure can be expressed as:

$$\sigma_p' = \gamma z K_p$$ (7.14)

The pressure distribution, total force, and point of application are shown in Figure 7.11(b).

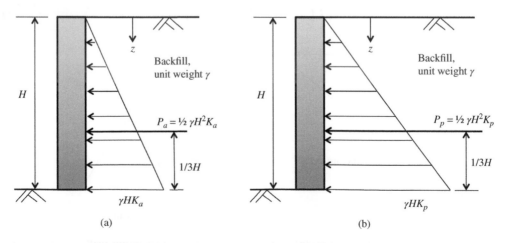

(a)   (b)

Fig. 7.11   Rankine earth pressures for cohesionless backfill. (a) Active earth pressure, (b) passive earth pressure.

Pressure distribution diagram with cohesive backfill (c' ≠ 0)

- *Active earth pressure*

  When $c' \neq 0$ and there is no groundwater in the backfill, the Rankine active earth pressure can be expressed as:

  $$\sigma'_a = \gamma z K_a - 2c' \sqrt{K_a} \tag{7.15}$$

  When $z = 0$, $\sigma'_a = -2c' \sqrt{K_a}$, indicating tension at the top of the retaining wall. Because of the soil's minimum tensile strength, a tensile crack can develop at the top portion of the retaining wall where tension (negative pressure) exists. The active earth pressure distribution is shown in Figure 7.12. The length of the tensile crack can be calculated using:

  $$\sigma'_a = \gamma \overline{z} K_a - 2c' \sqrt{K_a} = 0 \tag{7.16}$$

  So:

  $$\overline{z} = \frac{2c'}{\gamma \sqrt{K_a}} \tag{7.17}$$

  When a tensile crack is developed, the soil and the retaining wall within the tensile crack are no longer in contact. So the total earth force per unit length of the wall is equal to the area of the triangle ABC (Figure 7.12).

  $$P_a = \frac{1}{2}(\gamma H K_a - 2c' \sqrt{K_a})(H - \overline{z}) \tag{7.18}$$

- *Passive earth pressure*

  When $c' \neq 0$ and there is no groundwater in the backfill, the Rankine passive earth pressure can be expressed as:

  $$\sigma'_p = \gamma z K_p + 2c' \sqrt{K_p} \tag{7.19}$$

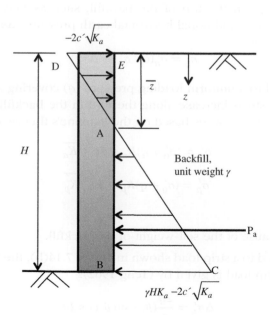

Fig. 7.12   Rankine active earth pressure for cohesive backfill.

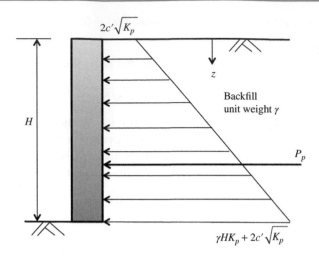

Fig. 7.13   Rankine passive earth pressure for cohesive backfill.

The passive earth pressure distribution is shown in Figure 7.13. The total resultant earth force per unit length of the wall is:

$$P_p = \frac{1}{2}\left(\gamma H K_p + 4c'\sqrt{K_p}\right)H \tag{7.20}$$

Lateral earth pressure with surcharge on backfill

When there is a surcharge on the top of the backfill, such as because of buildings, roads, railways, the surcharge causes additional horizontal earth pressure ($\Delta\sigma_h$) at various depths. The total lateral earth pressure is:

$$\sigma'_h = \sigma'_a(\text{or } \sigma'_p) + \Delta\sigma'_h \tag{7.21}$$

If a backfill is subjected to a uniform loading pressure (q) covering a large area, as shown in Figure 14(a), the vertical stress increase along the wall in the backfill is assumed to be q. So, the active and passive earth pressures based on the Rankine's theory are:

$$\sigma'_a = (\sigma'_0 + q)K_a - 2c'\sqrt{K_a} \tag{7.22}$$

$$\sigma'_p = (\sigma'_0 + q)K_p + 2c\sqrt{K_p} \tag{7.23}$$

where:

σ'_0 = effective stress because of the self-weight of the backfill.

If the backfill is subjected to a strip load shown in Figure 7.14(b), the horizontal earth pressure increase following the strip load is given by (Teng 1962):

$$\Delta\sigma'_h = \frac{2q}{\pi}(\beta - \sin\beta\ \cos 2\alpha) \tag{7.24}$$

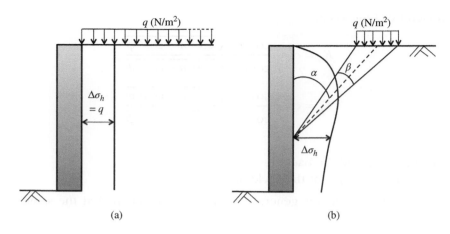

Fig. 7.14 Lateral earth pressure increase because of surcharge on the backfill. (a) Uniform surcharge covering large area, (b) strip load.

If the surcharge is a point load, line load, or uniform rectangular load, the calculations of $\Delta\sigma_h$ can follow the recommendations in the Naval Facilities Engineering Command Design Manual 7.02 (NAVFAC 1986).

Lateral earth pressure with inclined backfill

If the backfill is inclined as shown in Figure 7.15, the calculations of active and passive earth pressures still follow Equations (7.9) and (7.12), except that the lateral earth pressure coefficients are different. For granular backfill, the Rankine active and passive earth pressure coefficients

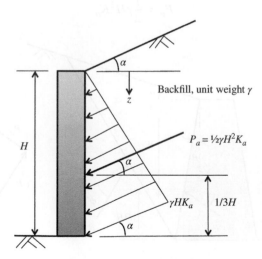

Fig. 7.15 Rankine lateral earth pressure with inclined backfill (active case).

can be expressed by (Teng 1962):

$$K_a = \cos \alpha \frac{\cos \alpha - \sqrt{\cos^2 \alpha - \cos^2 \phi}}{\cos \alpha + \sqrt{\cos^2 \alpha - \cos^2 \phi}} \tag{7.25}$$

$$K_p = \cos \alpha \frac{\cos \alpha + \sqrt{\cos^2 \alpha - \cos^2 \phi}}{\cos \alpha - \sqrt{\cos^2 \alpha - \cos^2 \phi}} \tag{7.26}$$

where:

α = the inclination angle of the backfill,
ϕ = the internal friction angle of the backfill.

The direction of the pressures is generally assumed to be inclined at the same angle of the backfill inclination.

7.2.3 Coulomb's theory

In 1776, Charles-Augustin de Coulomb, a French physicist, presented the Coulomb's theory to determine the active and passive earth forces on a retaining wall. The Coulomb's theory considered the inclination of the backfill, the inclination of the back of the wall, and the friction between the backfill and the retaining wall. It is important to note that the Coulomb's theory applies only to cohesionless soil ($c = 0$). The development of the Coulomb's theory is based on force equilibrium as shown in Figure 7.16 (active case) and Figure 7.17 (passive case). Therefore, the Coulomb's theory provides only the active and passive earth resultant forces; it does not provide earth pressure distribution with depth, although "earth pressure" is still conventionally used.

Coulomb active earth force

The Coulomb active earth force per unit length of a wall is:

$$P_a = \frac{1}{2}\gamma H^2 K_a \tag{7.27}$$

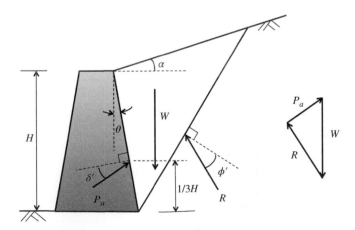

Fig. 7.16   Coulomb's active case of lateral earth "pressure."

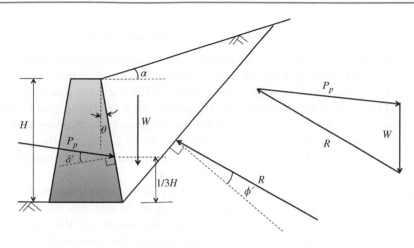

Fig. 7.17   Coulomb's passive case of lateral earth "pressure."

where K_a is the Coulomb active earth pressure coefficient:

$$K_a = \frac{\cos^2(\phi'-\theta)}{\cos^2\theta\cos(\delta'+\theta)\left[1+\sqrt{\dfrac{\sin(\phi'+\delta')\sin(\phi'-\alpha)}{\cos(\theta+\delta')\cos(\theta-\alpha)}}\right]^2} \tag{7.28}$$

where:

ϕ' = effective internal friction angle of the backfill,
θ = inclination angle of the back of the wall,
δ' = effective external friction angle between the backfill and the wall,
α = inclination angle of the backfill.

The point of application of P_a is at $H/3$ from the bottom of the wall.

Coulomb passive earth force

The Coulomb passive earth force per unit length of a wall is:

$$P_p = \frac{1}{2}\gamma H^2 K_p \tag{7.29}$$

where K_p is the Coulomb passive earth pressure coefficient:

$$K_p = \frac{\cos^2(\phi'+\theta)}{\cos^2\theta\cos(\theta-\delta')\left[1-\sqrt{\dfrac{\sin(\phi'+\delta')\sin(\phi'+\alpha)}{\cos(\theta-\delta')\cos(\theta-\alpha)}}\right]^2} \tag{7.30}$$

where the angles ϕ', θ, δ', and α are the same as defined in the Coulomb's active case.

The point of application of P_p is at $H/3$ from the bottom of the wall. It is noted that the direction of P_p is different from that of P_a.

7.3 Conventional retaining wall design

There are generally two major types of retaining walls: conventional retaining walls and MSE walls. The conventional retaining walls are constructed using (reinforced) concrete and include gravity, semigravity, cantilever retaining walls (Figure 7.18). Gravity retaining walls have no reinforcement and use self-weight for stability. Therefore, they are bulky and heavy and cannot be very tall. Semigravity walls have minimum reinforcement and are slimmer than gravity walls. But they are still relatively bulky and heavy. Cantilever walls contain sufficient reinforcement and can be very tall. Figure 7.19 shows a cantilever retaining wall in construction. The MSE walls are made of reinforced soil and are described in Chapter 8.

The conventional terminologies used in rigid conventional retaining walls are shown in Figure 7.20. The design of rigid conventional retaining walls are the internal stability (reinforcement) and the external stability. The external stability design are the checks against three failure modes: (1) overturning about the toe, (2) sliding along the base, and (3) bearing capacity failure of the foundation soil.

A typical cantilever retaining wall is shown in Figure 7.21. The earth pressure is active pressure. When designing a retaining wall, the soil on the footing, indicated by areas ④ and ⑤, is considered as part of the retaining wall. So the earth pressure acting on the vertical profile of AB is calculated. The backfill is typically granular soils, so $c'_1 = 0$. Using the Rankine's theory, the active earth pressure distribution is shown in Figure 7.21. The pressure is inclined at the same inclination of the backfill. The total resultant earth force per unit length of the wall is:

$$P_a = \frac{1}{2}\gamma_1(H')^2 K_a \tag{7.31}$$

where K_a is calculated using Equation (7.8) or (7.25).

The active earth force has horizontal and vertical components:

$$P_\mathrm{H} = P_a \cos \alpha \tag{7.32}$$

$$P_\mathrm{V} = P_a \sin \alpha \tag{7.33}$$

7.3.1 Factor of safety against overturning

The factor of safety against overturning is expressed by:

$$\mathrm{FS}_{\text{overturn}} = \frac{\sum M_R}{\sum M_O} \geq 2.0 \tag{7.34}$$

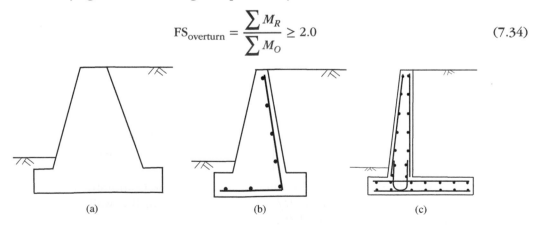

Fig. 7.18 Conventional retaining walls. (a) Gravity wall, (b) semi-gravity wall, (c) cantilever wall.

Fig. 7.19 Conventional retaining wall construction process (Highway 41, Fresno, California, USA, 2010).

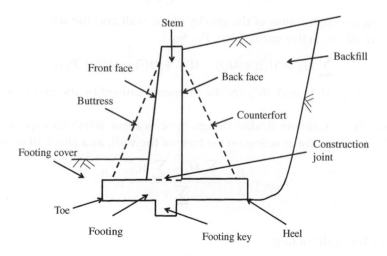

Fig. 7.20 Terms used in rigid conventional cantilever retaining wall.

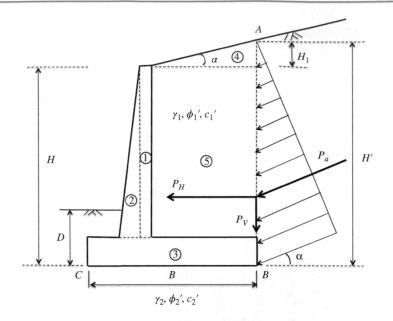

Fig. 7.21 A typical cantilever retaining wall configuration.

where:

$\sum M_R$ = sum of the resisting moments to the overturning, about the toe C.

$\sum M_O$ = sum of the overturning moments, about the toe C.

The overturning moment is caused only by the horizontal component of the active earth force, P_H.

$$\sum M_O = P_H \left(\frac{H'}{3} \right) \tag{7.35}$$

The resisting moments are because of the gravity of the wall and the soil on the footing and the vertical component of the active earth force, P_V. So:

$$\sum M_R = M① + M② + M③ + M④ + M⑤ + P_V B \tag{7.36}$$

where: $M①$, $M②$, $M③$, $M④$, and $M⑤$ are the moments caused by the gravity of areas ① to ⑤, respectively.

Some guidelines (e.g., Caltrans Bridge Design Specification 2004) also specify the maximum eccentricity of the resultant force acting on the base of the wall, as a check of overturning failure:

$$e_{max} = \frac{B}{2} - \frac{\sum M_R - \sum M_O}{\sum V} \leq \frac{B}{6} \tag{7.37}$$

where:

B = width of the wall footing,

$\sum V$ = total vertical force on the footing, namely, the gravity of the concrete wall, the backfill on the footing, and P_V.

Note that oftentimes a wall footing is embedded with shallow embedment depth (D) at the toe of the wall. When the backfill pushes the wall, passive earth pressure develops at the toe and contributes to the resistance to overturning and sliding. To be conservative, the passive earth pressure at the toe may be ignored, to account for the cases when the footing cover may not be present during construction or in service.

Alternatively, a limit state design approached can be used to explicitly consider the equilibrium of the structure. This means that the value of the design effect of the destabilizing actions (disturbing moments) are less than or equal to the value of the design effects of the stabilizing actions, such that:

$$E_{dst,d} \leq E_{stb,d} + T_d \tag{7.38}$$

where:

$E_{dst,d}$ is the design effect of the destabilizing actions (disturbing moments), such as from lateral earth pressure;

$E_{stb,d}$ design effects of the stabilizing actions, such as from gravity;

T_d is the contribution through the resistance of the ground around the structure.

7.3.2 Factor of safety against sliding

The factor of safety against sliding is expressed by:

$$FS_{slide} = \frac{\sum F_R}{\sum F_S} = \frac{\left(\sum V\right)\tan\delta + B \cdot c_a'}{P_H} \geq 1.5 \tag{7.39}$$

where:

$\sum F_S$ = total sliding force = P_H,

$\sum F_R$ = the sum of resisting forces to the sliding because of the friction and adhesion between the wall footing and the foundation soil,

δ = external friction angle between the concrete footing and the foundation soil,

c_a' = adhesion between the concrete footing and the foundation soil.

It is typically assumed that:

$$\delta = \frac{2}{3}\phi_2' \tag{7.40}$$

and

$$c_a' = \frac{2}{3}c_2' \tag{7.41}$$

The passive earth pressure at the toe of the wall may be ignored.

Alternatively, within the context of limit state design (and using partial factors of safety) it shall be verified that:

$$E_d \leq R_{V,d}\tan\delta + T_d \tag{7.42}$$

where:

E_d is the design effect of the actions because of loads imposed on the foundation by the retaining wall (which are also affected by various partial factors of safety);

Chapter 7

$R_{V,d} \tan \delta$ is the design resistance against sliding produced by the sum of permanent, favorable design vertical actions ($R_{V,d}$) and the interface angle of shear resistance between the soil and the structure (δ);

T_d is the passive earth pressure at the toe of the wall that, as stated before, it can be ignored for most cases.

7.3.3 Factor of safety of bearing capacity

The factor of safety against bearing capacity failure is expressed by:

$$FS_{bearing} = \frac{q_{ult}}{p_{max}} \geq 3.0 \qquad (7.43)$$

where:

q_{ult} = ultimate bearing capacity of the foundation soil,
p_{max} = maximum pressure at the wall footing.

The minimum and maximum pressures exerted by the wall at the bottom of the footing are:

$$p_{\substack{max \\ min}} = \frac{\sum V}{B}\left(1 \pm \frac{6e}{B}\right) \qquad (7.44)$$

The eccentricity (e) is calculated by Equation (7.37). It is important that e is less than $B/6$, so that $p_{min} > 0$. Otherwise, negative pressure (tension) will develop at the bottom of the footing, causing the separation between the footing and the soil beneath, and the footing is only partially supported on the soil, as shown in Figure 7.22.

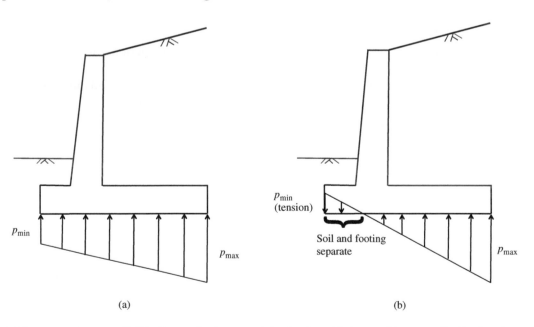

Fig. 7.22 Eccentric stress distribution at the bottom of the footing. (a) Acceptable stress distribution ($e < B/6$), (b) unacceptable stress distribution ($e > B/6$).

The ultimate bearing capacity of the wall footing can be calculated using the Meyerhof equation:

$$q_{ult} = c_2' N_c F_{cd} F_{ci} + q N_q F_{qd} F_{qi} + \frac{1}{2} \gamma_2 B' N_\gamma F_{\gamma d} F_{\gamma i}$$ (7.45)

where:

c_2' = effective cohesion of the foundation soil (note: not of the backfill);

N_c, N_q, N_γ = bearing capacity factors that are based on the effective friction angle of the foundation soil (ϕ_2') and can be found from Table 3.3 in Chapter 3,

q = effective stress at the bottom of the foundation, $q = \gamma_2 D$,

γ_2 = unit weight of the footing cover at the toe,

D = depth of footing cover at the toe,

B' = $B - 2e$.

Depth factors:

$$F_{qd} = 1 + 2 \tan \phi_2' (1 - \sin \phi_2')^2 \left(\frac{D}{B'} \right)$$ (7.46)

$$F_{cd} = F_{qd} - \frac{1 - F_{qd}}{N_c \tan \phi_2'}$$ (7.47)

$$F_{\gamma d} = 1$$ (7.48)

Inclination factors:

$$F_{ci} = F_{qi} = \left(1 - \frac{\psi^\circ}{90^\circ} \right)^2$$ (7.49)

$$F_{\gamma i} = \left(1 - \frac{\psi}{\phi_2'} \right)^2$$ (7.50)

where:

$$\psi = \tan^{-1} \left(\frac{P_H}{\sum V} \right) \quad \text{(in degrees)}$$ (7.51)

If the calculated factors of safety for overturning, sliding, or bearing capacity are less than the allowable values, the retaining wall should be redesigned, such as by increasing the footing width (B), adding a footing key, reducing the wall height (H), using alternative lightweight backfill, or changing wall type (such as from gravity wall to cantilever wall, from conventional rigid wall to MSE wall).

Alternatively, a limit state design approach combined with partial factors of safety can be used as discussed in Chapter 3 for the bearing capacity of shallow foundations. The design resistance V_d (= $q_{ult} A$, where A is the area of the foundation and q_{ult} is the ultimate bearing capacity) is used to compare with the design effect of the actions E_d because of loads imposed on the foundation by the retaining wall (which are also affected by various partial factors of safety).

$$V_d \geq E_d$$ (7.52)

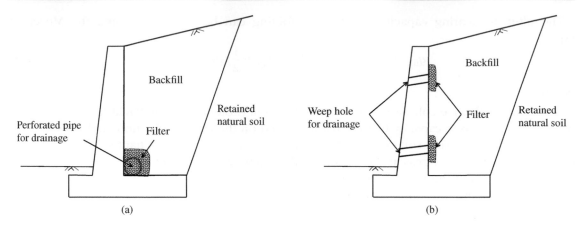

Fig. 7.23 Drainage for retaining walls. (a) Drainage using perforated pipe, installed along the back of the wall, (b) drainage using weep holes, installed in the wall stem.

7.3.4 Retaining wall drainage

Drainage is extremely important for retaining walls. Groundwater behind the wall can exert large lateral hydrostatic pressure on the wall and can cause catastrophic failure of the wall. An effective drainage system can prevent hydrostatic pressure buildup behind the wall. There are various types of drainage systems: some employ conventional pipe drains; some use combinations of geosynthetics. Figure 7.23 illustrates two common types of drainages for cantilever retaining walls. The filter around a perforated pipe or behind a weep hole protects the drainage channel from clogging. The drainage and filter designs are described in Chapter 6.

Sample Problem 7.1: **External stability of cantilever retaining wall**

A conventional cantilever retaining wall is shown below (Figure 7.24). Check the external stability of the retaining wall. Make appropriate assumptions if needed.

Solution: Using working stress design and total factor of safety:

The dimensions and the active earth force are shown in Figure 7.25.

$$H' = H_1 + H = 0.42 + 6 = 6.42 \text{ m}$$

The Rankine active earth pressure coefficient:

$$K_a = \tan^2\left(45 - \frac{\phi'_1}{2}\right) = \tan^2\left(45 - \frac{30}{2}\right) = 0.333$$

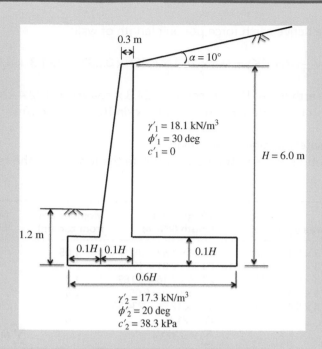

Fig. 7.24   Sample Problem: cantilever retaining wall.

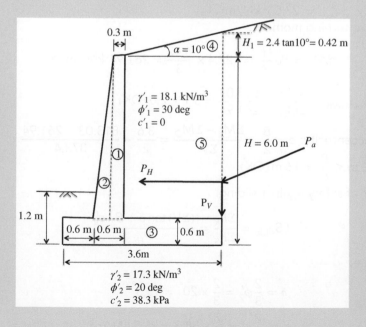

Fig. 7.25   Configuration and dimensions of a cantilever retaining wall in sample problem 7.1.

The Rankine active earth force per unit length of wall:

$$P_a = \frac{1}{2}\gamma_1(H')^2 K_a = 0.5 \times 18.1 \times 6.42^2 \times 0.333 = 124.3 \ \text{kN/m}$$

Horizontal earth force: $P_H = P_a \cos\alpha = 124.3 \times \cos 10° = 122.4 \ \text{kN/m}$
Vertical earth force: $P_V = P_a \sin\alpha = 124.3 \times \sin 10° = 21.6 \ \text{kN/m}$

1. Factor of safety against overturning
   The following table is prepared to calculate the forces and the resisting moments.

Section no.	Area (m^2)	Weight/unit length (kN/m)	Moment arm from toe (m)	Moment (kN·m/m)
1	$0.3 \times 5.4 = 1.62$	$23.56 \times 1.62 = 38.17$	1.05	40.08
2	$0.5 \times 0.3 \times 5.4 = 0.81$	$23.56 \times 0.81 = 19.08$	0.8	15.26
3	$3.6 \times 0.6 = 2.16$	$23.56 \times 2.16 = 50.89$	1.8	91.60
4	$0.5 \times 2.4 \times 0.42 = 0.50$	$18.1 \times 0.50 = 9.05$	2.8	25.34
5	$2.4 \times 5.4 = 12.96$	$18.1 \times 12.96 = 234.58$	2.4	562.99
		$P_V = 21.6$	3.6	77.76
		$\sum V = 373.4$		$\sum M_R = 813.03$

Note: $\gamma_{concrete} = 23.56 \ \text{kN/m}^3$.
The overturning moment:

$$\Sigma M_O = P_H \frac{H'}{3} = 122.4 \times \frac{6.42}{3} = 261.94 \ \text{kN} \cdot \text{m/m}$$

So: $\text{FS}_{overturn} = \dfrac{\Sigma M_R}{\Sigma M_O} = \dfrac{813.03}{261.94} = 3.1 \geq 2.0 \ (\text{OK})$

And eccentricity: $e = \dfrac{B}{2} - \dfrac{\Sigma M_R - \Sigma M_O}{\Sigma V} = \dfrac{3.6}{2} - \dfrac{813.03 - 261.94}{373.4}$

$= 0.32 \ \text{m} \leq \dfrac{B}{6} = 0.6 \ \text{m} \ (\text{OK})$

2. Factor of safety against sliding

$$\text{FS}_{slide} = \frac{\Sigma F_R}{\Sigma F_S} = \frac{(\Sigma V)\tan\delta + B \cdot c'_a}{P_H}$$

Assume:

$$\delta = \frac{2}{3}\phi'_2 = \frac{2}{3} \times 20° = 13.3°$$

$$c'_a = \frac{2}{3}c'_2 = \frac{2}{3} \times 38.3 = 25.5 \ \text{kN/m}^2$$

So: $FS_{slide} = \dfrac{\left(\sum V\right) \tan\delta + B \cdot c_a'}{P_H} = \dfrac{373.4 \times \tan 13.3° + 3.6 \times 25.5}{124.3}$

$= 1.45 < 1.5$ (not OK)

If considering the passive earth force at the toe: $P_p = 162.15$ kN/m,

then: $FS_{slide} = \dfrac{\left(\sum V\right) \tan\delta + B \cdot c_a' + P_p}{P_H} = 2.72 > 1.5$ (OK)

3. Factor of safety for bearing capacity

The maximum and minimum pressures at the bottom of the footing are:

$p_{\substack{max \\ min}} = \dfrac{\sum V}{B}\left(1 \pm \dfrac{6e}{B}\right) = \dfrac{373.4}{3.6}\left(1 \pm \dfrac{6 \times 0.32}{3.6}\right) = \dfrac{159.04}{48.40}$ kN/m^2

The ultimate bearing capacity of the wall footing:

$$q_{ult} = c_2' N_c F_{cd} F_{ci} + q N_q F_{qd} F_{qi} + \frac{1}{2}\gamma_2 B' N_\gamma F_{\gamma d} F_{\gamma i}$$

For $\phi_2' = 20°$, find in Table 3.3: $N_c = 14.83$, $N_q = 6.4$, $N_\gamma = 5.39$.

$$q = \gamma_2 D = 17.3 \times 1.2 = 20.76 \text{ kN/m}^2$$

$$B' = B - 2e = 3.6 - 2 \times 0.32 = 2.96 \text{ m}$$

Depth factors:

$$F_{qd} = 1 + 2\tan\phi_2'(1 - \sin\phi_2')^2\left(\dfrac{D}{B'}\right)$$

$$= 1 + 2 \times \tan 20°(1 - \sin 20°)^2\left(\dfrac{1.2}{2.96}\right) = 1.130$$

$$F_{cd} = F_{qd} - \dfrac{1 - F_{qd}}{N_c \tan\phi_2'} = 1.130 - \dfrac{1 - 1.130}{14.83 \times \tan 20°} = 1.154$$

$$F_{\gamma d} = 1$$

Inclination factors:

$$\psi = \tan^{-1}\left(\dfrac{P_H}{\sum V}\right) = \tan^{-1}\left(\dfrac{122.4}{373.4}\right) = 18.9°$$

$$F_{ci} = F_{qi} = \left(1 - \dfrac{\psi°}{90°}\right)^2 = \left(1 - \dfrac{18.9°}{90°}\right)^2 = 0.624$$

$$F_{\gamma i} = \left(1 - \dfrac{\psi}{\phi_2'}\right)^2 = \left(1 - \dfrac{18.9°}{20°}\right)^2 = 0.003 \approx 0$$

So: $q_{ult} = 38.3 \times 14.83 \times 1.154 \times 0.652 + 20.76 \times 6.4 \times 1.130 \times 0.624$
$+ 0.5 \times 17.3 \times 2.96 \times 5.39 \times 1 \times 0 = 521.0$ kN/m^2

$$FS_{bearing} = \dfrac{q_{ult}}{p_{max}} = \dfrac{521.04}{159.04} = 3.2 > 3.0 \qquad \text{(OK)}$$

Alternative solution using limit state design and partial factors of safety:

The dimensions and the active earth force are shown in Figure 7.25.

$$H' = H_1 + H = 0.42 + 6.0 = 6.42 \text{ m}$$

Assumed partial factors of safety:

$$\gamma_{\phi'} = 1.00, \gamma_{\gamma'} = 1.00, \gamma_{G;fav} = 0.90, \gamma_{G;unfav} = 1.10$$

Design geotechnical parameters: $\phi'_d = \tan^{-1}\left(\dfrac{\tan \phi'_k}{\gamma_{\phi'}}\right) = \tan^{-1}\left(\dfrac{\tan 30}{1.0}\right) = 30°$

$$\gamma'_d = \frac{\gamma'_k}{\gamma'_\gamma} = \frac{18.1}{1.00} = 18.1 \text{ kN/m}^3$$

The Rankine active earth pressure coefficient:

$$K_a = \tan^2\left(45 - \frac{\phi'_d}{2}\right) = \tan^2\left(45 - \frac{30}{2}\right) = 0.333$$

Design actions:
The self-weight of the wall is a permanent, favorable action ($G_{w,d}$). The design actions because of self-weight can therefore be calculated as

$$G_{W,d} = \text{Area} \times \gamma \times \gamma_{G,fav}$$

where $\gamma_{concrete} = 24$ kN/m^3, a value commonly used in Europe. Detailed calculations are illustrated below:

Section no.	Area (m²)	Weight/unit length (kN/m)	Design action/unit length (kN/m)
1	$0.3 \times 5.4 = 1.62$	$24 \times 1.62 = 38.9$	$38.9 \times 0.9 = 35.0$
2	$0.5 \times 0.3 \times 5.4 = 0.81$	$24 \times 0.81 = 19.4$	$19.4 \times 0.9 = 17.5$
3	$3.6 \times 0.6 = 2.16$	$24 \times 2.16 = 51.8$	$51.8 \times 0.9 = 46.6$
4	$0.5 \times 2.4 \times 0.42 = 0.5$	$18.1 \times 0.50 = 9.1$	$9.1 \times 0.9 = 8.1$
5	$2.4 \times 5.4 = 12.96$	$18.1 \times 12.96 = 234.6$	$234.6 \times 0.9 = 211.1$

The active earth pressure per unit length of wall is a permanent, unfavorable action:

$$P_a = \frac{1}{2}\gamma'_1(H')^2 K_a \gamma_{G,unfav} = 0.5 \times 18.1 \times 6.42^2 \times 0.333 \times 1.10 = 136.6 \text{ kN/m}$$

Horizontal earth force: $P_H = P_a \cos \alpha = 136.6 \times \cos 10 = 134.5$ kN/m
Vertical earth force: $P_V = P_a \sin \alpha = 136.6 \times \sin 10 = 23.7$ kN/m

1. Verification of overturning limit state
   The following table is prepared to calculate the design effect of the stabilizing actions.

No.	Area (m^2)	Weight/unit length (kN/m)	Design action/unit length (kN/m)	Moment arm from toe (m)	Moment (kN·m/m)
1	$0.3 \times 5.4 = 1.62$	$24 \times 1.62 = 38.9$	$38.9 \times 0.9 = 35.0$	1.05	36.75
2	$0.5 \times 0.3 \times 5.4 = 0.81$	$24 \times 0.81 = 19.4$	$19.4 \times 0.9 = 17.5$	0.80	14.00
3	$3.6 \times 0.6 = 2.16$	$24 \times 2.16 = 51.8$	$51.8 \times 0.9 = 46.6$	1.80	83.88
4	$0.5 \times 2.4 \times 0.42 = 0.5$	$18.1 \times 0.50 = 9.1$	$9.1 \times 0.9 = 8.1$	2.80	22.68
5	$2.4 \times 5.4 = 12.96$	$18.1 \times 12.96 = 234.6$	$234.6 \times 0.9 = 211.1$	2.40	506.6
			$P_V = 23.7$	3.6	85.32
		Total stabilizing moment ($E_{stb,d}$)			749.23

Note that $\gamma_{concrete}$ is 24 kN/m^3. It is also important to highlight that P_V is a stabilizing action in the strictest sense, which would imply that it should be factored using a $\gamma_{G,fav}$ partial factor of safety. In the table above, the value of P_V has been obtained using $\gamma_{G,unfav}$ instead. The argument behind this assumption is that actions/forces produced by the same source should be factored in the same way (i.e. the single source principle).
The overturning moment:

$$\sum M_O = E_{dst;d} = P_H \frac{H'}{3} = 134.5 \times \frac{6.42}{3} = 287.3 \text{ kN} \cdot \text{m/m}$$

Assuming that T_d is small, it can be verified that $E_{dst,d} \le E_{stb,d} + T_d$ hence the limit state against overturning is satisfied.

2. Limit state verification against sliding
   It is assumed here that the same partial factors of safety are applicable, even if that might not be the case as required in BS EN 1997–1:2004. The design effect of the action is the horizontal component of the active earth pressure. This is a permanent, unfavorable action such that:

$$P_a = \frac{1}{2}\gamma_1'(H')^2 K_a \gamma_{G,unfav} = 0.5 \times 18.1 \times 6.42^2 \times 0.333 \times 1.10$$

$$= 136.6 \text{ kN/m}$$

and

$$E_d = P_H = P_a \cos \alpha = 136.6 \times \cos 10 = 134.5 \text{ kN/m}$$

The design resistance is given by $R_{V,d} \tan \delta$ where $\delta = 30°$ assuming that the wall is made of reinforced concrete and $R_{V,d}$ is the sum of permanent, favorable vertical actions as calculated in the following table:

No.	Area (m^2)	Weight/unit length (kN/m)	Design action/unit length (kN/m)
1	$0.3 \times 5.4 = 1.62$	$24 \times 1.62 = 38.9$	$38.9 \times 0.9 = 35.0$
2	$0.5 \times 0.3 \times 5.4 = 0.81$	$24 \times 0.81 = 19.4$	$19.4 \times 0.9 = 17.5$
3	$3.6 \times 0.6 = 2.16$	$24 \times 2.16 = 51.8$	$51.8 \times 0.9 = 46.6$
4	$0.5 \times 2.4 \times 0.42 = 0.5$	$18.1 \times 0.50 = 9.1$	$9.1 \times 0.9 = 8.1$
5	$2.4 \times 5.4 = 12.96$	$18.1 \times 12.96 = 234.6$	$234.6 \times 0.9 = 211.1$
			$P_V = 23.7$
		TOTAL ($R_{V,d}$)	342.0

Hence, assuming once again that T_d is small ...

$$E_d = 134.5 \leq R_{V,d} \tan 30 + T_d = 342.0 \tan 10 = 60.3$$

The limit state against sliding is satisfied

3. Verification for bearing resistance

As the calculations here follow the same principles explained for shallow foundations in Chapter 3, readers are referred to Sample Problem 3.1 for a typical calculation example.

7.4   Sheet pile wall design

7.4.1   Failure modes

There are two basic types of steel sheet pile walls: cantilever walls and anchored walls. There are three types of failure modes of sheet pile walls (USACE 1994), as explained below.

Deep-seated failure: This is a rotational failure of the entire soil mass retained by a sheet pile wall, as shown in Figure 7.26. The failure mode is independent of the structural characteristics of the wall and the anchor. This type of failure should be assessed using conventional slope stability analysis. Usually, the failure cannot be remedied by increasing the depth of penetration or by repositioning the anchor. The only remedy is to change the geometry of the retaining material or improve the soil's strength.

Rotational failure because of inadequate pile penetration: Lateral earth and/or water pressure on the sheet pile can cause the rigid wall to rotate about the point O, as shown in Figure 7.27. This type of failure can be prevented by adequate penetration of the piling in a cantilever wall or by a proper combination of penetration and anchor in an anchored wall.

Structural failure: Overstress of the piling or the anchor can break the structural component, as shown in Figures 7.28 and 7.29. Adequate structural design can prevent this type of failure.

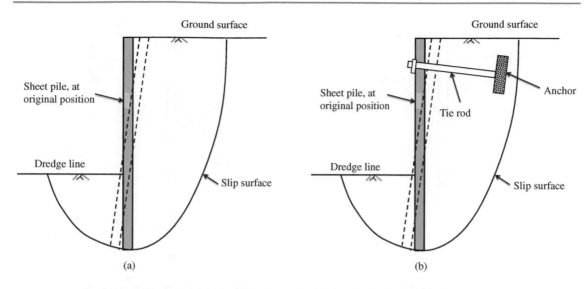

Fig. 7.26 Deep-seated failure of sheet pile walls. (a) Cantilever wall, (b) anchored wall.

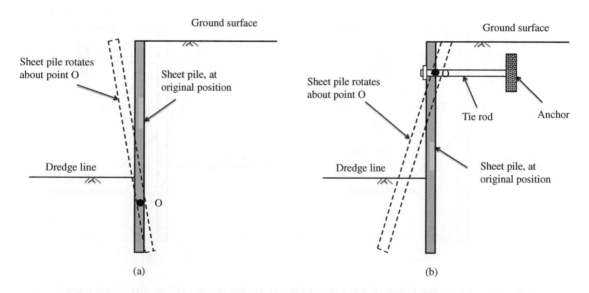

Fig. 7.27 Rotational failure because of inadequate penetration. (a) Cantilever wall, (b) anchored wall.

7.4.2 Preliminary data for the design

The following preliminary data should be established before the design of sheet pile walls (USACE 1994).

- Elevation at the top of the sheet piling.
- The ground surface profile that covers a distance of at least 10 times of the exposed wall height, on both sides of the wall.
- The subsoil profile on both sides of the wall, namely, cohesion, friction angle, and unit weight.

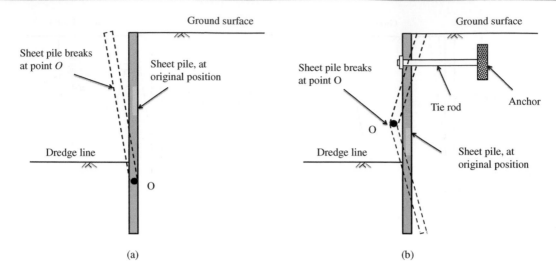

Fig. 7.28   Flexural failure of sheet pile walls. (a) Cantilever wall, (b) anchored wall.

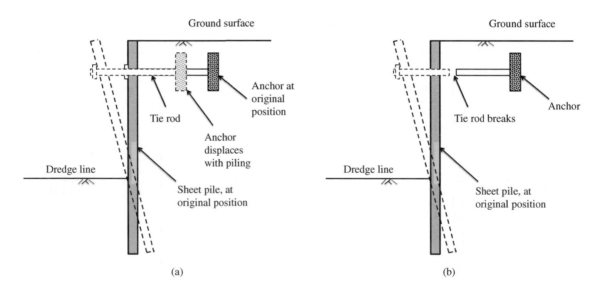

Fig. 7.29   Anchorage failure of sheet pile walls. (a) Anchor passive failure, (b) tie rod failure.

- Water elevation on both sides of the wall and seepage characteristics.
- Magnitudes and locations of surface charges, on both sides of the wall.
- Magnitudes and locations of external loads that are applied directly to the wall.

7.4.3   Design of cantilever walls penetrating cohesionless soils

The following approach of designing cantilever walls penetrating sand is based on the Steel Sheet Piling Design Manual (United States Steel 1984) and Principles of Foundation Engineering (Das 2011).

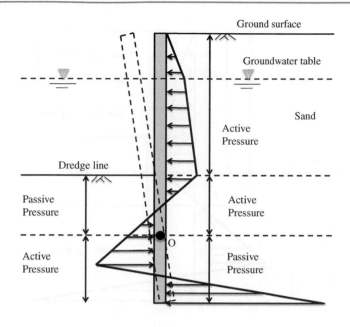

Fig. 7.30 Cantilever sheet pile penetrating sand with groundwater table.

Figure 7.30 illustrates a cantilever sheet pile penetrating sand with groundwater table in the sand. The relative movements of the sheet pile and a simplified lateral earth pressure distribution are shown in the figure. The design of a sheet pile are:

(a) embedment depth,
(b) maximum moment in the sheet pile,
(c) required section modulus.

To explain the design approach, Figure 7.31 is used to show the variables that will be determined in the design. The design approach is as follows.

The following parameters should be first determined: L_1, L_2, soil's unit weight, and strength parameters on both sides of the piling.

The active earth pressure based on the Rankine's theory at the groundwater table is:

$$\sigma_1' = \gamma L_1 K_a \tag{7.53}$$

The active earth pressure at the dredge line is:

$$\sigma_2' = (\gamma L_1 + \gamma' L_2) K_a \tag{7.54}$$

The active earth pressure below the dredge line and above the rotational point O (on the right side of the sheet pile) (Refer Figure 7.30) is:

$$\sigma_a' = [\gamma L_1 + \gamma' L_2 + \gamma'(z - L_1 - L_2)] K_a \tag{7.55}$$

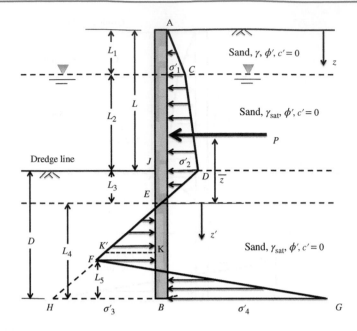

Fig. 7.31   Design graph and parameters for a cantilever sheet pile penetrating sand.

The passive earth pressure below the dredge line and above the rotational point O (on the left side of the sheet pile) (Refer Figure 7.30) is:

$$\sigma'_p = \gamma'(z - L_1 - L_2)K_p \tag{7.56}$$

Note: the hydrostatic pressures on both sides of the pile are equal and cancel each other.
The net lateral earth pressure below the dredge line above the rotational point O is:

$$\sigma' = \sigma'_a - \sigma'_p = [\gamma L_1 + \gamma' L_2 + \gamma'(z - L_1 - L_2)]K_a - \gamma'(z - L_1 - L_2)K_p$$
$$= (\gamma L_1 + \gamma' L_2)K_a - \gamma'(z - L_1 - L_2)(K_p - K_a)$$
$$= \sigma'_2 - \gamma'(z - L)(K_p - K_a) \tag{7.57}$$

L_3 can be determined where the net lateral pressure is zero above the rotational point O:

$$\sigma' = \sigma'_2 - \gamma' L_3 (K_p - K_a) = 0 \tag{7.58}$$

So:

$$L_3 = \frac{\sigma'_2}{\gamma'(K_p - K_a)} \tag{7.59}$$

The total resultant earth force for the active earth pressure zone ACDE is:

$$P = \frac{1}{2}\sigma'_1 L_1 + \frac{1}{2}(\sigma'_1 + \sigma'_2)L_2 + \frac{1}{2}\sigma'_2 L_3 \tag{7.60}$$

Calculate the moment arm $(\bar{z})$ of P about point E, using moment equilibrium:

$$P \cdot \bar{z} = \frac{1}{2}\sigma'_1 L_1 \left(\frac{1}{3}L_1 + L_2 + L_3\right) + \sigma'_1 L_2 \left(\frac{L_2}{2} + L_3\right) + \frac{1}{2}(\sigma'_2 - \sigma'_1)L_2 \left(\frac{L_2}{3} + L_3\right) + \frac{1}{2}\sigma'_2 L_3 \left(\frac{2}{3}L_3\right)$$

(7.61)

Calculate the length of HB (σ'_3) using the similar triangles of DJE and HBE:

$$\frac{\sigma'_2}{L_3} = \frac{\sigma'_3}{L_4}$$

(7.62)

So:

$$\sigma'_3 = \frac{\sigma'_2}{L_3}L_4 = \frac{(\gamma L_1 + \gamma' L_2)K_a}{L_3}L_4$$

(7.63)

where L_4 is unknown.

At the bottom of the sheet pile, the pile moves to the right, so the passive earth pressure occurs on the right side of the sheet pile:

$$\sigma'_p = [\gamma L_1 + \gamma'(L_2 + L_3 + L_4)]K_p$$

(7.64)

The active earth pressure occurs on the left side of the sheet pile:

$$\sigma'_a = \gamma'(L_3 + L_4)K_a$$

(7.65)

So, the net earth pressure at the bottom of the sheet pile is:

$$\sigma'_4 = \sigma'_p - \sigma'_a = [\gamma L_1 + \gamma'(L_2 + L_3 + L_4)]K_p - \gamma'(L_3 + L_4)K_a$$

(7.66)

where L_4 is unknown.

The stability of the sheet pile wall requires force equilibrium and moment equilibrium.

For force equilibrium:

$$\Sigma(\text{horizontal forces on the sheet pile}) = 0$$

(7.67)

So: area of ACDE + area of FHG−area of EHB = 0
So:

$$P + \frac{1}{2}(\sigma'_3 + \sigma'_4)L_5 - \frac{1}{2}\sigma'_3 L_4 = 0$$

(7.68)

L_5 can be solved:

$$L_5 = \frac{\sigma'_3 L_4 - 2P}{\sigma'_3 + \sigma'_4}$$

(7.69)

where L_4 is unknown.

For moment equilibrium:

$$\Sigma(\text{moments of the horizontal forces on the sheet pile about any point}) = 0$$

(7.70)

Take moments about the point B:

$$P(\bar{z} + L_4) + \frac{1}{2}(\sigma'_3 + \sigma'_4)L_5 \left(\frac{L_5}{3}\right) - \frac{1}{2}\sigma'_3 L_4 \left(\frac{L_4}{3}\right) = 0$$

(7.71)

where only L_4 is unknown.

Use trial and error by using different L_4 values and find L_4 to make the total moments zero in Equation (7.71).

The theoretical penetration depth is:

$$D = L_3 + L_4 \tag{7.72}$$

The maximum moment occurs at a cross section with zero shear force. This location is denoted as K (Figure 7.31). So: area of ACDE = area of EKK'.

For a new axis z' starting from point E, use similar triangles:

$$\frac{\sigma_2'}{L_3} = \frac{KK'}{z'} \tag{7.73}$$

From Equation (7.59):

$$KK' = \frac{\sigma_2'}{L_3}z' = \gamma'(K_p - K_a)z' \tag{7.74}$$

At the cross section with zero shear:

$$P = \frac{1}{2}(z')^2\gamma'(K_p - K_a) \tag{7.75}$$

so:

$$z' = \sqrt{\frac{2P}{\gamma'(K_p - K_a)}} \tag{7.76}$$

The maximum moment is at point K:

$$M_{max} = P(\bar{z} + z') - \frac{1}{2}\gamma'(K_p - K_a)(z')^2 \left(\frac{z'}{3}\right) \tag{7.77}$$

Given the allowable flexural stress of the sheet pile, σ_{all}, the section modulus of the sheet pile per unit length is:

$$S = \frac{M_{max}}{\sigma_{all}} \tag{7.78}$$

Factor of safety for sheet pile stability:

Two common approaches have been adopted by different agencies to introduce the factor of safety into the design process. A desirable method recommended by the US Army Corps of Engineers (USACE 1994) and the United States Steel (1984) is to apply a factor of safety (strength reduction factor) to the soil strength parameters ϕ and c. Because passive pressures are less likely to be fully developed than active pressures on the retaining side (USACE 1994), the current practice is to evaluate passive pressures using the "effective" values of ϕ and c:

$$\tan(\phi_{eff}) = \frac{\tan\phi}{FS_p} \tag{7.79}$$

$$c_{eff} = \frac{c}{FS_p} \tag{7.80}$$

where: FS_p = factor of safety for passive pressure and typically 1.5. ϕ_{eff} is used in calculating K_p. When calculating active pressure coefficient, K_a, ϕ is used without strength reduction. The penetration depth calculated using a strength reduction is the design penetration depth.

Another approach is to add 20–40% to the theoretical penetration depth (*D*) without using strength reduction (Das 2011, USS 1984).

Alternatively, a limit state design approach using partial factors of safety can also be used in conjunction with the method proposed by the Steel Sheet Piling Design Manual (United States Steel 1984) and Principles of Foundation Engineering (Das 2011). However, care must be taken in the selection of the values for these partial factors of safety. Sample Problem 7.2 is solved using these three different approaches, and more details are provided as well for the limit state design approach.

Sample Problem 7.2: Design of a cantilever wall penetrating cohesionless soil

A cantilever sheet pile wall penetrating granular soil is shown in Figure 7.32. Determine:

(a) The depth of the embedment, *D*, using a factor of safety of 1.5 for strength reduction.
(b) The maximum moment on the sheet pile.
(c) Given the allowable flexural stress of the sheet pile, $\sigma_{all} = 150$ MN/m², determine the section modulus of the sheet pile per unit length.

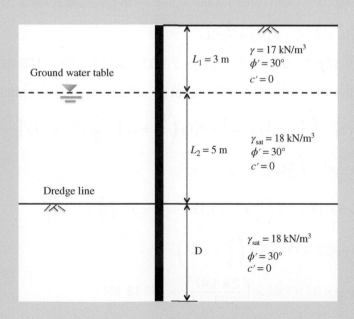

Ground water table

$L_1 = 3$ m

$\gamma = 17$ kN/m³
$\phi' = 30°$
$c' = 0$

$L_2 = 5$ m

$\gamma_{sat} = 18$ kN/m³
$\phi' = 30°$
$c' = 0$

Dredge line

D

$\gamma_{sat} = 18$ kN/m³
$\phi' = 30°$
$c' = 0$

Fig. 7.32   Sample problem of sheet pile design in sand.

Solution:

The "effective" internal friction angle considering the strength reduction is:

$$\tan(\phi_{\text{eff}}) = \frac{\tan\phi}{FS_p} = \frac{\tan 30°}{1.5} = 0.385$$

Find: $\phi_{\text{eff}} = 21°$

Rankine active and passive earth pressure coefficients are:

$$K_a = \tan^2\left(45 - \frac{\phi}{2}\right) = \tan^2\left(45 - \frac{30}{2}\right) = 0.333$$

$$K_p = \tan^2\left(45 + \frac{\phi_{\text{eff}}}{2}\right) = \tan^2\left(45 + \frac{21}{2}\right) = 2.117$$

$$\gamma' = \gamma_{\text{sat}} - \gamma_w = 18 - 9.81 = 8.19 \text{ kN/m}^3$$

$$\sigma'_1 = \gamma L_1 K_a = 17 \times 3 \times 0.333 = 17 \text{ kN/m}^2$$

$$\sigma'_2 = (\gamma L_1 + \gamma' L_2)K_a = (17 \times 3 + 8.19 \times 5) \times 0.333 = 30.6 \text{ kN/m}^2$$

$$L_3 = \frac{\sigma'_2}{\gamma'(K_p - K_a)} = \frac{30.6}{8.19 \times (2.117 - 0.333)} = 2.09 \text{ m}$$

The total resultant earth force for the active earth pressure zone ACDE is:

$$P = \frac{1}{2}\sigma'_1 L_1 + \frac{1}{2}(\sigma'_1 + \sigma'_2)L_2 + \frac{1}{2}\sigma'_2 L_3$$

$$= 0.5 \times 17 \times 3 + 0.5 \times (17 + 30.6) \times 5 + 0.5 \times 30.6 \times 2.09 = 176.48 \text{ kN/m}$$

Calculate the moment arm ($\bar{z}$) of P about point E, using moment equilibrium:

$$P \cdot \bar{z} = \frac{1}{2}\sigma'_1 L_1\left(\frac{1}{3}L_1 + L_2 + L_3\right) + \sigma'_1 L_2\left(\frac{L_2}{2} + L_3\right) + \frac{1}{2}(\sigma'_2 - \sigma'_1)L_2\left(\frac{L_3}{3} + L_3\right)$$

$$+ \frac{1}{2}\sigma'_2 L_3\left(\frac{2}{3}L_3\right)$$

$$= 0.5 \times 17 \times 3 \times \left(\frac{3}{3} + 5 + 1.47\right) + 17 \times 5 \times \left(\frac{5}{2} + 1.47\right)$$

$$+ \frac{1}{2}(32 - 17) \times 5 \times \left(\frac{5}{3} + 1.47\right)$$

$$+ \frac{1}{2} \times 32 \times 1.47 \times \left(\frac{2 \times 1.47}{3}\right) = 757.12 \text{ kN}$$

Find: $\bar{z} = 3.05$m

$$\sigma'_3 = \frac{\sigma'_2}{L_3}L_4 = \frac{(\gamma L_1 + \gamma' L_2)K_a}{L_3}L_4 = \frac{(17 \times 3 + 8.19 \times 5) \times 0.333}{1.47}L_4 = 21.77L_4$$

The net earth pressure at the bottom of the sheet pile:

$$\sigma_4' = \sigma_p' - \sigma_a' = [\gamma L_1 + \gamma'(L_2 + L_3 + L_4)]K_p - \gamma'(L_3 + L_4)K_a$$

$$= [17 \times 3 + 8.19 \times (5 + 2.09 + L_4)] \times 2.117 - 8.19 \times (2.09 + L_4) \times 0.333$$

$$= 14.6L_4 + 225.2$$

Using force equilibrium:

$$\Sigma(\text{horizontal forces on the sheet pile}) = 0$$

$$P + \frac{1}{2}(\sigma_3' + \sigma_4')L_5 - \frac{1}{2}\sigma_3'L_4 = 0$$

Solve: $L_5 = \dfrac{\sigma_3'L_4 - 2P}{\sigma_3' + \sigma_4'} = \dfrac{14.6L_4^2 - 2 \times 176.48}{14.6L_4 + (14.6L_4 + 225.2)} = \dfrac{14.6L_4^2 - 353.0}{29.2L_4 + 225.2}$

Using moment equilibrium:

$$\Sigma(\text{moments of the horizontal forces on the sheet pile about point B}) = 0$$

$$P(\bar{z} + L_4) + \frac{1}{2}(\sigma_3' + \sigma_4')L_5\left(\frac{L_5}{3}\right) - \frac{1}{2}\sigma_3'L_4\left(\frac{L_4}{3}\right) = 0$$

$$176.48 \times (4.36 + L_4) + \frac{1}{2}(14.65L_4 + 14.60L_4 + 225.2) \times \frac{1}{3} \times \left(\frac{14.6L_4^2 - 353.0}{29.2L_4 + 225.2}\right)^2$$

$$-\frac{1}{2} \times 14.65 \times \left(\frac{L_4^3}{3}\right) = 0$$

Using trial and error by plugging different L_4 values, find: $L_4 = 11.15$ m
   The penetration depth: $D = L_3 + L_4 = 2.09 + 11.15 = 13.24$ m
   To find the maximum moment:

$$z' = \sqrt{\frac{2P}{\gamma'(K_p - K_a)}} = \sqrt{\frac{2 \times 176.48}{8.19 \times (2.117 - 0.333)}} = 4.91 \text{ m}$$

$$M_{max} = P(\bar{z} + z') - \frac{1}{2}\gamma'(K_p - K_a)(z')^2\left(\frac{z'}{3}\right)$$

$$= 176.48 \times (4.36 + 4.91) - \frac{1}{2} \times 8.19 \times (2.117 - 0.333) \times \left(\frac{4.91^3}{3}\right)$$

$$= 1347.7 \text{ kN} \cdot \text{m/m}$$

The section modulus of the sheet pile per unit length is:

$$S = \frac{M_{max}}{\sigma_{all}} = \frac{1347.7 \text{ kN} \cdot \text{m/m}}{150 \times 10^3 \text{ kN/m}^2} = 9.0 \times 10^{-3} \text{ m}^3/\text{m of wall}$$

The earth pressure distribution is shown in Figure 7.33.

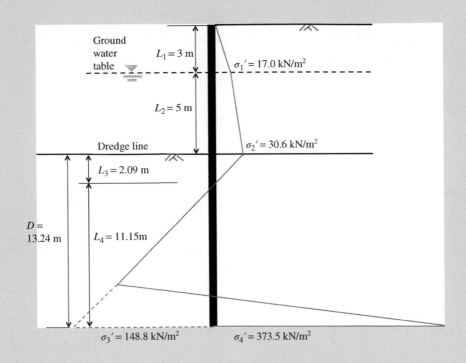

Fig. 7.33   Sample problem solution for sheet pile design in sand.

As an alternative, if the strength reduction is not used, the theoretical penetration depth would be 10.35 m. With 25% increase, the design penetration depth would be 12.94 m and the maximum moment would be 1076.3 kN·m/m.

Alternative solution using limit state design and partial factors of safety:

First allowance is made in the depth of excavation considering future unplanned excavations and/or possible scour. This allowance (Δa) is equal to 10% of the distance between the lowest support and the excavation level (but limited to a maximum of 0.5 m), as suggested in BS EN 1997–1:2004. Hence

$$\Delta a = 0.10(8) = 0.8 > 0.5 \rightarrow 0.5 \text{ m}$$

In terms of this example, this means that the dredge line is lowered and that $L_2 = 5.5$ m

The following partial factors of safety will be assumed in the calculations that follow:

$$\gamma_{G,dst} = 1.35, \quad \gamma_{\phi'} = 1.00, \quad \gamma_{\gamma} = 1.00$$

Design values of geotechnical parameters:

$$\phi'_d = \tan^{-1}\left(\frac{\tan\phi'_k}{\gamma_{\phi'}}\right) = \tan^{-1}\left(\frac{\tan 30}{1.00}\right) = 30$$

$$\gamma_d = \frac{\gamma_k}{\gamma_\gamma} = \frac{17}{1.00} = 17.0 \ kN/m^3$$

$$\gamma'_d = \frac{\gamma_{Ik}}{\gamma_\gamma} = \frac{18 - 9.81}{1.00} = 8.19 \ kN/m^3$$

Rankine active and passive earth pressure coefficients are:

$$K_a = \tan^2\left(45 - \frac{\phi'_d}{2}\right) = \tan^2\left(45 - \frac{30}{2}\right) = 0.333$$

$$K_p = \tan^2\left(45 + \frac{\phi'_d}{2}\right) = \tan^2\left(45 + \frac{30}{2}\right) = 3.0$$

The simplified diagram used in the preceding solution is also assumed for this solution. Although this is the subject of discussion, it is also assumed that hydrostatic pressures at both sides of the wall derive from the same source; hence the level of uncertainty should be the same and as consequence they are factored equally and cancel each other.

$$\sigma'_1 = (\gamma_d L_1)K_a = (17 \times 3) \times 0.33 = 17 \ kN/m^2$$

$$\sigma'_2 = (\gamma_d L_1 + \gamma'_d L_2)K_a = (17 \times 3 + 8.19 \times 5.5) \times 0.33 = 32.0 \ kN/m^2$$

$$L_3 = \frac{\sigma'_2}{\gamma'_d(K_p - K_a)} = \frac{32.0}{8.19 \times (3.00 - 0.333)} = 1.47 \ m$$

The total resultant earth force for the active earth pressure zone ACDE is considered to be a permanent, unfavorable action; hence:

$$P = \left(\frac{1}{2}\sigma'_1 L_1 + \frac{1}{2}(\sigma'_1 + \sigma'_2)L_2 + \frac{1}{2}\sigma'_2 L_3\right)\gamma_{G,dst}$$

$$P = (0.5 \times 17 \times 3 + 0.5 \times (17 + 32) \times 5.5 + 0.5 \times 32 \times 1.47) \times 1.35$$

$$P = 248.1 \ kN/m$$

Calculate the moment arm ($\bar{z}$) of P about point E, using moment equilibrium:

$$P \cdot \bar{z} = \frac{1}{2}\sigma'_1 L_1\left(\frac{1}{3}L_1 + L_2 + L_3\right) + \sigma'_1 L_2\left(\frac{L_2}{2} + L_3\right) + \frac{1}{2}(\sigma'_2 - \sigma'_1)L_2\left(\frac{L_3}{3} + L_3\right)$$

$$+ \frac{1}{2}\sigma'_2 L_3\left(\frac{2}{3}L_3\right)$$

$$= 0.5 \times 17 \times 3 \times \left(\frac{3}{3} + 5 + 1.47 \right) + 17 \times 5 \times \left(\frac{5}{2} + 1.47 \right)$$

$$+ \frac{1}{2}(32 - 1.47) \times 5 \times \left(\frac{5}{3} + 1.47 \right)$$

$$+ \frac{1}{2} \times 32 \times 1.47 \times \left(\frac{2 \times 1.47}{3} \right) = 757.12 \ \text{kN}$$

Find: $\bar{z} = 3.05$ m

$$\sigma_3' = \frac{\sigma_2'}{L_3} L_4 = \frac{(\gamma L_1 + \gamma' L_2) K_a}{L_3} L_4 = \frac{(17 \times 3 + 8.19 \times 5) \times 0.333}{1.47} L_4 = 21.77 L_4$$

The net earth pressure at the bottom of the sheet pile:

$$\sigma_4' = \sigma_p' - \sigma_a' = [\gamma L_1 + \gamma'(L_2 + L_3 + L_4)] K_p - \gamma'(L_3 + L_4) K_a$$

$$\sigma_4' = [17 \times 3 + 8.19 \times (5.5 + 1.47 + L_4)] \times 3.00 - 8.19(1.47 + L_4)0.333$$

$$\sigma_4' = 320.2 + 21.8 L_4$$

There is controversy on whether passive earth pressures should be treated as actions or resistances. The choice made will of course affect the values of the partial factors of safety. It is assumed here that the passive thrust is derived from the same source of the active thrust. Hence passive pressures are also treated as permanent, unfavorable actions.

Using force equilibrium:

$$\Sigma(\text{horizontal forces on the sheet pile}) = 0$$

$$P + \left[\frac{1}{2} \left(\sigma_3' + \sigma_4' \right) L_5 - \frac{1}{2} \sigma_3' L_4 \right] \gamma_{G,\text{unfav}} = 0$$

Solve: $L_5 = \dfrac{\frac{1}{2} \sigma_3' L_4 \gamma_{G,\text{unfav}} - P}{\frac{1}{2}(\sigma_3' + \sigma_4') \gamma_{G,\text{unfav}}} = \dfrac{0.5 \times 21.77 L_4{}^2 \times 1.35 - 248.1}{0.5 \times (21.77 L_4 + (320.2 + 21.8 L_4)) \times 1.35}$

$$= \frac{14.7 L_4^2 - 248.1}{29.4 L_4 + 216.1}$$

Using moment equilibrium:

$$\Sigma(\text{moments of the horizontal forces on the sheet pile about point B}) = 0$$

$$P(\bar{z} + L_4) + \frac{1}{2}(\sigma_3' + \sigma_4') L_5 \left(\frac{L_5}{3} \right) - \frac{1}{2} \sigma_3' L_4 \left(\frac{L_4}{3} \right) = 0$$

$$248.1 \times (3.05 + L_4) + 0.5(21.77 L_4 + 320.2 + 21.8 L_4) \times 0.333$$

$$\times \left(\frac{14.7 L_4^2 - 248.1}{29.4 L_4 + 216.1} \right)^2 - 0.5 \times 21.77 L_4{}^3 \times 0.333 = 0$$

Solving by trial and error → $L_4 = 10.57$ m
The penetration depth: $D = \Delta a + L_3 + L_4 = 0.5 + 1.47 + 10.57 = 12.54$ m
To find the maximum moment and the section modulus the same equations described in the previous solutions can be used.

7.4.4 Design of cantilever walls penetrating cohesive soils

There are two cases of cantilever sheet pile walls penetrating cohesive soils: (1) the sheet pile wall is entirely in clay; (2) the sheet pile wall is driven in clay and backfilled with sand above the dredge line. The two cases result in different earth pressure development. In this session, the design of the second case is discussed.

The relative wall movements for a sheet pile penetrating clay are the same as shown in Figure 7.30. Figure 7.34 shows the typical earth pressure distribution and the design parameters for a sheet pile penetrating clay with sand backfill. Above the dredge line, the active earth pressure distribution is the same as in a sheet pile penetrating sand.

The active earth pressure at the groundwater table is:

$$\sigma_1' = \gamma L_1 K_a \tag{7.81}$$

The active earth pressure at the dredge line is:

$$\sigma_2' = (\gamma L_1 + \gamma_{sand}' L_2) K_a \tag{7.82}$$

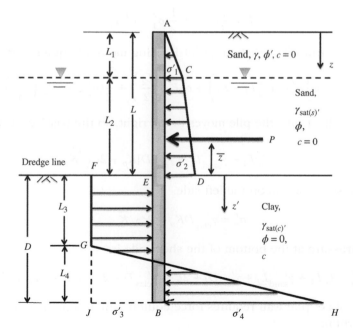

Fig. 7.34 Design graph for cantilever sheet pile penetrating undrained clay with sand backfill.

Below the dredge line, the soil is clay in undrained condition, so $\phi = 0$. Therefore,

$$K_a = K_p = 1.0 \tag{7.83}$$

The active earth pressure below the dredge line and above the rotational point O (on the right side of the sheet pile) is:

$$\sigma'_a = [\gamma L_1 + \gamma'_{sand}L_2 + \gamma'_{clay}(z - L_1 - L_2)]K_a - 2c\sqrt{K_a} \tag{7.84}$$

where: $\gamma'_{sand} = \gamma_{sat(sand)} - \gamma_w$

The passive earth pressure below the dredge line and above the rotational point O (on the left side of the sheet pile) is:

$$\sigma'_p = \gamma'_{clay}(z - L_1 - L_2)K_p + 2c\sqrt{K_p} \tag{7.85}$$

where: $\gamma'_{clay} = \gamma_{sat(clay)} - \gamma_w$

Note: the hydrostatic pressures on both sides of the pile are equal and cancel each other.

The net lateral pressure below the dredge line and above the rotational point O is:

$$\sigma'_3 = \sigma'_p - \sigma'_a = 4c - (\gamma L_1 + \gamma'_{sand}L_2) \tag{7.86}$$

If

$$\gamma_1 L_1 + \gamma'_{sand}L_2 > 4c \tag{7.87}$$

there will be no net passive earth pressure to balance the active pressure above the dredge line, and the sheet pile will fail. So the critical height can be calculated using Equation (7.87).

The total resultant earth force for the active earth pressure zone ACDE is:

$$P = \frac{1}{2}\sigma'_1 L_1 + \frac{1}{2}(\sigma'_1 + \sigma'_2)L_2 \tag{7.88}$$

Calculate the moment arm ($\bar{z}$) of P about point E, using moment equilibrium:

$$P \cdot \bar{z} = \frac{1}{2}\sigma'_1 L_1\left(\frac{1}{3}L_1 + L_2\right) + \sigma'_1 L_2\left(\frac{L_2}{2}\right) + \frac{1}{2}(\sigma'_2 - \sigma'_1)L_2\left(\frac{L_2}{3}\right) \tag{7.89}$$

At the bottom of the sheet pile, the pile moves to the right, so the passive earth pressure occurs on the right side:

$$\sigma'_p = [\gamma L_1 + \gamma'_{sand}L_2 + \gamma'_{clay}D]K_p + 2c\sqrt{K_p} \tag{7.90}$$

The active earth pressure occurs on the left side:

$$\sigma'_a = \gamma'_{clay}DK_a - 2c\sqrt{K_a} \tag{7.91}$$

So, the net earth pressure at the bottom of the sheet pile is:

$$\sigma'_4 = \sigma'_p - \sigma'_a = (\gamma L_1 + \gamma'_{sand}L_2 + \gamma'_{clay}D) + 2c - (\gamma'_{clay}D - 2c) = 4c + \gamma L_1 + \gamma'_{sand}L_2 \tag{7.92}$$

The stability of the sheet pile wall requires force equilibrium and moment equilibrium.

For force equilibrium:

$$\Sigma(\text{horizontal forces on the sheet pile}) = 0 \tag{7.93}$$

So, area of ACDE + area of JHG − area of FEBJ = 0

$$P + \frac{1}{2}(\sigma_3' + \sigma_4')L_4 - \sigma_3'D = 0 \tag{7.94}$$

Solve for D

$$D = \frac{P + 4cL_4}{4c - \gamma L_1 - \gamma_{sand}'L_2} \tag{7.95}$$

For moment equilibrium:

$$\Sigma(\text{moments of the horizontal forces on the sheet pile about any point}) = 0 \tag{7.96}$$

Take moments about point B:

$$P(\bar{z} + D) - \sigma_3'\frac{D^2}{2} + \frac{1}{2}(\sigma_3' + \sigma_4')L_4\left(\frac{L_4}{3}\right) = 0 \tag{7.97}$$

Substitute D in Equation (7.97) with Equation (7.95); L_4 is the only unknown.

Use trial and error by using different L_4 values and find L_4 to satisfy Equation (7.97). Then D can be determined using Equation (7.95).

The maximum moment occurs at a cross section with zero shear force. Use a new axis z' starting at the dredge line, at the location of zero shear force:

$$P = \sigma_3'z' \tag{7.98}$$

So:

$$z' = \frac{P}{\sigma_3'} = \frac{P}{4c - (\gamma L_1 + \gamma_{sand}'L_2)} \tag{7.99}$$

The maximum moment is:

$$M_{max} = P(\bar{z} + z') - \frac{1}{2}\sigma_3'z'^2 \tag{7.100}$$

The section modulus can be determined using Equation (7.78).

Factor of safety for sheet pile stability:

The factor of safety for stability follows the same approach as in the design of a sheet pile penetrating sand. Strength reduction is used in evaluating passive pressures using "effective" values of ϕ and c. As $\phi = 0$ for undrained clay, only the cohesion is reduced:

$$c_{eff} = \frac{c}{FS_p} \tag{7.101}$$

where:

FS_p = factor of safety for passive pressure and typically 1.5. c_{eff} should be used in the design.

Note: The long-term condition for sheet piling in clays must also be considered, because of the time-dependent changes in ϕ and c. The analysis should be carried out using effective stress parameters ϕ' and c' obtained from consolidated-drained triaxial tests, or from consolidated-undrained tests triaxial in which pore pressure measurements are made. Limited experimental data indicated that the long-term values of c are quite small; for design purpose, c may be conservatively taken as zero. The final value of ϕ is usually between 20 and 30 degrees (United States Steel 1984).

Chapter 7

As demonstrated in Sample Problem 7.2, a limit state design approach using partial factors of safety can also be used for the design of cantilever walls penetrating cohesive soils. Clearly, the pressure diagram in Figure 7.34 differs slightly from that in Figure 7.31. However, for such approach the same equations can be used and should differ only in the use of partial factors of safety in the determination of design geotechnical parameters and the actions used in the force and momentum equilibrium equations.

Sample Problem 7.3: Design of a cantilever wall penetrating cohesive soil

A cantilever sheet pile wall penetrating clay with sand backfill is shown in Figure 7.35. The subsoil profile is similar to Figure 7.33, except that the soil below the dredge line is saturated clay. Determine:

(a) The depth of the embedment, D, using a factor of safety of 1.5 for passive pressure.
(b) The maximum moment on the sheet pile.

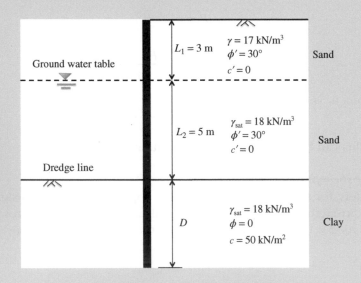

Fig. 7.35 Sample problem of sheet pile design in clay

Solution:

Strength reduction for clay: $c_{eff} = \dfrac{c}{FS_p} = \dfrac{50}{1.5} = 33.3 \text{ kN/m}^2$

$$K_a = \tan^2\left(45 - \frac{\phi}{2}\right) = \tan^2\left(45 - \frac{30}{2}\right) = 0.333$$

$$\gamma'_{sand} = \gamma_{sat(sand)} - \gamma_w = 18 - 9.81 = 8.19 \text{ kN/m}^3$$

$$\sigma'_1 = \gamma L_1 K_a = 17 \times 3 \times 0.333 = 17 \text{ kN/m}^2$$

$$\sigma'_2 = (\gamma L_1 + \gamma'_{sand} L_2) K_a = (17 \times 3 + 8.19 \times 5) \times 0.333 = 30.6 \text{ kN/m}^2$$

$$\sigma'_3 = \sigma'_p - \sigma'_a = 4c_{eff} - (\gamma L_1 + \gamma'_{sand} L_2) = 4 \times 33.3 - (17 \times 3 + 8.19 \times 5)$$

$$= 41.25 \text{ kN/m}^2$$

The total resultant earth force for the active earth pressure zone ACDE is:

$$P = \frac{1}{2}\sigma'_1 L_1 + \frac{1}{2}(\sigma'_1 + \sigma'_2)L_2 = 0.5 \times 17 \times 3 + 0.5 \times (17 + 30.6) \times 5 = 144.5 \text{ kN/m}$$

Calculate the moment arm ($\bar{z}$) of P about point E, using moment equilibrium:

$$P \cdot \bar{z} = \frac{1}{2}\sigma'_1 L_1 \left(\frac{1}{3}L_1 + L_2\right) + \sigma'_1 L_2 \left(\frac{L_2}{2}\right) + \frac{1}{2}(\sigma'_2 - \sigma'_1)L_2 \left(\frac{L_2}{3}\right)$$

$$144.5\bar{z} = 0.5 \times 17 \times 3 \times \left(\frac{3}{3} + 5\right) + 17 \times 5 \times \frac{5}{2} + 0.5 \times (30.6 - 17) \times \left(\frac{5}{3}\right)$$

Find: $\bar{z} = 2.61$ m

The net earth pressure at the bottom of the sheet pile is:

$$\sigma'_4 = 4c_{eff} + \gamma L_1 + \gamma_{sand} L_2 = 4 \times 33.3 + 17 \times 3 + (18 - 9.81) \times 5 = 241.25 \text{ kN/m}^2$$

For force equilibrium:

$$p + \frac{1}{2}(\sigma'_3 + \sigma'_4)L_4 - \sigma'_3 D = 0$$

Solve for D:

$$D = \frac{P + 4c_{eff}L_4}{4c_{eff} - \gamma L_1 - \gamma'_{sand}L_2} = \frac{144.5 + 4 \times 33.3 \times L_4}{4 \times 33.3 - 17 \times 3 - 8.19 \times 5} = 3.50 + 3.23L_4$$

For moment equilibrium:

$$\Sigma(\text{moments of the horizontal forces on the sheet pile about B}) = 0$$

$$P(\bar{z} + D) - \sigma'_3\frac{D^2}{2} + \frac{1}{2}(\sigma'_3 + \sigma'_4)L_4 \left(\frac{L_4}{3}\right) = 0$$

Substitute D in the above Equation; L_4 is the only unknown.
Using trial and error with different L_4 values, find $L_4 = 1.94$ m
So: $D = 9.77$ m ≈ 10.0 m
The maximum moment is:

$$M_{max} = P(\bar{z} + z') - \frac{1}{2}\sigma'_3 z'^2$$

where $z' = \dfrac{P}{\sigma'_3} = \dfrac{P}{4c_{eff} - \gamma L_1 - \gamma'_{sand} L_2} = \dfrac{144.5}{4 \times 33.3 - 17 \times 3 - 8.19 \times 5} = 3.50$ m

So: $M_{max} = 144.5 \times (2.61 + 3.50) - 0.5 \times 41.25 \times 3.50^2 = 630$ kN·m/m

The pressure distribution and the design parameters are shown in Figure 7.36.

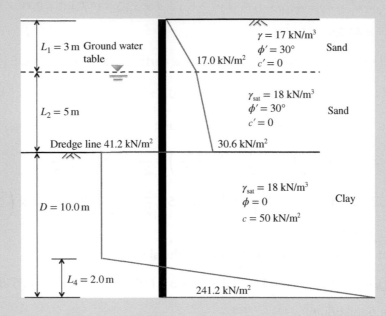

Fig. 7.36   Sample problem solution for sheet pile design in clay with sand backfill.

7.5   Soil nail wall design

Soil nailing is the reinforcement of existing walls and slopes by installing closely spaced steel bars (i.e., soil nails) into the soil. Figure 7.37 shows a typical cross section and the basic elements of a soil nail wall. Soil nail walls have been successfully used in the temporary or permanent soil retaining applications, namely, roadway cuts and widening, repairs or reconstructions of existing retaining structures, and excavations in urban environment. This section introduces the basics of soil nail walls and the preliminary design. Detailed guidelines can be referenced in the FHWA manual "Soil Nail Walls" (Lazarte et al. 2003), "BS 8006–2:2011 – Code of practice for strengthened/reinforced soils, Part 2: Soil nail design," and in "BS EN 14490:2010 – Execution of special geotechnical works – Soil nailing." Note that the standard BS EN 1997–1:2004 widely discussed in previous chapters explicitly excludes soil nail design.

There are various types of soil nail walls, based on the nail installation techniques:

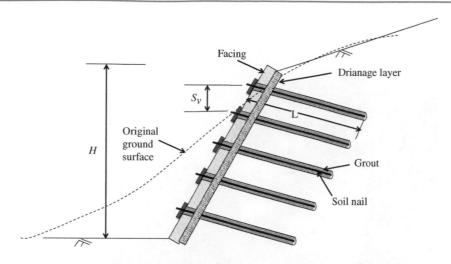

Fig. 7.37 Typical cross section and basic elements of a soil nail wall.

- *Drilled and grouted soil nails*: Holes of 100–200 mm (4–8 inch) diameters are first drilled, and steel bars are placed in the holes and the holes are grouted. The holes are typically 1.5 m apart. Grouted soil nails are used for temporary and permanent applications.
- *Driven soil nails*: The nails are relatively small in diameter (19–25 mm or 0.75–1.0 inch) and are mechanically driven into the soil. Grout is not used. This type of installation is fast, but it cannot provide good corrosion protection.
- *Hollow bar soil nails*: The nails are hollow and grout is simultaneously injected through the hollow bar with the drilling. This soil nail type allows for a faster installation than drilled and grouted soil nails, and it can provide some level of corrosion protection. This method is commonly used as a temporary retaining structure.
- *Jet-grouted soil nails*: Jet grouting is used to cut the soil; after the holes are drilled, steel bars are installed using the vibropercussion through the grouted holes. The grout provides corrosion protection.
- *Launched soil nails*: In this new technique, soil nails are launched into the soil at a high speed using a firing mechanism involving compressed air. The installation is fast, but it is difficult to control the soil penetration depth of nails, particularly when the subsoil contains cobbles. Figure 7.38 shows a soil nail launcher installing soils nails to stabilize coastal bluffs.

The design and analysis of soil nail walls should consider two limiting conditions and other design aspects as follows:

(A) Strength limit states, namely:
  - External failure modes:
    a. global stability failure
    b. sliding stability failure
    c. bearing capacity failure
  - Internal failure modes:
    d. nail-soil pullout failure
    e. bar-grout pullout failure

Fig. 7.38 Soil nails are used to stabilize the bluff at Pebble Beach Drive, Crescent City, California. (photo courtesy of Soil Nail Launcher, Inc.)

    f. nail tensile failure

    g. nail bending and shear failure

- Facing failure modes:

    h. facing flexure failure

    i. facing punching shear failure

    j. head-stud failure

(B) Service limit states

- Excessive wall deformation.

(C) Other design considerations:

- Drainage behind the wall.
- Corrosion protection of the soil nails.
- Frost protection.
- Support of dead load of temporary facing.

Figure 7.39 shows the strength failure modes of soil nail walls. Nail length, diameter, and spacing typically control the external and internal stabilities of a soil nail wall. The external stability analysis of soil nail walls is presented as follows; it is based on the recommendations of the FHWA manual "Soil Nail Walls" (Lazarte et al. 2003).

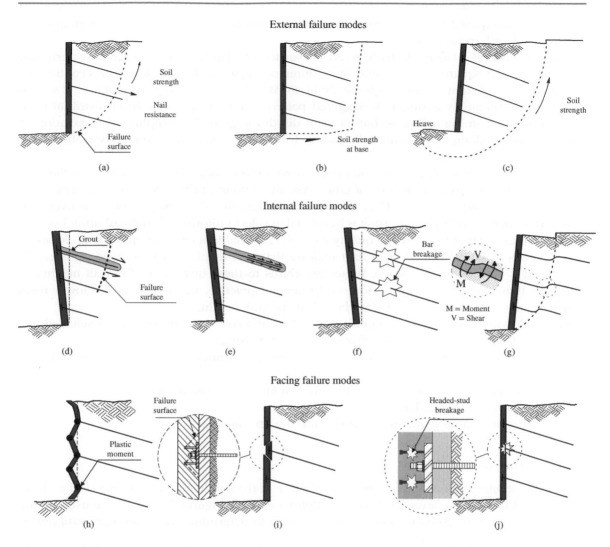

Fig. 7.39 Failure modes of soil nail walls (after Lazarte et al. 2003). (a) Global stability failure, (b) sliding stability failure, (c) bearing failure (basal heave), (d) nail–soil pullout failure, (e) bar–grout pullout failure, (f) nail tensile failure, (g) nail bending and/or shear failure, (h) facing flexure failure, (i) facing punching shear failure, (j) headed-stud failure.

7.5.1   Initial design parameters and conditions

The following initial parameters should be first determined and used in the stability design of soil nail walls.

- *Wall layout*: The layout refers to the wall height (H), the length of the wall, and the inclination of the wall face (the typical range is from 0° to 10°).
- *Soil nail vertical (S_V) and horizontal (S_H) spacing*: S_V is typically the same as S_H. The nail spacing ranges from 1.25 to 2 m (4–6.5 ft) for conventional drilled and grouted soil nails with a preferred routine value of 1.5 m (5 ft). A reduced spacing for driven nails (as low as 0.5 m

or 1.5 ft) is required because driven nails develop bond strengths that are lower than those for drilled and grouted nails.

- *Soil nail patterns on face*: The patterns are (1) square, (2) staggered in a triangular pattern, and (3) irregular. A square pattern results in a column of aligned soil nails and enables continuous and easy installation of geocomposite drain strips behind the facing. In practice, a square pattern is commonly adopted. A staggered pattern results in more uniform distribution of earth pressure in the soil nails, but its main disadvantage is the complicated installation of geocomposite drain strips behind the facing. Irregular nail spacing is project-specific where reduced spacing is needed.
- *Soil nail inclination*: The inclination ranges from 10 to 20 degrees with a typical inclination of 15 degrees to ensure easy flow of grout from the bottom of the hole to the nail head.
- *Soil nail length distributions*: They are (1) uniform length, when the potential for excessive wall deformation is not a concern; it is beneficial to select uniform length distribution because it simplifies the construction and quality control; (2) variable length, when the wall deformation needs to be controlled; field data indicate that wall displacements can be significantly reduced if the nail lengths in the upper two-thirds to three quarters of the wall height are greater than those in the lower portion. In general practice, nail length in the lower rows should never be shorter than $0.5H$, where H is the wall height.
- *Soil nail materials*: Appropriate grade of steel for the soil nail bars should be selected; for most applications, Grade 420 MPa (Grade 60) steel is used.
- *Soil properties*: The soil properties and strata are determined on the basis of the subsoil exploration.
- *Drilling methods*: They are based on the site condition and contractors.
- *Factor of safety*: Refer to Table 7.1.
- *Loads*: The surcharge on the retained soil and the seismic load are determined.

7.5.2   Global stability failure

The global stability failure refers to the slope failure of the soil behind a retaining wall and can be analyzed using the conventional slope stability methods. Figures 7.40–7.46 are used by the FHWA (Lazarte et al. 2003) to conduct initial design to determine the preliminary *nail length*

Table 7.1   Minimum recommended factors of safety for the stability design of soil nail walls, using the allowable stress design (ASD) method, based on the FHWA manual "soil nail walls" (Lazarte et al. 2003).

Failure modes	Resisting components	Symbol	Minimum recommended factors of safety under static loads	
			Temporary structure	Permanent structure
External stability	Global stability (long term)	FS_G	1.35	1.5
	Global stability (excavation)	FS_G	1.2–1.3	
	Sliding	FS_{SL}	1.3	1.5
	Bearing capacity	FS_{BC}	2.5	3.0
Internal stability	Pullout resistance	FS_P	2.0	
	Nail bar tensile strength	FS_T	1.8	

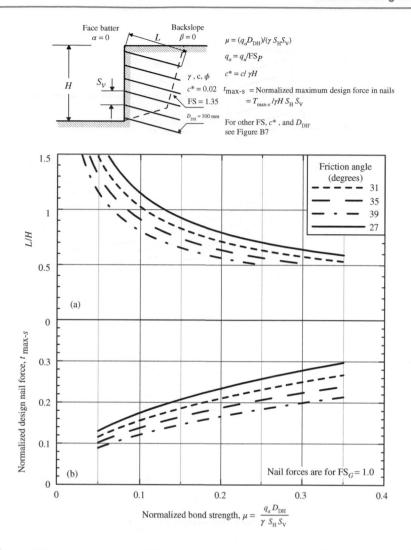

Fig. 7.40 Preliminary soil nail wall design: batter 0°, back slope 0°. (After Lazarte et al. 2003.)

and *maximum tensile forces*. The charts were developed using soil nail wall design computer programs and were based on the following assumptions:

- Homogeneous soil
- No surcharge
- No seismic forces
- Uniform length, spacing, and inclination of soil nails
- No groundwater.

The following initial parameters should be determined first to use the charts:

- Wall face inclination (face batter) with respect to vertical direction, α
- Inclination angle of the back slope with respect to horizontal direction, β

Chapter 7

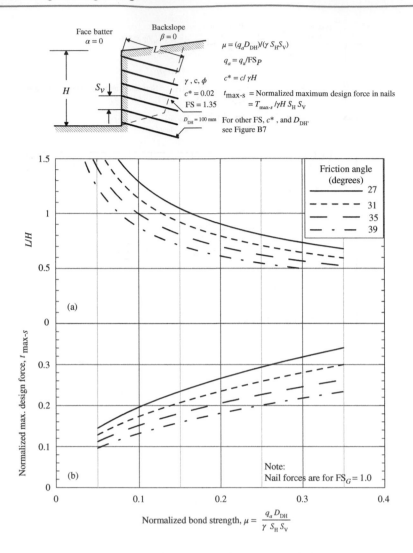

Fig. 7.41 Preliminary soil nail wall design: batter 0°, back slope 10°. (After Lazarte et al. 2003.)

- Effective internal friction angle of the soil, ϕ'
- Unit weight of the soil, γ
- Horizontal (S_V) and vertical (S_H) spacing, assume $S_V = S_H$
- Global factor of safety for external stability, $FS_G = 1.35$
- Factor of safety of pullout resistance, FS_P
- Ultimate bond strength (ultimate pullout resistance), q_u
- Effective diameter of the drilled hole, D_{DH}

Figures 7.40–7.46 use the normalized allowable pullout resistance:

$$\mu = \frac{q_u D_{DH}}{FS_p \gamma S_H S_V} \tag{7.102}$$

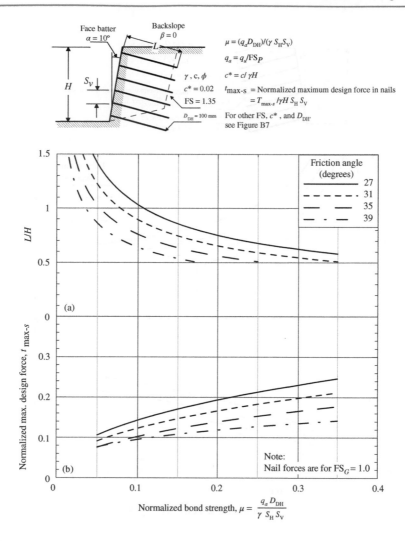

Fig. 7.42 Preliminary soil nail wall design: batter 10°, back slope 0°. (After Lazarte et al. 2003.)

The required tensile strength of the soil nail is expressed using the normalized maximum design tensile force, t_{max-s} (unitless):

$$t_{max-s} = \frac{T_{max-s}}{\gamma \cdot H \cdot S_H \cdot S_V} \tag{7.103}$$

where: T_{max-s} is the maximum design nail force, which can be calculated from Equation (7.103) once t_{max-s} is determined from Figures 7.40–7.46. Equation (7.103) is based on $FS_p = 1.0$. Figures 7.40–7.46 were developed on the basis of the fixed parameters shown in Table 7.2.

The normalized nail length (L/H) and the maximum force (T_{max-s}) depend on the global factor of safety FS_G, drillhole diameter D_{DH}, and the soil's cohesion c. The effect of wall height (6–24 m) on the L/H and T_{max-s} is not significant (Lazarte et al. 2003). If the face batter, back slope angle, and soil's effective friction angle are different from the ones used in the charts, interpolation

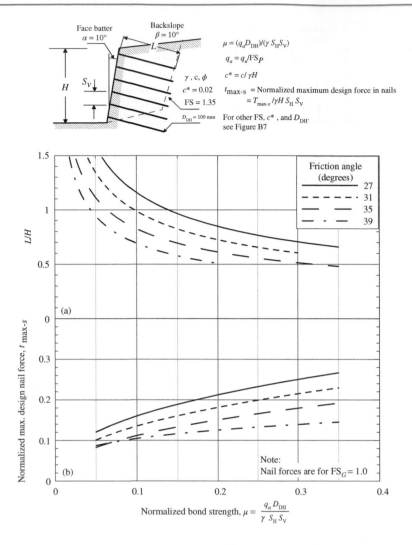

Fig. 7.43 Preliminary soil nail wall design: batter 10°, back slope 10°. (After Lazarte et al. 2003.)

between the values in the charts can be used. If the drillhole diameter, soil's cohesion, and the global factor of safety are not the ones used in the charts, correction factors, C_{1L}, C_{2L}, C_{3L}, are used following Figure 7.46.

7.5.3 Sliding failure

Active earth pressure can be mobilized to push the reinforced wall to slide along the base or slightly below the base if a weak seam below the base is present. The factor of safety against sliding can be calculated using the same approach as in rigid cantilever retaining wall design. In the analysis, the reinforced soil portion is examined as one rigid block: section CDEF as shown in Figure 7.47. The factor of safety is defined by:

$$FS_{slide} = \frac{\Sigma R}{\Sigma D} \geq 1.5 \qquad (7.104)$$

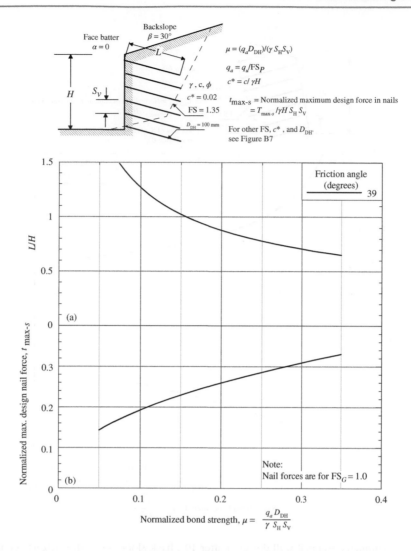

Fig. 7.44 Preliminary soil nail wall design: batter 0°, back slope 30°. (After Lazarte et al. 2003.)

where

ΣR = total horizontal resisting force to sliding, per unit length of wall,
ΣD = total horizontal driving force that causes sliding, per unit length of wall.

$$\Sigma R = c_b B + (\Sigma V)\tan\phi_b \tag{7.105}$$

where

c_b = soil's cohesion at the base,
B = width of the base of the reinforced soil portion (Figure 7.47),
ΣV = total vertical force on section CDEF, per unit length of wall,
ϕ_b = soil's friction angle at the base.

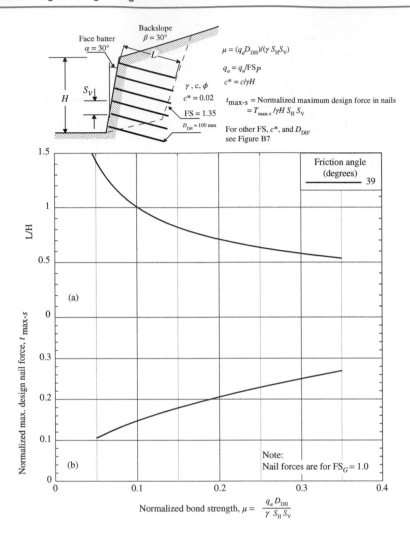

Fig. 7.45   Preliminary soil nail wall design: batter 10°, back slope 30°. (After Lazarte et al. 2003.)

$$\Sigma V = W + Q_D + P_a \sin \beta \qquad (7.106)$$

where

W = total weight of section CDEF, per unit length of wall,

Q_D = total dead surcharge on section CDEF, per unit length of wall,

P_a = total active force acting on the vertical profile EF, per unit length of wall. It can be calculated using Equation (7.31).

And

$$\Sigma D = P_a \cos \beta \qquad (7.107)$$

It is noted that section CDEF is analyzed as one rigid wall, and P_a acts on the vertical profile of EF. Therefore, the face batter (face inclination α) does not affect the calculation of the active

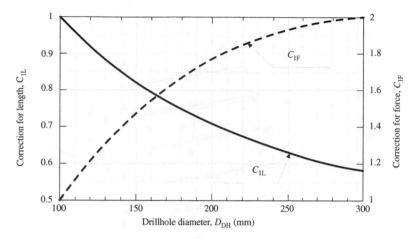

(a) Correction for drillhole diameter

(b) Correction for different values of normalized cohesion($c^* = c/\gamma H$)

$$C_{2L} = -4.0\,c^* + 1.09 \geq 0.85$$

$$C_{2F} = -4.0\,c^* + 1.09 \geq 0.85$$

(c) Correction for different global factors of safety

$$C_{3L} = 0.52\,\text{FS} + 0.30 \geq 1.0$$

Fig. 7.46 Correction factors for preliminary soil nail wall design. (After Lazarte et al. 2003.)

Table 7.2 Fixed parameters used to develop the initial design charts in Figures 7.40–7.46. (Lazarte et al. 2003).

Parameters	Fixed values used for design charts
Global factor of safety, FS_G	1.35
Nail horizontal spacing, S_H	1.5 m (or 5 ft)
Nail vertical spacing, S_V	1.5 m (or 5 ft)
Nail inclination	15°
Drillhole diameter, D_{DH}	100 mm (or 4 inch)
Unit weight of soil, γ	18.9 kN/m³ (or 120 pcf)
Cohesion of soil, c	5 kN/m² (or 100 psf)
Wall height, H	12 m (40 ft)

earth pressure. If the factor of safety is less than 1.5, longer soil nails may be used to increase the width B, or the surcharge on the reinforced soil section may be reduced or removed if possible, to increase the factor of safety.

7.5.4 Bearing capacity failure

As shown in Figure 7.48, when a soil nail wall is used to retain a vertical excavation in fine-grained, soft soil, the unbalanced load between the excavated soil and the retained soil

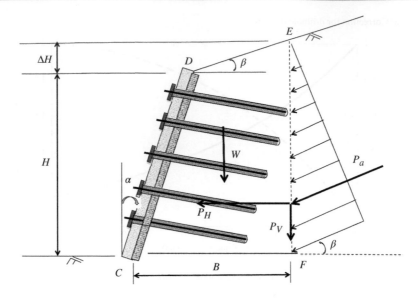

Fig. 7.47   Sliding analysis of soil nail wall.

can cause the bottom of the excavation to heave and trigger a bearing capacity failure. Lazarte et al. (2003) recommended the factor of safety against heaving be analyzed using the Terzaghi's equation (Terzaghi et al. 1996).

$$\text{FS}_H = \frac{S_u N_c}{H_{\text{eq}} \left(\gamma - \frac{S_u}{B'} \right)} \tag{7.108}$$

where

S_u = undrained shear strength of the foundation soil,
γ = unit weight of the foundation soil,
N_c = bearing capacity factor (Figure 7.48c),
H_{eq} = equivalent wall height = $H + \Delta H$,
H = height of the soil nail wall,
ΔH = height of equivalent overburden stress on top of the retained soil,
B' = width of influence,

$$B' = \frac{B_e}{\sqrt{2}} \tag{7.109}$$

where

B_e = width of excavation.

When a strong soil deposit underlies the soft soil beneath the excavation at a depth of $D_B < 0.71 B_e$, as shown in Figure 7.48b, D_B should replace B' in Equation (7.108).

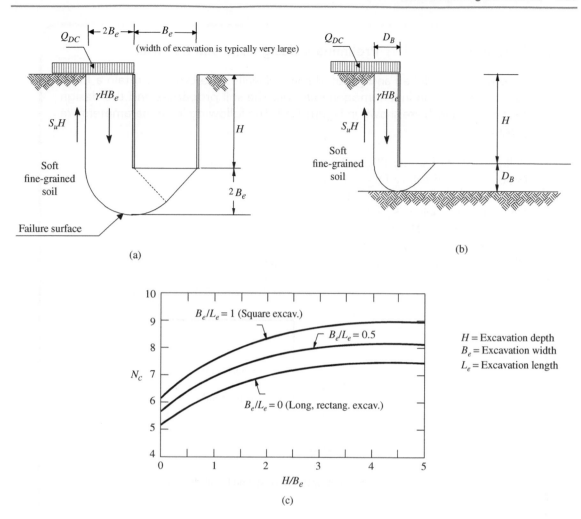

Fig. 7.48 Bearing capacity (heave) analysis of soil nail wall, (After Lazarte et al. 2003.)

In Figure 4.48c: for wide excavations, $H/B_e = 0$; for long walls, $B_e/L = 0$, and $N_c = 5.14$. The factor of safety FS_H uses the values in Table 7.1.

Limit state design of soil nails requires verification for local or general rotational failure (i.e., using the method of slices described in Chapter 5), translational failure through the soil nail or the facing, and local over stressing of nails, among others. Just as in any other geotechnical structure, serviceability limit states (e.g., related to soil movement and settlement) should also be verified. Limit state design procedures for soil nails are, however, based on a "trial-and-error" approach that uses partial factors of safety on the geotechnical parameters and the actions imposed on the soil nail wall. Extensive explanations are included in "BS EN 14490:2010 – Execution of special geotechnical works – Soil nailing."

Sample Problem 7.4: Preliminary design of a soil nail wall

A conventional drilled and grouted soil nail wall is preferred to retain a long, vertical excavation in a homogeneous soil. The soil properties and the design excavation depth are shown in Figure 7.49. The following initial parameters are given:

Pullout factor of safety, $FS_p = 2.0$
Global factor of safety, $FS_G = 1.35$
Nail horizontal spacing, $S_H = 1.5$ m
Nail vertical spacing, $S_V = 1.5$ m
Nail inclination = 15 degrees
Drillhole diameter, $D_{DH} = 100$ mm
Ultimate pullout resistance, $q_u = 120$ kN/m²

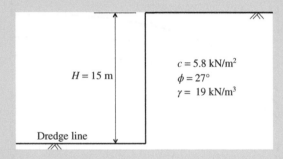

$c = 5.8$ kN/m²
$\phi = 27°$
$\gamma = 19$ kN/m³

$H = 15$ m

Dredge line

Fig. 7.49 Sample problem of soil nail wall design.

Determine:

1. The preliminary nail length and maximum design nail tensile force to satisfy the global stability.
2. The factor of safety against sliding.
3. The factor of safety against heaving of the foundation soil.

Solution:

1. The normalized allowable pullout resistance:

$$\mu = \frac{q_u D_{DH}}{FS_p \gamma S_H S_V} = \frac{120 \times 0.1}{2.0 \times 19 \times 1.5 \times 1.5} = 0.14$$

The soil nail wall configuration in Figure 7.49 shows the face batter = 0 and the back slope = 0.

$$c^* = \frac{c}{\gamma H} = \frac{5.8}{19 \times 15} = 0.02 \text{(No correction is needed on } c^* \text{)}$$

Using Figure 7.40 and given the friction angle of the soil is 27°, find:
$L/H = 0.96$
So, the soil nail length: $L = 0.96 \times 15 = 14.4 \text{ m} \approx 15.0 \text{ m}$
Using Figure 7.40 and $\mu = 0.14$, find $t_{max-s} = 0.2$.
So, the maximum design nail tensile force is:

$$T_{max-s} = t_{max-s} \cdot \gamma \cdot H \cdot S_H \cdot S_V = 0.2 \times 19 \times 15 \times 1.5 \times 1.5 = 128 \text{ kN}$$

2. Check for sliding. The configuration of the soil nail wall is shown in Figure 7.50.

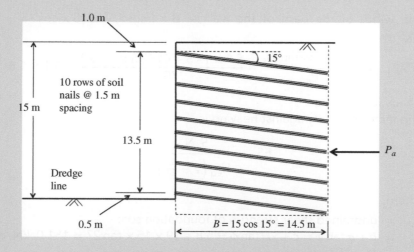

Fig. 7.50 Configuration of the soil nail wall for Sample Problem 7.4.

To simplify the calculation, assume the reinforced soil block slides along the dredge line.
The factor of safety for sliding is:

$$FS_{slide} = \frac{\Sigma R}{\Sigma D} = \frac{c_b B + (\Sigma V) \tan \varphi_b}{P_a}$$

The total vertical force is:

$$\Sigma V = W + Q_D + P_a \sin \beta = W = 19 \times 15 \times 14.5 = 4132.5 \text{ kN/m}$$

Using Equation (7.18), the total earth force per unit length of the wall is:

$$P_a = \frac{1}{2}\left(\gamma H K_a - 2c'\sqrt{K_a}\right)(H - \bar{z})$$

where:

$$K_a = \tan^2\left(45 - \frac{\phi}{2}\right) = \tan^2\left(45 - \frac{27}{2}\right) = 0.376$$

$$\bar{z} = \frac{2c}{\gamma\sqrt{K_a}} = \frac{2 \times 5.8}{19 \times \sqrt{0.376}} = 1.0 \text{ m}$$

$$P_a = \frac{1}{2}\left(\gamma H K_a - 2c'\sqrt{K_a}\right)(H - \bar{z})$$

$$= \frac{1}{2}(19 \times 15 \times 0.376 - 2 \times 5.8 \times \sqrt{0.376})(15 - 1)$$

$$= 700.3 \text{ kN/m}$$

The width of the base of the wall is: $B = L\cos 15° = 15 \times \cos 15° = 14.5$ m

$$FS_{slide} = \frac{c_b B + (\Sigma V)\tan\phi_b}{P_a} = \frac{5.8 \times 14.5 + 4132.5 \times \tan 27}{700.3} = 3.1 > 1.5$$

3. Check for heaving
   The factor of safety against heaving is

$$FS_H = \frac{S_u N_c}{H_{eq}\left(\gamma - \frac{S_u}{B'}\right)}$$

where:

S_u = undrained shear strength of foundation soil:
S_u = $c + \sigma\tan\phi = c + \gamma H\tan\phi = 5.8 + 19 \times 15 \times \tan 27 = 151.0$ kN/m²
H_{eq} = $H = 15.0$ m

Assume wide excavation, B_e and B' are very large, so: $S_u/B' = 0$

So: $FS_H = \dfrac{S_u N_c}{H_{eq}\gamma}$

Using Figure 7.48c, and given $H/B_e = 0$, and $B_e/L_e = 0$ (long excavation), find:

$$N_c = 5.14$$

$$FS_H = \frac{S_u N_c}{H_{eq}\gamma} = \frac{151.0 \times 5.14}{15 \times 19} = 2.72$$

For temporary excavation: $FS_H > 2.5$, OK.
For permanent excavation: FS_H is less than 3.0, which is not OK. Foundation soil improvement or reducing the excavation height may be needed.

Homework Problems

1. Figure 7.51 shows a simplified retaining wall with horizontal backfill. Assume the soil pressure behind the retaining wall is at-rest pressure, and the backfill is normally consolidated. Determine:
   (1) The at-rest soil pressure distribution along the depth of the retaining wall.
   (2) The total resultant force (viz, the hydrostatic force) per unit length of the wall.
   (3) The location of the point of application of the resultant force.

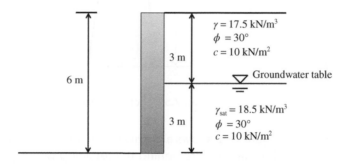

$\gamma = 17.5$ kN/m^3
$\phi = 30°$
$c = 10$ kN/m^2

Groundwater table

$\gamma_{sat} = 18.5$ kN/m^3
$\phi = 30°$
$c = 10$ kN/m^2

6 m
3 m
3 m

Fig. 7.51 Backfill profile behind a retaining wall.

2. Figure 7.51 shows a simplified retaining wall with horizontal backfill. Assume the soil pressure behind the retaining wall is active pressure. Determine:
   (1) The active soil pressure distribution along the depth of the retaining wall.
   (2) The total resultant force (viz, the hydrostatic force) per unit length of the wall, before a tensile crack is developed.
   (3) The location of the point of application of the resultant force.
3. Figure 7.51 shows a simplified retaining wall with horizontal backfill. Assume the soil pressure behind the retaining wall is passive pressure. Determine:
   (1) The passive soil pressure distribution along the depth of the retaining wall.
   (2) The total resultant force (viz, the hydrostatic force) per unit length of the wall.
   (3) The location of the point of application of the resultant force.
4. Figure 7.52 shows a gravity retaining wall with inclined backfill. Assume the soil pressure behind the retaining wall is active pressure

and the groundwater is not present in the backfill. The external friction angle between the granular backfill and the concrete wall is 28°. Determine:
(1) The total resultant soil force per unit length of the wall.
(2) The location and direction of the resultant force.

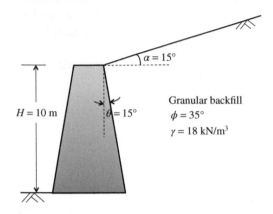

Fig. 7.52   Retaining wall and backfill configuration.

5. Figure 7.52 shows a gravity retaining wall with inclined backfill. Assume the soil pressure behind the retaining wall is passive pressure and the groundwater is not present in the backfill. The external friction angle between the granular backfill and the concrete wall is 28°. Determine:
   (1) The total resultant soil force per unit length of the wall.
   (2) The location and direction of the resultant force.

6. A cantilever retaining wall is shown in Figure 7.53. The backfill is granular soil. The wall geometry and the characteristics of the backfill and foundation soil are shown in the figure. Weep holes are installed at the bottom of the retaining wall to drain excessive water. The weep holes are clogged and water table rises to the top of the backfill. Determine the external stability of the retaining wall, in terms of overturning, sliding, and bearing capacity. Make appropriate assumptions if needed.

7. A gravity concrete retaining wall is shown in Figure 7.54. The properties of the backfill and the foundation soil are also shown in the figure. The external friction angle between the concrete and the foundation soil is assumed to be two-third of the internal friction angle of the foundation soil. Determine the external stability of the retaining wall, in terms of overturning, sliding, and bearing capacity. Make appropriate assumptions if needed.

8. A cantilever sheet pile wall penetrating granular soil is shown in Figure 7.55. Groundwater is not present.

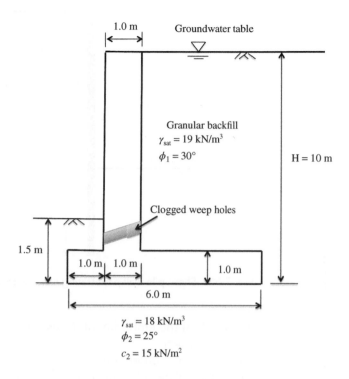

1.0 m

Groundwater table

Granular backfill
$\gamma_{sat} = 19$ kN/m³
$\phi_1 = 30°$

H = 10 m

Clogged weep holes

1.5 m

1.0 m | 1.0 m

1.0 m

6.0 m

$\gamma_{sat} = 18$ kN/m³
$\phi_2 = 25°$
$c_2 = 15$ kN/m²

Fig. 7.53 Cantilever retaining wall.

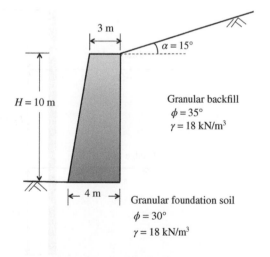

3 m

$\alpha = 15°$

H = 10 m

Granular backfill
$\phi = 35°$
$\gamma = 18$ kN/m³

4 m

Granular foundation soil
$\phi = 30°$
$\gamma = 18$ kN/m³

Fig. 7.54 Gravity retaining wall.

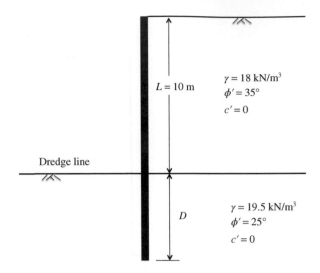

Fig. 7.55 Sheet pile wall penetrating sand, without groundwater table.

Determine:

(1) The depth of the embedment, *D*, using a factor of safety of 1.5 for strength reduction.
(2) The maximum moment on the sheet pile.
(3) Given the allowable flexural stress of the sheet pile, $\sigma_{all} = 150$ MN/m², determine the section modulus of the sheet pile per unit length.

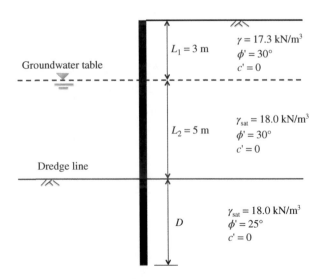

Fig. 7.56 Sheet pile wall penetrating sand, with groundwater table.

9. A cantilever sheet pile wall penetrating granular soil is shown in Figure 7.56. Determine:

(1) The depth of the embedment, D, using a factor of safety of 1.5 for strength reduction.
(2) The maximum moment on the sheet pile.
(3) Given the allowable flexural stress of the sheet pile, $\sigma_{all} = 148$ kN/m², determine the section modulus of the sheet pile per unit length.

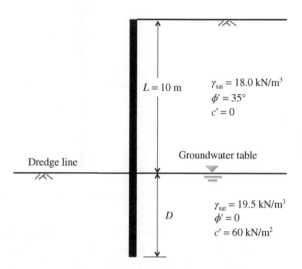

$L = 10$ m

$\gamma_{sat} = 18.0$ kN/m³
$\phi' = 35°$
$c' = 0$

Groundwater table

Dredge line

$\gamma_{sat} = 19.5$ kN/m³
$\phi' = 0$
$c' = 60$ kN/m²

D

Fig. 7.57   Sheet pile wall penetrating undrained clay, with the groundwater table at the dredge line.

10. A cantilever sheet pile wall is driven into undrained clay and then back-filled with granular soil. The groundwater table is at the dredge line. The sheet pile and the soil characteristics are shown in Figure 7.57. Determine:
    (1) The depth of the embedment, D, using a factor of safety of 1.5 for strength reduction.
    (2) The maximum moment on the sheet pile.
    (3) Given the allowable flexural stress of the sheet pile, $\sigma_{all} = 150$ MN/m², determine the section modulus of the sheet pile per unit length.
11. A cantilever sheet pile wall is driven into undrained clay and then back-filled with granular soil. The sheet piling, the groundwater table, and the soil characteristics are shown in Figure 7.58. Determine:
    (1) The depth of the embedment, D, using a factor of safety of 1.5 for strength reduction.
    (2) The maximum moment on the sheet pile.

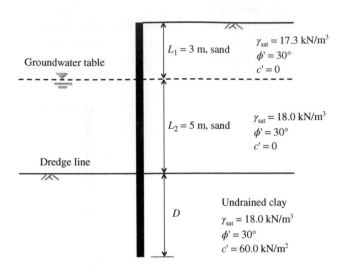

Fig. 7.58   Sheet pile wall penetrating undrained clay.

12. A soil nail wall is illustrated in Figure 7.59. Analyze its global stability. Make appropriate assumptions if needed.
13. A conventional drilled and grouted soil nail wall will be designed to retain a long, vertical excavation in a silty sand. The soil properties and

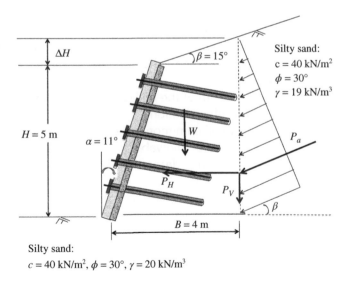

Silty sand:
$c = 40$ kN/m^2, $\phi = 30°$, $\gamma = 20$ kN/m^3

Fig. 7.59   Soil nail wall configuration for problem 12.

the design excavation depth are shown in Figure 7.60. The following initial parameters are also given:

Pullout factor of safety, $FS_p = 2.0$
Global factor of safety, $FS_G = 1.35$
Nail horizontal spacing, $S_H = 1.5\,m$
Nail vertical spacing, $S_V = 1.5\,m$
Nail inclination = 15 degrees
Drillhole diameter, $D_{DH} = 10\,cm$
Ultimate pullout resistance, $q_u = 120\,kN/m^2$

Determine:

(1) The preliminary nail length and maximum design nail tensile force to satisfy the global stability.
(2) The factor of safety against sliding.
(3) The factor of safety against heaving of the foundation soil.

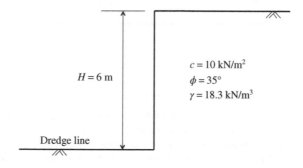

Fig. 7.60   Soil nail wall design for problem 13.

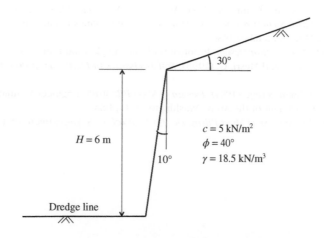

Fig. 7.61   Soil nail wall design for problem 14.

Chapter 7

14. A hollow bar soil nail wall is used as a temporary soil retention structure. The retained soil and its properties are shown in Figure 7.61. The following design parameters are also selected for this project:

Pullout factor of safety, $FS_p = 2.0$
Global factor of safety, $FS_G = 1.4$
Nail horizontal spacing, $S_H = 1.5\,m$
Nail vertical spacing, $S_V = 1.5\,m$
Nail inclination $= 15$ degrees
Drillhole diameter, $D_{DH} = 200\,mm$
Ultimate pullout resistance, $q_u = 120\,kN/m^2$

Determine:

(1) The preliminary nail length and maximum design nail tensile force to satisfy the global stability.
(2) The factor of safety against sliding.
(3) The factor of safety against the heaving of the foundation soil.

References

California Department of Transportation (Caltrans) (2004). *Bridge Design Specification*, Division of Engineering Services, California Department of Transportation. Sacramento, CA, USA.

Das, B. (2011). *Principles of Foundation Engineering*. Cengage Learning, Stamford, CT.

Lazarte, C.A., Elias, V., Espinoza, R.D., and Sabatini, P. (2003). Geotechnical Engineering Circular No. 7 *Soil Nail Walls*. Report FHWA0-IF-03-017, Federal Highway Administration, Washington, D.C., March 2003.

Mayne, P.W., and Kulhawy, F.H. (1982). "K_0-OCR relationships in soil." *ASCE Journal of Geotechnical Engineering*, Vol. 108, GT6, pp 851–872.

Naval Facilities Engineering Command (NAVFAC) (1986). *Naval Facilities Engineering Command Design Manual 7.02 Foundations & Earth Structures,* NAVFAC DM-7.02, Naval Facilities Engineering Command, 200 Stovall Street, Alexandria, Virginia 22322. September 1986.

Teng, W.C. (1962). *Foundation Design*, Prentice-Hall, Inc., Englewood Cliffs, NJ.

Terzaghi, K., Peck, R.B., and Mesri, G. (1996). *Soil Mechanics in Engineering Practice,* 3rd Edition, John Wiley & Sons, New York, NY.

U.S. Army Corps of Engineering, (1994). *Design of Sheet Pile Walls*, Engineer Manual, EM 1110-2-2504, U.S. Army Corps of Engineers, Department of the Army. Washington D.C., USA.

United States Steel (1984). *Steel Sheet Piling Design Manual*. U. S. Department of Transportation/FHWA. Pittsburg, PA, USA.

Chapter 8

Introduction to Geosynthetics Design

8.1 Geosynthetics types and characteristics

Geosynthetics are a variety of man-made polymeric products that are used in a wide array of civil engineering applications. Their primary functions are separation, reinforcement, filtration, drainage, containment, and protection. The major types of geosynthetics are shown in Figure 8.1 and include the following.

The polymers that make the geosynthetics are derived from polymerization, a process that connects monomers (molecular compounds) into three-dimensional chains (Figure 8.2). The major types of polymers used in the manufacturing of geosynthetics are polyethylene (PE), polypropylene (PP), polyvinyl chloride (PVC), polyester (PET), polyamide (PA), and polystyrene (PS).

There are six main functions of geosynthetics (Koerner 2005; Holtz 2001):

1. Separation: to separate two dissimilar materials to retain or improve the functionality and integrity of the two materials, for example, the aggregate base material and the subgrade natural soil of roadways.
2. Reinforcement: to improve the strength of the system using the tension of the geosynthetics.
3. Filtration: to allow adequate liquid flow, limit soil loss across the plane of the geosynthetics, and maintain the service life of the application.
4. Drainage: to allow sufficient liquid flow.
5. Containment: as an impermeable barrier, to prevent or limit the flow of liquid or gas across the plane of the geosynthetics.
6. Protection: as a cover, to protect the underlying soils.

Each of the geosynthetics is described as follows.

1. *Geotextiles*. Geotextiles are the most popular geosynthetics and are used in various civil engineering applications with the aforementioned six primary functions. There are three major types of geotextiles based on the manufacturing processes: woven, nonwoven (Figure 8.1(a)), and knitted. The woven geotextiles are made on conventional textile weaving machinery using a wide variety of fabric weaves. The nonwoven geotextiles are needle-punched, resin-bonded, and heat-bonded geotextiles. In the needle-punched geotextiles, the mechanical bonding is achieved by introducing a fibrous web into a machine equipped with barbed needles that punch and reorient the fibers. In resin-bonded geotextiles,

Geotechnical Engineering Design, First Edition. Ming Xiao.
© 2015 John Wiley & Sons, Ltd. Published 2015 by John Wiley & Sons, Ltd.
Companion Website: www.wiley.com/go/Xiao

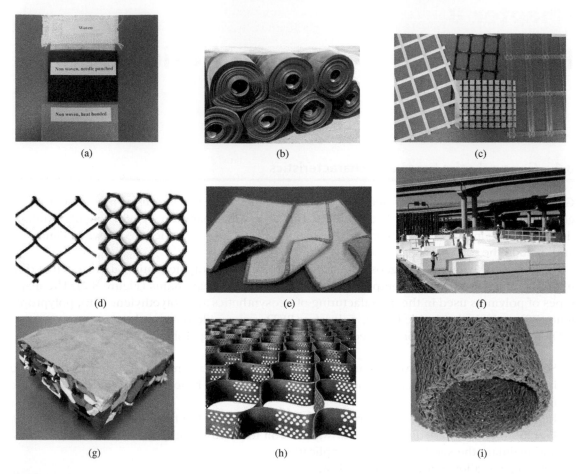

Fig. 8.1 Families of geosynthetics. (a) Geotextile, (b) geomembrane, (c) geogrid, (d) geonet, (e) geosynthetic clay liner (GCL), (f) geofoam. (Photo courtesy of *Geosynthetics* Magazine and ACH Foam Technologies), (g) geocomposite, (h) geocell, (i) geopipe.

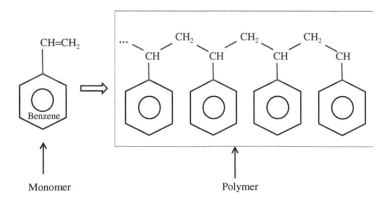

Fig. 8.2 Example of polymerization.

the bonds are formed by spraying or impregnating the fibrous web with an acrylic resin. The heat-bonded (also called melt-bonded) geotextiles are made by melting and compressing the fibers. Generally, woven geotextiles exhibit high tensile strength, high modulus, and low strain. Nonwoven, needle-punched geotextiles have high permeability because of the high porosity and conformability following their high elongation characteristics. Nonwoven heat-bonded geotextiles typically have high modulus and high conformability. Depending on the manufacturing processes, knitted geotextiles can offer high tensile strength and elasticity.

2. *Geomembranes*. Because of their low equivalent diffusion permeability of 1×10^{-11}cm/sec to 1×10^{-14}cm/sec (Koerner 2005), geomembranes are used primarily as impermeable barriers to contain solids or liquids. Depending on the applications, the rigidity, thickness, and surface texture of geomembranes can be controlled during manufacturing. As liquid containments, the primary concerns of geomembrane applications are puncture failure and poor sealing of two joining geomembranes. Many modern techniques have been developed to prevent and detect these types of failures.

3. *Geogrids*. The primary function of geogrids is reinforcement. The rigid longitudinal and transverse ribs of geogrids are perpendicular to each other. There are uniaxial and biaxial geogrids, depending on the direction of the required tensile strength.

4. *Geonets*. The primary function of geonets is drainage. Although they are gridlike materials and not weak, the orientations and configurations of the ribs of geonets are different from geogrids (Figures 8.1(c) and (d)). Geonets are exclusively used as in-plane drains.

5. *Geosynthetic clay liners (GCL)*. GCLs consist of a thin layer of bentonite contained between geotextiles and/or geomembranes on both sides, and they serve as impermeable barriers. The highly expansive bentonite may close punctures on hydration and make GCLs advantageous over geomembranes. GCLs have similar thickness to geomembranes and are easy to install. Therefore, GCLs are used to replace compacted clay liners or geomembranes.

6. *Geofoams*. Geofoams are lightweight foams with apparent density of $10-30\,\text{kg/m}^3$. They are typically used as lightweight backfills to alleviate overburden pressure or as compressible inclusions (buffers) to reduce the stresses on soil or structures.

7. *Geocomposites*. Geocomposites comprise various combinations of geosynthetics to achieve multiple functions simultaneously. For instance, Figure 8.1(g) shows a composite of geofoam and nonwoven, heat-bonded geotextile – the geotextile functions as a filter, and the porous geofoam is a lightweight and compressible drain.

8. *Geocells*. Geocells are three-dimensional, expandable panels made from high-density polyethylene (HDPE), polyester, or another polymer materials. The cells can be filled with gravels, sands, or fine soils to form a permeable or impermeable structure, depending on the applications.

9. *Geopipes*. Geopipes are plastic and sometimes permeable pipes for underground transmission of water, wastewater, gas, oil, or other liquids. Geopipes can also incorporate other functions such as filtration, as shown in Figure 8.1(i). Geopipes can replace traditional steel, cast iron, or concrete pipelines.

The polymer types, main functions, and examples of applications of the nine types of geosynthetics are summarized in Table 8.1.

Table 8.1   Summary of geosynthetics.

Type of geosynthetics	Polymer types	Main functions	Examples of applications
Geotextile	PP, PET, PE, PA	Separation, reinforcement, filtration, drainage, containment, protection	Retaining walls, filters and drains, pavement subgrade, slope stabilization
Geomembrane	HDPE, LDPE, PP	Containment	Landfill covers and bottom liners, retention pond liners, facings of dams
Geogrid	HDPE, PP	Reinforcement, protection	Pavement subgrade, railroad ballasts, slope stabilization, retaining walls
Geonet	PE	Drainage	Drainage in retaining walls, slopes, landfills, embankments, and dams
GCL	Combination of geotextile and geomembrane	Containment	Landfill covers and bottom liners, retention pond liners
Geofoam	Expanded polystyrene	Separation	Lightweight roadway fills, retaining wall/abutment backfills
Geocomposite	Combination of geosynthetics for combined functions	Separation, reinforcement, filtration, drainage, containment, protection	Landfill liners and drains, filter/drain/reinforcement for retaining walls and embankments
Geocell	HDPE, PET	Reinforcement, filtration, protection	Erosion control, slope protection, retaining walls, ground stabilization, temporary flood walls
Geopipe	PVC, HDPE, PP	Drainage	Roadway drains, various pipelines

8.2   Design of mechanically stabilized Earth walls using geosynthetics

Mechanically stabilized earth (MSE) walls use geosynthetics or metal strips as reinforcements that are embedded in granular backfills. The reinforcement and the soil together form a wall or embankment as shown in Figure 8.3.

Compared with the conventional (reinforced) concrete walls, the MSE walls have the following advantages and disadvantages:

Advantages:

- Increased internal integrity because of the geosynthetics' tensile strength and the friction between the soil and the reinforcement.
- Increased shear resistance to resist slope failure.
- Rapid construction.
- Flexible wall system accommodating large differential settlement.
- Suited for seismic regions.

(a)   (b)

Fig. 8.3 Examples of construction of MSE wall and abutment. (a) MSE wall, (b) MSE bridge abutment. (Photo courtesy of Tensar International.)

Disadvantages:

- Require large base width.
- Not applicable to locations requiring future access to underground utilities.
- Susceptible to damages during construction.
- Corrosion of metallic strips.

MSE walls are constructed in layers. During the installation, the geosynthetics should be orientated such that the direction with higher tensile strength is perpendicular to the wall face to maximize the reinforcement and prevent failure. MSE walls typically have fasciae that cover the wall faces. The fasciae prevent ultraviolet damage to the geosynthetics, provide esthetics, or are used to anchor the reinforcements. The fasciae can be interlocking precast concrete modular blocks or precast concrete panels that are fixed to the finished wall faces (Figure 8.4). Welded wire panels, gabion baskets, and treated timber facings can also serve as MSE fasciae. Generally, the fasciae are not considered to contribute to the internal or external stability during the design. Figure 8.5 illustrates three typical MSE wall configurations that use geosynthetics as reinforcements.

Figure 8.6 illustrates an example of geogrid reinforcement in an MSE wall section. In this example, two types of geogrids are used: uniaxial and biaxial. The uniaxial geogrids provide design tensile strength in one direction; the biaxial geogrids provide design tensile strength in two directions and are used to wrap the backfill. In this figure, the uniaxial geogrids are placed at the bottom of each layer to the design length and provide the primary reinforcement; the biaxial geogrids (labeled as BX1120) wraparound each layer and function as containment (cover).

Designing MSE walls consists of internal stability and external stability designs. Given the wall height, soil properties, and surcharges on the top of the wall, the internal stability design determines the following parameters:

- Vertical spacing of each layer and the number of layers.
- Embedment length of the geosynthetics or metal strips in each layer.

(a)   (b)

Fig. 8.4 Examples of MSE wall fasciae. (a) Modular blocks (photo courtesy of Norman Retaining Walls), (b) modular panels. (Photo courtesy of the Reinforced Earth Company.)

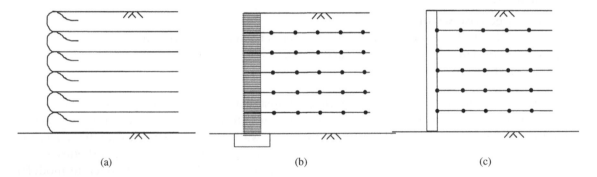

(a)   (b)   (c)

Fig. 8.5 Typical MSE wall configurations. (a) Geotextile wraparound facing, (b) geogrid reinforcement with modular block masonry units, (c) geogrid reinforcement with full-height precast panel.

• Overlap length of the wraparound section of the geosynthetics in each layer, in the case that the geosynthetics rely on overlap rather than modular facing to anchor the geosynthetics.

The external stability design follows the design of conventional retaining walls, namely:

• Overturning about the toe.
• Sliding of the base.
• Bearing capacity of the foundation soil.

8.2.1 Design procedures of geosynthetic MSE walls

Design approach stipulated in the Eurocode

Design of "BS EN 14475:2006 – Execution of special geotechnical works – Reinforced fill" and "BS 8006–1:2010 – Code of practice for strengthened/reinforced soils and other fills" deal with

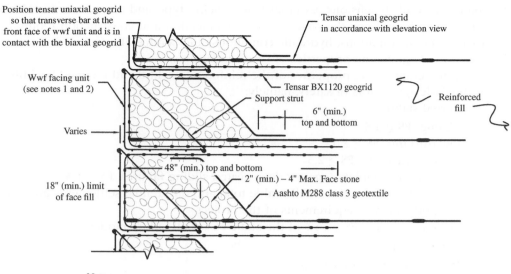

Position tensar uniaxial geogrid
so that transverse bar at the
front face of wwf unit and is in
contact with the biaxial geogrid

Tensar uniaxial geogrid
in accordance with elevation view

Wwf facing unit
(see notes 1 and 2)

Tensar BX1120 geogrid

Support strut

Reinforced
fill

6" (min.)
top and bottom

Varies

48" (min.) top and bottom

2" (min.) – 4" Max. Face stone

18" (min.) limit
of face fill

Aashto M288 class 3 geotextile

Notes:

1. See welded wire form (wwf) facing unit detail
   for facing material and dimensions.

2. All facing units shall be fabricated from
   galvanized steel.

Welded wire form facing detail

Not to scale

Fig. 8.6   Example of geogrid reinforcement in MSE walls. (Photo courtesy of Tensar International.)

the execution of reinforced fill in the United Kingdom. They include specific details regarding drainage, construction stages, type, and configuration of reinforcement, among many other related topics. Currently, existing design procedures for reinforced fill are not compatible with the limit state design approaches discussed in previous chapters. The values of partial factors of safety have not been calibrated. Hence BS EN 1997–1:2004 cannot be used in the design of reinforced fills. However, BS EN 14475:2006 adopts a limit state design approach in which the design strength should be greater or equal than the design load. Within this context, ultimate limit states related to local and global failure that may or may not include the failure of reinforcing elements need to be considered.

Design approach used in the United States

In the United States, the design of geosynthetically reinforced walls commonly follows the "Geosynthetic Design and Construction Guidelines Reference Manual" (FHWA 2008). The design steps described here follow the recommendations in the "Geosynthetic Design and Construction Guidelines Reference Manual" (FHWA 2008) and "Designing with Geosynthetics" (Koerner 2005).

Step 1: Determine wall dimensions, external loads, facing type and connections, and vertical spacing requirement on the basis of facing connection and construction requirements, environmental conditions, hydraulic conditions, and service life period.

Step 2: Determine engineering properties of the foundation soil.

Step 3: Determine engineering properties of the reinforced backfill and retained backfill.

Step 4: Establish design factors of safety, namely:

  A. External stability:

    ◦ Sliding: FS $\geq$ 1.5
    ◦ Overturning: FS $\geq$ 2.0
    ◦ Bearing capacity: FS $\geq$ 2.5
    ◦ Deep-seated stability (against deep-seated sliding failure): FS $\geq$ 1.3
    ◦ Settlement: the maximum allowable total and differential settlements are based on the performance requirements of the project.
    ◦ Seismic stability: FS $\geq$ 75% of the static FS for all failure modes.

  B. Internal stability:

    ◦ Pullout resistance: FS $\geq$ 1.5
    ◦ Pullout resistance under seismic resistance: FS $\geq$ 1.1
    ◦ Minimum embedment length: 1 m (3 ft)
    ◦ Determine the *allowable tensile strength* of the reinforcement. The connection between the facing and the reinforcement should be considered, as the connection may limit the design tensile strength.

The allowable tensile strength ($T_{allowable}$) is obtained using the laboratory-derived pulling test and considering the specific applications shown in Table 8.2 (Koerner 2005).

$$T_{allowable} = T_{ult}\left(\frac{1}{\Pi RF}\right) = T_{ult}\left(\frac{1}{RF_{ID} \times RF_{CR} \times RF_{CBD}}\right) \tag{8.1}$$

where:

$T_{allowable}$ = allowable tensile strength,
T_{ult} = ultimate tensile strength, obtained from lab testing,
RF_{ID} = reduction factor for installation damage,
RF_{CR} = reduction factor for creep,
RF_{CBD} = reduction factor for chemical and biological degradation, also called durability reduction factor,
ΠRF = cumulative value of reduction factors.

The reduction factors can be obtained from Table 8.2.

Step 5: Determine wall embedment depth

To refer to Figure 8.7, the minimum embedment depth at the front of the wall, H_1, depends on the inclination of the ground slope in front of the wall and the wall height, H.

Slope in front of the wall	Minimum H_1
Horizontal, wall	$H/20$
Horizontal, abutment	$H/10$
3H : 1V	$H/10$
2H : 1V	$H/7$
1.5H : 1V	$H/5$

The minimum H_1 is 0.5 m (15 in). If the wall is founded on a slope, a minimum horizontal bench of 1.2 m (4 ft) wide should be provided in front of the wall.

Step 6: Internal stability design, including the following four major steps:

(1) Calculate the lateral earth pressure, σ_h.

If there is no external strut holding the wall, the lateral earth pressure is typically active pressure. The Rankine's theory is used to calculate the earth pressure distribution along the height of the wall. The calculations follow the procedure in Sections 7.2 and 7.3. The lateral (horizontal) soil pressure, σ_h, can be caused by three factors: backfill, surcharge on backfill, and live load:

$$\sigma_h = \sigma_{hs} + \sigma_{hq} + \sigma_{hl} \tag{8.2}$$

where

σ_{hs} = horizontal soil pressure caused by the backfill (soil),
σ_{hq} = horizontal soil pressure caused by surcharge on the backfill, q,
σ_{hl} = horizontal soil pressure caused by live load.

(2) Calculate the vertical spacing, S_v.

The maximum vertical spacing should be large enough to mobilize the allowable tensile strength of the geosynthetics. On the basis of the force equilibrium, the resultant lateral earth force acting on the vertical spacing, S_v, should be balanced by the tensile force provided by the geosynthetics on the same vertical spacing. It is assumed that the tensile strength of each layer of the geosynthetics is divided evenly to resist the lateral soil force from the top and the bottom of the geosynthetics layer, as shown in Figure 8.8.

Assuming σ_h is constant throughout a short distance of S_v, the resultant lateral soil force per unit length of the wall is $\sigma_h \cdot S_v$. It is counterbalanced by the factored tensile strength of the geosynthetics per unit length of the wall, $T_{allowable}/FS_{wall}$, where $T_{allowable}$ is the allowable tensile strength of the geosynthetics, and FS_{wall} is the design factor of safety of the wall. Therefore:

$$\sigma_h \cdot S_v. = \frac{T_{allowable}}{FS_{wall}} \tag{8.3}$$

As σ_h, FS_{wall}, and $T_{allowable}$ are known, the vertical spacing (S_v) at different elevations can be determined.

(3) Calculate the effective length, L_e, and the nonreacting length, L_{NR}, of the geosynthetics (Figure 8.8).

The nonreacting length refers to the portions of the geosynthetics that are in the failure zone and therefore do not provide resistance to the shear failure. The effective length is the length of geosynthetics that extends beyond the theoretical failure plane and therefore provides resistance. The resistance is caused by adhesion and friction that the geosynthetics transfer to the soil. The adhesion and friction are provided on both sides of each geosynthetics layer, and the maximum resistance a geosynthetics layer can provide is its allowable tensile strength. Therefore, the effective length can be solved from the following equation:

$$T_{allowable} = \text{adhesion} + \text{friction} = 2 \times (c_a + \sigma_v \tan \delta) \times L_e \tag{8.4}$$

Chapter 8

So:

$$L_e = \frac{T_{\text{allowable}}}{2 \times (c_a + \sigma_v \tan \delta)} \tag{8.5}$$

where:

c_a = adhesion between the geosynthetics and the soil,
δ = external friction angle between the geosynthetics and the soil,
δ_v = vertical effective stress of the backfill.

If $L_e < 1.0$ m, use $L_{e(\text{min})} = 1.0$ m.

The nonreacting length (L_{NR}) at each reinforcement elevation (z) can simply be solved using trigonometry:

$$L_{\text{NR}} = (H - z) \cdot \tan\left(45° - \frac{\phi}{2}\right) \tag{8.6}$$

The total horizontal length of a geosynthetic reinforcement layer is:

$$L = L_{\text{NR}} + L_e \tag{8.7}$$

(4) Calculate the overlap length, L_0, if the geosynthetics rely on the overlap rather than the modular facing for anchoring (Figure 8.5a). The overlap of geosynthetics ensures the anchorage of the reinforcement without pullout failure. The overlap is given by the following equation following a similar approach as in Equation (8.4), except that in each layer the adhesion and friction are provided by the four sides of the geosynthetic reinforcement (sides A, B, C, and D in Figure 8.9). Therefore:

$$L_0 = \frac{T_{\text{allowable}}}{4 \times (c_a + \sigma_v \tan \delta)} \tag{8.8}$$

If $L_0 < 1.0$ m, use $L_0 = 1.0$ m.

Step 7: External stability design follows the same procedure as outlined in Section 7.3, namely:
- Overturning resistance, check with and without surcharge.
- Sliding resistance, check with and without surcharge.
- Bearing capacity of the foundation.
- Deep-seated (overall) stability.
- Seismic analysis.

Step 8: Settlement analysis.
- Estimate total and differential settlements along the wall alignment, and the differential settlement from the back to the front backfills, using conventional settlement analyses.
- Compare estimated differential settlement along the wall alignment to distortion limits of potential facings.

Step 9: Internal and surficial drainage design.

The detailed approach of MSE wall design is illustrated by the following sample problem.

Table 8.2   Recommended strength reduction factor values (after Koerner 2005).

Applications/ functions	Ranges of reduction factors		
	Installation damage	Creep	Chemical/biologi-cal degradation
Separation	1.1 ~ 2.5	1.5 ~ 2.5	1.0 ~ 1.5
Cushioning	1.1 ~ 2.0	1.2 ~ 1.5	1.0 ~ 2.0
Unpaved roads	1.1 ~ 2.0	1.5 ~ 2.5	1.0 ~ 1.5
Walls	1.1 ~ 2.0	2.0 ~ 4.0	1.0 ~ 1.5
Embankments	1.1 ~ 2.0	2.0 ~ 3.5	1.0 ~ 1.5
Bearing and foundations	1.1 ~ 2.0	2.0 ~ 4.0	1.0 ~ 1.5
Slope stabilization	1.1 ~ 1.5	2.0 ~ 3.0	1.0 ~ 1.5
Pavement overlays	1.1 ~ 1.5	1.0 ~ 2.0	1.0 ~ 1.5
Railroads (filter/separation)	1.1 ~ 3.0	1.0 ~ 1.5	1.5 ~ 2.0
Flexible forms	1.1 ~ 1.5	1.5 ~ 3.0	1.0 ~ 1.5
Silt fences	1.1 ~ 1.5	1.5 ~ 2.5	1.0 ~ 1.5

Chapter 8

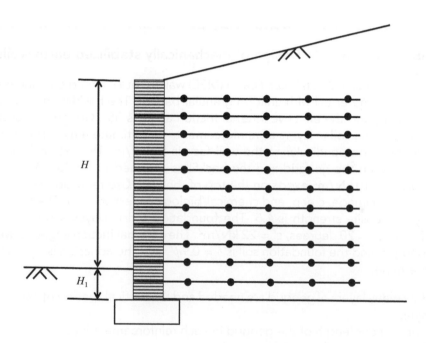

Fig. 8.7   Embedment depth of MSE wall.

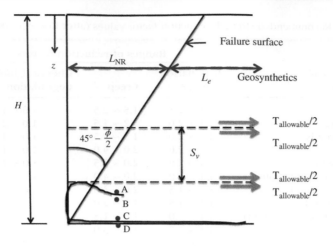

Fig. 8.8 Illustration of reinforcement of MSE retaining wall.

Sample Problem 8.1: Design of mechanically stabilized earth walls

Design a mechanically stabilized earth (MSE) wall to retain a road embankment. The wall is 5-meter tall with uniform granular backfill. The backfill's unit weight is 20 kN/m³, and the effective internal friction angle is 35°. Geogrid is used to wraparound the soil in layers as reinforcement. The ultimate tensile strength in the direction perpendicular to the wall face is 50 kN/m. The external friction angle between the geogrid and the backfill is assumed to be 35°. A uniform surcharge of 10 kN/m² exerts on the top of the wall. Use reduction factors for installation damage, creep, and chemical/biological degradation. The factor of safety for tensile strength is 1.5. The foundation soil is clayey sand with $\gamma = 18$ kN/m³, $\phi' = 20$ degrees, $c' = 22$ kN/m². The external friction angle between the geogrid and the foundation soil is $\delta = 0.9\phi'$, and the adhesion is $c_a = 0.9c'$.
Determine:

1. How many layers of geogrid are needed and the vertical spacing of each layer.
2. Embedment length of the geogrid in each reinforcement layer.
3. Overlap length of geogrid in each layer.
4. The external stability in terms of overturning, sliding, and bearing capacity of the MSE wall.

Solution:

Step 1: Determine wall dimension, external loads.
Step 2: Determine engineering properties of the foundation soil.
Step 3: Determine engineering properties of the reinforced backfill and retained backfill.

The wall dimensions, external loads, and soil engineering properties in the first three steps are shown in Figure 8.9.

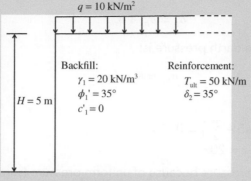

$q = 10 \text{ kN/m}^2$

$H = 5 \text{ m}$

Backfill:
$\gamma_1 = 20 \text{ kN/m}^3$
$\phi_1' = 35°$
$c_1' = 0$

Reinforcement:
$T_{ult} = 50 \text{ kN/m}$
$\delta_2 = 35°$

Foundation soil:
$\gamma_2 = 18 \text{ kN/m}^3$
$\phi_2' = 20°$
$c_2' = 22 \text{ kN/m}^2$

Reinforcement:
$\delta_2 = 0.9 \, \phi_2' = 0.9 \times 20 = 18°$
$c_a = 0.9 \, c_2' = 0.9 \times 22 = 20 \text{ kN/m}^2$

Fig. 8.9 Given parameters for the MSE wall.

Step 4: The following design factors of safety are required:
   A. External stability:
   - Sliding: FS ≥ 1.5
   - Overturning: FS ≥ 2.0
   - Bearing capacity: FS ≥ 2.5
   B. Internal stability:
   - Pullout resistance: FS ≥ 1.5
   - Minimum embedment length: 3 ft (1m)
   - The allowable tensile strength of the reinforcement:

$$T_{allowable} = \frac{T_{ult}}{RF_{ID} \cdot RF_{CR} \cdot RF_{CBD}}$$

From Table 8.2 for the wall application: $RF_{ID} = 1.2$; $RF_{CR} = 2.5$; $RF_{CBD} = 1.2$.

So: $T_{allowable} = \frac{T_{ult}}{RF_{ID} \cdot RF_{CR} \cdot RF_{CBD}} = \frac{50}{1.2 \times 2.5 \times 1.2} = 13.9 \text{ kN/m}$

Step 5: Determine wall embedment depth
The minimum embedment depth at the front of the wall is:

$$H_1 = H/20 = 0.25 \text{ m}$$

Use the minimum depth of 0.5 m.

Step 6: Internal stability design, namely, the following four major steps:

(1) Calculate the lateral earth pressure, σ_h.
Only the lateral earth pressures caused by the backfill and surcharge are considered.

$$\sigma_h = \sigma_{hs} + \sigma_{hq}$$

Rankine active earth pressure is:

$$\sigma_{hs} = K_a \sigma_0'$$

where:

$$K_a = \tan^2 \left(45 - \frac{35}{2}\right) = 0.271$$
$$\sigma_0' = \sigma_0' = \gamma_1 z = 20z$$

Lateral earth pressure because of uniform pressure (surcharge) of infinite area:

$$\sigma_{hq} = K_a q = 2.71 \ \text{kN/m}^2$$

So: $\sigma_h = \sigma_{hs} + \sigma_{hq} = 0.271 \times 20z + 2.71 = 5.42z + 2.71$

(2) Calculate the vertical spacing, S_v, given FS = 1.5

$$S_v = \frac{T_{\text{allowable}}}{\text{FS} \cdot \sigma_h} = \frac{13.9}{1.5 \times (5.42z + 2.71)} = \frac{9.26}{5.42z + 2.71}$$

Note:
- To be conservative, the chosen vertical spacing should be ≤ the calculated vertical spacing;
- The calculated reinforcement length varies with depth. For easy construction, the reinforcement lengths can be grouped with one length in each group. The selected length should be ≥ the calculated length.

$z = 5$ m (bottom of the wall): $S_v = 0.31$ m, use 0.30 m.
After five layers with $S_v = 0.30$ m, $z = 3.5$ m : $S_v = 0.43$ m, use 0.40 m.
After five layers with $S_v = 0.40$ m, $z = 1.5$ m : $S_v = 0.85$ m, use 0.50 m.
From the bottom of the wall:
Five layers with S_v of 0.30 m + 5 layers with S_v of 0.40 m + 3 layers with S_v of 0.50 m = 1.5 + 2.0 + 1.5 = 5.0 m.

(3) Calculate the effective length, L_e, and the nonreacting length, L_{NR}, of the geogrid.

$$L_e = \frac{T_{\text{allowable}}}{2(c_a + \sigma_v \tan \delta)}$$

As $c = 0$ for backfill, $c_a = 0$.

$$\delta = 35°$$

$\sigma_v = \gamma z = 20z$ (do not consider surcharge for conservative design)

So:

$$L_e = \frac{13.9}{2(0 + 20z \cdot \tan 35)} = \frac{0.496}{z}$$

The nonreacting length:

$$L_{NR} = (H - z) \cdot \tan\left(45 - \frac{\phi_1'}{2}\right) = (5 - z)\tan\left(45 - \frac{35}{2}\right) = 2.60 - 0.52z$$

Total embedment length: $L = L_{NR} + L_e$

(4) Overlap length:

$$L_0 = \frac{T_{allowable}}{4 \times (c_a + \sigma_v \tan \delta_1)} = \frac{0.248}{z}$$

$z_{min} = 0.5\,\text{m}$, so $L_0 = 0.496\,\text{m}$.

Use minimum $L_0 = 1.0$ m

The following stability is developed to calculate the embedment length of each layer (Table 8.3).

The configuration of the MSE wall and the reinforcement are shown in Figure 8.10.

Table 8.3 Calculation of embedment lengths of reinforcement.

Layer no.	Depth, z (m)	Spacing, S_v (m)	L_e (m)	$L_{e(min)}$ (m)	L_{NR} (m)	L_{cal} (m)	L_{used} (m)
13	0.5	0.5	0.99	1.0	2.34	3.34	Use 4.0m
12	1.0	0.5	0.50	1.0	2.08	3.08	
11	1.5	0.5	0.33	1.0	1.82	2.82	
10	1.9	0.4	0.26	1.0	1.61	2.61	Use 3.0
9	2.3	0.4	0.22	1.0	1.40	2.40	
8	2.7	0.4	0.18	1.0	1.20	2.20	
7	3.1	0.4	0.16	1.0	0.99	1.99	
6	3.5	0.4	0.14	1.0	0.78	1.78	
5	3.8	0.3	0.13	1.0	0.62	1.62	Use 2.0m
4	4.1	0.3	0.12	1.0	0.47	1.47	
3	4.4	0.3	0.11	1.0	0.31	1.31	
2	4.7	0.3	0.11	1.0	0.16	1.16	
1	5.0	0.3	0.10	1.0	0	1.0	

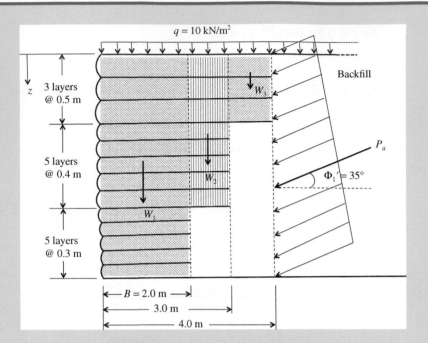

Fig. 8.10 Geogrid reinforcement configuration of the MSE wall.

Step 7: The following external stability is checked.
(1) Overturning resistance, check with and without surcharge.

$$FS_{OT} = \frac{\sum(\text{Resisting moments})}{\sum(\text{Overturning moments})}$$

The overturning moment is solely caused by the horizontal component of the active soil force, P_a.
As the active soil pressure is: $\sigma_h = \sigma_{hs} = \sigma_{hq} = 5.42z + 2.71$ the total resultant active soil force is:

$$P_a = \frac{1}{2}[2.71 + (5.42 \times 5 + 2.71)] \times 5 = 81.3 \text{ kN/m}$$

It is inclined at 35° (the effective internal friction angle of the backfill).

$$\sum(\text{Overturning moments}) = P_a \cos\phi'_1 \frac{H}{3} = 81.3 \times \cos 35° \times \frac{5}{3}$$
$$= 111.0 \text{ kN} \cdot \text{m/m}$$

The reinforced soil of the MSE wall (the hatched section in Figure 8.10) acts as one entity. Therefore, the resisting moments are the weight of the reinforced soil section, which can be divided into three sections, as shown in Figure 8.10.

Calculate the reinforced section weights per meter length of the wall (the unit weight of backfill is 20 kN/m^3):

$$W_1 = 2.0 \times 5.0 \times 20 = 200 \ kN/m$$
$$W_2 = 1.0 \times 3.5 \times 20 = 70 \ kN/m$$
$$W_3 = 1.0 \times 1.5 \times 20 = 30 \ kN/m$$

So:

Σ(Resisting moments)

= moments caused by the reinforced soil weight

+ moment caused by vertical component of the

lateral soil force + moment caused by the surcharge

$= [200 \times 1.0 + 70 \times 12.5 + 30 \times 3.5] + [81.3 \times \sin 35° \times 4]$

$+ [10 \times 4.0 \times 2.0] = 746.5 \ kN \cdot m/m$

So: $FS_{OT} = \frac{746.5}{111.0} = 6.72 > 2.0$, acceptable

If without surcharge, $FS_{OT} = 6.86$.

(2) Sliding resistance between the first layer of reinforcement and the underlying foundation soil. The contact width is $B = 2.0$ m. The external friction angle between the geogrid and the foundation soil is $\delta_2 = 0.9$, $\phi' = 18°$.

Consider surcharge:

$$FS_S = \frac{\Sigma(\text{Resisting forces})}{\Sigma(\text{Sliding forces})} = \frac{[c_a + \sigma_v \tan \delta_2]B}{P_a \cos \phi_1'}$$

The vertical stress at the bottom of the wall:

$$\sigma = \gamma H + q + \frac{P_a \sin \phi_1'}{B}$$
$$= 20 \times 5 + 10 + \frac{81.3 \times \sin 35°}{2}$$
$$= 133.3 \ kN/m^2$$

Σ(Resisting forces) $= [c_a + \sigma_v \tan \delta_2]B = (20 + 133.3 \times \tan 18°)$

$\times 2 = 126.6 \ kN/m$

Σ(Sliding forces) $= P_a \cos \phi_1' = 81.3 \times \cos 35° = 66.6 \ kN/m$

$$FS_S = \frac{126.6}{66.6} = 1.90 > 1.5 \ \text{Acceptable}$$

When without surcharge, FS = 2.13.

If FS < 1.5, the reinforcement width at the base, B, should be increased and the external stability should be rechecked.

(3) Bearing capacity of the foundation.
The entire soil wall exerts even pressure on the foundation soil; the pressure is not eccentric. Use Terzaghi's ultimate bearing capacity equation for wall footing:

$$q_{ult} = c_2'N_c + qN_q + \frac{1}{2}\gamma_2 BN_\gamma$$

Assume the wall sits on ground surface, so $q = 0$.
From $\phi\prime_2 = 20°$, find $N_c = 17.69, N_\gamma = 3.64$
So: $q_{ult} = 22 \times 17.69 + 0 + 0.5 \times 18 \times 2 \times 3.64 = 454.7$ kN/m^2.
From the calculation of FS$_s$, the even pressure exerted by the MSE wall is:

$$\sigma_v = 133.3 \text{ kN/m}^2.$$

$$FS_{bearing} = \frac{q_{ult}}{\sigma_v} = \frac{454.7}{133.3} = 3.4 > 2.0 \quad \text{Acceptable.}$$

Step 8: Settlement analysis.
The settlement analysis is omitted in this sample problem.
Step 9: Internal and surficial drainage design.
The drainage design is omitted in this sample problem.

8.3   Design of reinforced soil slopes

The designs of geosynthetically reinforced walls and slopes both use the limit equilibrium method, that is, the slope is at the onset of failure. The applications of reinforced soil slopes (RSSs) are highway embankments, levees, landslide repairs, residential developments, commercial parks, and landfills. RSS structures are internally stabilized fill slopes; they are constructed using alternating layers of compacted soil and reinforcement. An RSS is different from an MSE wall – slopes steeper than 70 degrees are defined as walls, and lateral earth pressure procedures apply (FHWA 2001, Koerner 2005).

Figure 8.11 shows an example of a RSS using geogrid. Figure 8.12 illustrates the typical layout of an RSS. The typical components of an RSS system are:

Retained soil	The soil that remains in place beyond the limits of excavation.
Reinforced soil	The soil that is placed and compacted in lifts with reinforcements to create the sloped structure.
Primary reinforcement	The geogrids or geotextiles that are placed horizontally in the slope to resist sliding.
Secondary reinforcement	The geogrids or geotextiles that locally stabilize the slope surface during and after slope construction.
Foundation	The foundation of the slope should be stable before the construction of a slope.

Fig. 8.11   Example of a reinforced soil slope in construction. (Photo courtesy of Tensar International.)

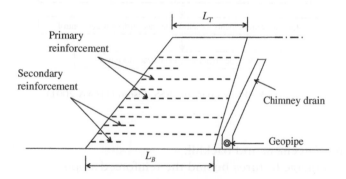

Fig. 8.12   Typical layout of a reinforced soil slope.

Subsurface drainage   The geosynthetic or granular drainage system should be installed in the slope to control, collect, and route groundwater seepage.

Surface erosion protection   The cover on the finished slope surface for erosion control.

The construction of RSS embankments is considerably simpler and consists of many of the elements outlined for MSE wall construction. They are summarized as follows:

- Site preparation.
- Construct subsurface drainage, if required.
- Place reinforcement layer.
- Place and compact backfill on reinforcement.

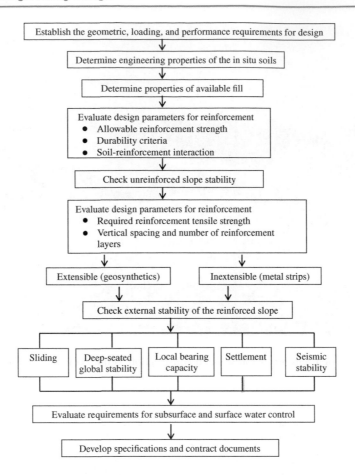

Fig. 8.13   Flow chart for RSS design (FHWA 2001).

- Construct face.
- Place additional reinforcement and backfill.
- Construct surface drainage features behind the reinforced slope

The design steps (given here) as summarized in Figure 8.13 for RSSs follow the Federal Highway Administration (FHWA) publications "Mechanically Stabilized Earth Walls and Reinforced Soil Slopes Design and Construction Guidelines" (FHWA 2001) and "Geosynthetic Design and Construction Guidelines Reference Manual" (FHWA 2008).

Step 1: Establish the geometric, loading, and performance requirements for design.
   (1) Geometric and loading requirements:
      ◦ Slope height, H.
      ◦ Slope angle, θ.
      ◦ Surcharge load, q.
      ◦ Temporary live load, Δq.
      ◦ Design seismic acceleration, a_{design}.

(2) Performance requirements:

The following factors of safety will be used in Steps 6, 7, and 8.

(a) Internal slope stability (for reinforcement design): FS $\geq$ 1.3

(b) External stability and settlement:

- Sliding: FS $\geq$ 1.3.
- Deep-seated failure (overall stability): FS $\geq$ 1.3.
- Local bearing failure: FS $\geq$ 1.3.
- Dynamic loading: FS $\geq$ 1.1.
- Postconstruction magnitude and time rate of settlement are based on project requirements.

Step 2: Determine the engineering properties of the in situ soils.

(1) The foundation and the retained soil profile (i.e., the soil beneath and behind the reinforced zone).

(2) Strength parameters for each soil layer of the retained soil and the foundation soil:

- Undrained c_u and ϕ_u or effective c' and ϕ'.
- Unit weight γ_{wet} and γ_{dry}.
- Consolidation parameters.
- Location of the groundwater table.
- For failure repair, identify the location of the previous failure surface and the cause of the failure.

Step 3: Determine the properties of reinforced fill.

(1) Gradation and plasticity index.

(2) Compaction results.

(3) Shear strength parameters: undrained c_u and ϕ_u or effective c' and ϕ'.

(4) Chemical composition of soil (pH).

Step 4: Evaluate design parameters for reinforcement.

(1) Allowable geosynthetics tensile strength:

$$T_{allowable} = T_{ult} \left(\frac{1}{FS \cdot \Pi RF} \right) = \left(\frac{1}{FS \cdot RF_{ID} \cdot RF_{CR} RF_{CBD}} \right) \tag{8.9}$$

where FS accounts for possible overstressing of the reinforcement without breakage. Table 8.2 shows the ranges of the three reduction factors for slope stabilization: $RF_{ID} = 1.1 \sim 1.5$, $RF_{CR} = 2.0 \sim 3.0$, and $RF_{CBD} = 1.0 \sim 1.5$. FHWA (2001) recommends $\Pi RF = 7$ for conservative and preliminary design. Leshchinsky (2002) recommended the following values for the reduction factors on the basis of the geosynthetic materials (Table 8.4):

(2) Soil-reinforcement interaction (pullout resistance)

- FS = 1.5 for granular soils
- FS = 2.0 for cohesive soils
- Minimum anchorage length, $L_e = 1$ m (3 ft)

Step 5: Check unreinforced slope stability.

(1) Perform stability analyses using the methods in Chapter 5 to determine the safety factors for potential failure surfaces. For the assumed failure surfaces, use circular and plane surfaces, and the failure surfaces should go through the toe, the face (at several elevations), and be deep seated below the toe. Checking for deep-seated

Table 8.4   Preliminary reduction factors for reinforced slopes (Leshchinsky 2002).

Polymer type	RF_{ID}	RF_{CR}	RF_{CBD}
Polyester	1.0 ~ 1.5	1.5 ~ 2.0	1.0 ~ 2.0
Polypropylene	1.0 ~ 1.5	3.0 ~ 5.0	1.0 ~ 2.0
Polyethylene	1.0 ~ 1.5	2.5 ~ 5.0	1.0 ~ 2.0
Polyvinyl alcohol	1.0 ~ 1.5	1.4 ~ 1.8	1.0~ 1.5

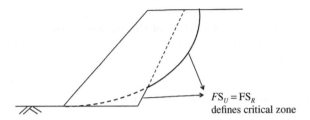

$FS_U = FS_R$
defines critical zone

Fig. 8.14 Critical zone defined by rotational and translational surfaces that meet the required safety factor (FS_u = Factors of safety for unreinforced slope; FS_R = required factor of safety).

slope failure is important. This analysis can be performed using slope stability software because of the large number of assumed failure surfaces.

(2) Determine the critical zone to be reinforced: Identify the failure surfaces whose safety factors are less than or equal to the required safety factor, and plot all these failure surfaces on the cross-section of the slope. The surfaces that just meet the required safety factor envelop the boundary of the critical zone to be reinforced (Figure 8.14).

(3) Critical failure surfaces extending below the toe of the slope are indications of deep foundation and edge bearing capacity problems that must be addressed prior to completing the design. For such cases, a more extensive foundation analysis is warranted, and foundation improvement measures should be considered.

Step 6: Design reinforcement to provide a stable slope.

(1) Calculate the total reinforcement tension, T_{s_max}, per unit length of the slope. The T_{s_max} is the maximum tension required to achieve the required factor of safety (FS_R) for all potential failure surfaces inside the critical zone that is identified in Step 5. For each potential failure circle, the resistance moment of unreinforced slope (M_R)+ the resistance moment provided by the reinforcement ($T_s \cdot D$) = required new resistance moment(M_{new}):

$$M_R + T_s \cdot D = M_{new} \tag{8.10}$$

and

$$M_R = FS_U \cdot M_D \tag{8.11}$$

$$M_{new} = FS_R \cdot M_D \tag{8.12}$$

and M_D = driving moment that causes slope failure.

The above relationships can be rearranged to obtain:

$$T_s = (\text{FS}_R - \text{FS}_U)\frac{M_D}{D}$$ (8.13)

where:

T_s = total tension provided by the reinforcement,
FS_R = required factor of safety to achieve stable slope,
FS_U = minimum factor of safety of unreinforced slope,
M_D = driving moment on the failure surfaces with FS_U,
D = length of moment arm of T_s to the center of the failure circle with FS_U,
   = radius of circle, R, for continuous, sheet type extensible reinforcement (such as geosynthetics) and continuous, sheet type inextensible reinforcement (such as wire mesh), which is based on the assumption that T_s acts tangentially to the circle,
   = vertical distance, Y, from T_s to the center of the failure circle, for discrete element, strip type reinforcement. Assume T_s acts at $H/3$ above slope base for preliminary design.

If there are more than one failure surfaces that have $\text{FS} = \text{FS}_U$, then multiple T_s will be obtained. From all calculated T_s, find the maximum T_{s_max}. Note, as FS_U, M_D, and D all vary with the failure surfaces, the minimum FS_U may not yield the maximum tension, T_{s_max}.

(2) Determine the distribution of the reinforcement:

   (a) For low slopes ($H \le 6\,\text{m}$), assume uniform distribution of reinforcement. Skip Step (b) and directly go to Step (3).

   (b) For high slops ($H > 6\,\text{m}$), divide the slope into two or three reinforcement zones of equal height.

   For two zones:

$$T_{\text{bottom}} = \frac{3}{4}T_{s\_max}$$ (8.14)

$$T_{\text{top}} = \frac{1}{4}T_{s\_max}$$ (8.15)

   For three zones:

$$T_{\text{bottom}} = \frac{1}{2}T_{s\_max}$$ (8.16)

$$T_{\text{middle}} = \frac{1}{3}T_{s\_max}$$ (8.17)

$$T_{\text{top}} = \frac{1}{6}T_{s\_max}$$ (8.18)

   In each zone, the tension is assumed to be distributed evenly among the reinforcement layers.

(3) Determine the reinforcement vertical spacing (S_v), the number of reinforcement layers (N), and the design tension of each reinforcement layer ($T_{\text{allowable}}$):

$$N = \frac{H_{\text{zone}}}{S_v}$$ (8.19)

$$T_{\text{design}} = \frac{T_{\text{zone}}}{N} \qquad (8.20)$$

where:

H_{zone} = thickness of each zone,

T_{zone} = the total tension for each zone, calculated from Step (2).

A typical vertical spacing of 400 mm (16 inch) can be used for face stability and compaction quality in initial design.

Secondary (or intermediate) reinforcement that is between the primary reinforcement is used to stabilize the slope surface during and after slope construction. Short lengths (1.2 to 2 m) of intermediate reinforcement into the fill from the face should be used.

For slopes flatter than 1H:1V (45°) and of well-graded soils, closer spaced reinforcement (no greater than 400 mm) may preclude wrapping the slope face. Wrapped faces are required for steeper slopes and uniformly graded soils to prevent face sloughing.

(4) Determine the required reinforcement lengths (L_e) at different elevations to provide adequate pullout resistance.

$$L_e = \frac{T_{\text{allowale}} \cdot \text{FS}}{2F^* \cdot \alpha \cdot \sigma_v' \cdot C \cdot R_c} \qquad (8.21)$$

where:

$T_{\text{allowable}}$ = design tension of each reinforcement layer,

FS = factor of safety for pullout resistance,

F^* = pullout friction factor (see explanation below),

α = scale effect correction factor to account for a nonlinear stress reduction over the embedded length of highly extensive reinforcement, based on laboratory data. For all metallic reinforcements, $\alpha = 1.0$; for geogrids, $\alpha = 0.8$; for geotextiles, $\alpha = 0.6$.

σ_v' = effective vertical stress at the reinforcement level in the reinforcement zone,

C = overall reinforcement surface area geometry factor, $C = 2$ for strip, grid, and sheet type reinforcement,

R_c = reinforcement coverage ratio,

$$R_c = \frac{b}{S_b} \qquad (8.22)$$

where

b = gross width of the reinforcement element,

S_b = center-to-center horizontal spacing between reinforcements, see Figure 8.15.

Evaluation of pullout friction factor, F^* (on the basis of SCDOT Geotechnical Design Manual, Appendix D, Reinforced Soil Slopes, 2010).

The pullout friction factor is the coefficient of friction between the soil and reinforcement. It can be obtained most accurately from laboratory or field pullout tests.

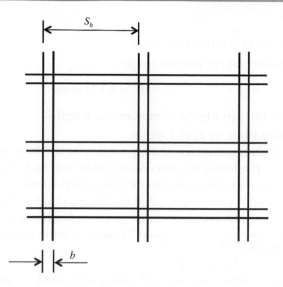

Fig. 8.15   Illustration of reinforcement coverage ratio.

It can also be derived from empirical or theoretical relationships.

$$F^* = \text{Passive   resistance} + \text{Frictional   resistance} = F_q \alpha_\beta + \tan \delta \qquad (8.23)$$

where:

F_q = the embedment (or surcharge) bearing capacity factor,
α_β = bearing factor for passive resistance which is based on the thickness per unit width of the bearing member,
δ = the soil-reinforcement friction angle.

In the absence of site-specific pullout test data, the following semiempirical relationships can be used for practical purposes.

° For steel ribbed reinforcement:
   $F^* = \tan \delta = 1.2 + \log C_u$, at the top of the slope, and maximum

$$F^* = 2.0 \qquad (8.24)$$

$$F^* = \tan \varphi_r, \text{at depth of } 6.1\,\text{m}\,(20 \text{ ft}) \text{ and below} \qquad (8.25)$$

where:

C_u = uniform coefficient of the reinforced backfill,
φ_r = reinforced backfill peak friction angle.

° For steel grid reinforcements with transverse spacing, $S_t \geq 6$ inch (0.13 m):

$$F^* = F_q \alpha_\beta = 40 \alpha_\beta = 40 \left(\frac{t}{2S_t} \right) = 20 \left(\frac{t}{S_t} \right), \text{at the top of the slope} \qquad (8.26)$$

$$F^* = F_q \alpha_\beta = 20 \alpha_\beta = 40 \left(\frac{t}{2S_t} \right) = 20 \left(\frac{t}{S_t} \right), \text{at depth of 6.1\,m\,(20ft) and below.}$$

$$\qquad (8.27)$$

where

t = thickness of the transverse bar.

∘ For geosynthetic sheet reinforcement:

$$F^* = 0.67 \tan \varphi_r \qquad (8.28)$$

A few notes on developing reinforcement lengths:

A. Minimum value of L_e is 1 meter.
B. The length required for sliding stability at the base will generally control the length of the lower reinforcement levels (see Step 7).
C. Lower-layer lengths must extend at least to the limits of the critical zone. Longer reinforcements may be needed to address deep-seated slope failure issues.
D. For ease of construction and inspection, lengthen some reinforcement layers to create two or three sections of equal reinforcement length.

Step 7: Check external stability (Figure 8.16).

(1) Sliding resistance (Figure 8.16a)

The horizontal sliding of the soil along any level of reinforcement within the slope should be checked. All soil strata and interface friction values should be considered. The back of a possible sliding wedge should be the back boundary of the reinforced zone, or be angled at $45 + \phi/2$ (ϕ = backfill/s internal friction angle), whichever is flatter. The sliding force is the horizontal component of the active soil force behind the sliding wedge, and the resisting force is the friction at the base of the slope. Adhesion between the slope and the foundation soil is not considered in the resistance.

$$FS_{slide} = \frac{Resisting\ force}{Sliding\ force} = \frac{(W + P_a \sin \phi_b) \tan \phi_{min}}{P_a \cos \phi_b} \geq 1.3 \qquad (8.29)$$

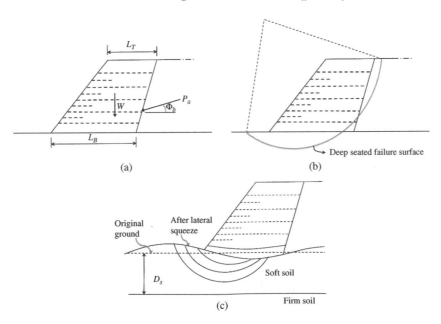

(a)

(b)

(c)

Fig. 8.16 External stability of reinforced soil slope. (a) Sliding, (b) Deep-seated slope failure, (c) Local bearing failure (lateral squeeze).

where:

W = weight of the potential sliding portion of the slope,

P_a = active soil force per unit length of the slope $= \frac{1}{2}\gamma_b H^2 K_a$,

ϕ_b = internal friction angle of the reinforced backfill; if drains/filters are placed between the reinforced slope and the native soil, then ϕ_b should be the friction angle of the reinforced slope and the drains/filters material,

γ_b = unit weight of the reinforced backfill,

ϕ_{min} = minimum friction angle between reinforced soil and the reinforcement or between the slope and the foundation soil,

H = height of the potential sliding portion of the slope,

K_a = active earth pressure coefficient.

(2) Deep-seated global stability (Figure 8.16b)

$$FS_{\text{deep-seated}} = \frac{\text{Total   resistance   moment}}{\text{Total   driving   moment}} \geq 1.3 \qquad (8.30)$$

If a deep-seated failure surface with FS < 1.3 extends beyond the reinforced section, the reinforcement needs to extend beyond the deep-seated failure surface, the toe of the new slope should be regraded, or the slope should be constructed at a flatter angle.

(3) Local bearing capacity at the toe (lateral squeeze) (Figure 8.16c)

If a weak soil layer exists beneath the slope to a limited depth of D_s, which is less than the slope width, b, the factor of safety can be calculated using (Silvestri 1983)

$$FS_{\text{squeeze}} = \frac{2c_u}{\gamma D_s \tan \beta} + \frac{4.14c_u}{H\gamma} \geq 1.3 \qquad (8.31)$$

where:

c_u = undrained cohesion of the soft soil layer,

γ = unit weight of the soil of the slope,

D_s = depth of the soft soil beneath the base of the slope,

β = slope angle,

H = slope height.

If $D_s > b$, then general slope (i.e., deep-seated) stability governs.

(4) Foundation settlement.

The magnitude and rate of total and differential foundation settlements should be calculated. The methodologies described in Chapter 3 can be used.

Step 8: Check seismic stability (refer to Chapter 9.3 for details).

Perform a pseudo-static analysis by adding a horizontal seismic force at the centroid of each slice using the method of slices. The horizontal seismic forces use a design seismic acceleration α_{design}, which takes a value of half of the seismic ground coefficient, α, according to the AASHTO Standard Specifications for Highway Bridges (Division 1A-Seismic Design, 6.4.3 Abutments). The dynamic factor of safety should be at least 1.1.

Step 9: Evaluate requirements for subsurface and surface water runoff control.

(1) Subsurface water control.

(a) Design of subsurface drainage features should address flow rate, filtration, placement, and outlet details.

(b) Drains are typically placed at the rear of the reinforced mass. Geocomposite drainage systems or conventional granular blanket and trench drains can be used.

(c) Lateral spacing of outlets is dictated by site geometry, estimated flow, and existing agency standards. Outlet design should address long-term performance and maintenance requirements.

(d) Geosynthetic drainage composites can be used in the subsurface drainage design. The FHWA Geosynthetic Design and Construction Guidelines (FHWA 1998) presented the criteria of geotextile permeability and filtration/clogging. Drainage composites should be designed with the consideration of:
  ∘ Geotextile filtration and clogging.
  ∘ Long-term compressive strength of polymeric core of the composites.
  ∘ Reduction of flow capacity because of intrusion of geotextile into the geocomposite core.
  ∘ Long-term inflow/outflow capacity.

(e) Slope stability analyses should account for interface shear strength along a geocomposite drain. The geocomposite and soil interface likely have a friction value that is lower than that of the soil. Thus, a potential failure surface may be induced along the interface.

(f) Geotextile reinforcements (primary and intermediate layers) must be more permeable than the reinforced fill material to prevent hydraulic buildup above the geotextile layers during precipitation.

(g) Where drainage is critical for maintaining slope stability, special emphasis on the design and construction of subsurface drainage features is recommended. Redundancy in the drainage system is also recommended for these cases.

(2) Surface water runoff.

(a) Surface water runoff should be collected above the reinforced slope and channeled or piped below the base of the slope.

(b) Wrapped faces and intermediate layers of secondary reinforcement may be required at the face of a reinforced slope to prevent local sloughing.

(c) Select a long-term facing system to prevent or minimize erosion because of rainfall and runoff on the slope face.

(d) The flow-induced tractive shear stress on the face of the reinforced slope can be calculated using:

$$\lambda = d \cdot \gamma_w \cdot s \tag{8.32}$$

where:

λ = tractive shear stress, kPa,
d = depth of water flow, m,
γ_w = unit weight of water, kN/m^3,
s = slope ratio (vertical:horizontal), m/m.

For $\lambda < 100$ kPa, consider vegetation with temporary or permanent erosion control mat.

For $\lambda > 100$ kPa, consider vegetation with permanent erosion control mat or other armor type systems.

The following example illustrates the design process of a RSS.

Sample Problem 8.2: Design of a reinforced soil slope

A new embankment will be constructed to support a new road. The desired embankment height is 10.00 m and the slope is 0.8H:1.0V. A geogrid with an ultimate tensile strength of 100 kN/m (ASTM D4595 width method) is desired for reinforcing the new slope. Assume the reinforcement coverage ratio (R_c) is 0.2. A uniform
surcharge of 12.5 kN/m^2 is to be used for the traffic loading. Available information indicates that the natural foundation soil and the retained soil both have a drained friction angle of 34°, effective cohesion of 12.5 kPa, and bulk unit weight of 19.0 kN/m^3. The granular backfill to be used in the reinforced section will have a minimum friction angle of 34° and dry unit weight of 19.0 kN/m^3. The reinforced slope design must have a minimum factor of safety of 1.5 for slope stability.

  Determine the number of layers, vertical spacing, and the total length of geogrid reinforcements that are required for the reinforced slope.

Solution:

Step 1: Establish the geometric, loading, and performance requirements for design.
  (1) Geometric and loading requirements:
    ◦ Slope height, $H = 10$ m. The surcharge on the top of the slope can be converted to equivalent height: $\Delta H = 12.5/19 = 0.66$ m
      The equivalent slope height is: $H' = 10.66$ m
    ◦ Slope angle, $\theta = \tan^{-1}(0.8/1) = 39°$
    ◦ Surcharge load, $q = 12.5$ kN/m^2
    ◦ Temporary live load, $\Delta q = 0$
    ◦ Design seismic acceleration, $a_{design} = 0$
  (2) Performance requirements:
    (a) Internal slope stability (for reinforcement design): FS = 1.5
    (b) External stability and settlement:
      ◦ Sliding: FS = 1.3
      ◦ Deep-seated failure (overall stability): FS = 1.3
      ◦ Local bearing failure: FS = 1.5
      ◦ Dynamic loading: N/A
      ◦ Postconstruction magnitude and time rate of settlement are based on project requirements: N/A

Step 2: Determine the engineering properties of the in situ soils.
The foundation soil's properties are:
- $c' = 12.5$ kN/m^2, $\phi' = 34°Z$
- Unit weight, $\gamma_{wet} = 19.0$ kN/m^3
- Assume the external friction angle between the geogrid and the foundation soil is 20°
- Consolidation parameters: N/A
- Location of the groundwater table: N/A. Assume no GWT effect.

Step 3: Determine the properties of reinforced fill.
- Gradation and plasticity index: granular soil, no plasticity.
- Compaction results: $\gamma_d = 19.0$ kN/m^3
- $c' = 0$, $\phi' = 34°$
- Chemical composition of soil (pH) : N/A.

Step 4: Evaluate design parameters for reinforcement.
(1) Allowable geosynthetics tensile strength:

$$T_{allowable} = T_{ult} \left(\frac{1}{FS \cdot \Pi RF} \right) = \frac{100}{1.5 \times 7} = 9.5 \text{ kN/m}$$

Use $\Pi RF = 7$ for conservative and preliminary design.
The geogrid reinforcement coverage ratio: $R_c = 0.2$.
(2) Soil-reinforcement interaction (pullout resistance)
- FS = 1.5 for granular soils (in this sample problem)
- FS = 2.0 for cohesive soils
- Minimum anchorage length, $L_e = 1$ m

Step 5: Check unreinforced slope stability.
Ordinary method of slices is used. A computer program is used to analyze multiple trial failure surfaces. The failure surfaces with FS = 1.3 are identified and they encompass the critical zone to be reinforced. Figure 8.17 shows the critical surface with FS = 1.3 that was obtained from the computer program. The center of the failure surface is (23.76 m, 27.50 m), the radius is 18.16 m.
The computer program yielded the minimum factor of safety of 0.9, indicating the slope will fail without reinforcement. Deep-seated failure in the slope was not identified by the computer program.

Step 6: Design reinforcement to provide a stable slope.
(1) Total tension provided by the reinforcement should be:
$T_s = (FS_R - FS_U)M_D/D$
where: $FS_R = 1.3$.

The computer program provided the following results for the critical failure surface with the minimum FS:

$$FS_U = 0.9$$

$$M_D = 615.35 \text{ kN} \cdot \text{m/m}$$

$$D = R = 18.16 \text{ m, as geogrid is used.}$$

So, $\quad T_s = T_{s\_max} = (FS_R - FS_U)\dfrac{M_D}{D} = (1.3 - 0.9) \times \dfrac{615.35}{18.16}$
$= 13.55 \text{ kN/m}$

(2) Determine the distribution of the reinforcement:
As $H = 10.66 \text{ m} > 6 \text{ m}$, divide the slope into two reinforcement zones of equal height.

$$T_{bottom} = \frac{3}{4}T_{s\_max} = \frac{3}{4} \times 13.55 = 10.16 \text{ kN/m}$$

$$T_{top} = \frac{1}{4}T_{s\_max} = \frac{1}{4} \times 13.55 = 3.39 \text{ kN/m}$$

(3) In each zone, the tension is assumed to be distributed evenly among the reinforcement layers. Assume vertical spacing $S_v = 0.4 \text{ m}$.
In the top zone:

$$H_{zone} = \frac{10.0}{2} = 5.0 \text{ m}$$

$$T_{top} = 3.39 \text{ kN/m}$$

Number of reinforcement layers:

$$N = \frac{H_{zone}}{S_v} = \frac{5}{0.4} = 12.5 \text{ layers}$$

Design tension of each reinforcement layer:

$$T_{design(top)} = \frac{T_{top}}{N} = \frac{3.39}{12.5} = 0.27 \text{ kN/m}$$

In the bottom zone:

$$H_{zone} = \frac{10.0}{2} = 5.0 \text{ m}$$

$$T_{bottom} = 10.16 \text{ kN/m}$$

Number of reinforcement layers:

$$N = \frac{H_{zone}}{S_v} = \frac{5}{0.4} = 12.5 \text{ layers}$$

Design tension of each reinforcement layer:

$$T_{design(bottom)} = \frac{T_{bottom}}{N} = \frac{10.16}{12.5} = 0.81 \ \text{kN/m}$$

Secondary reinforcement of 1.5-m long into the fill is selected to stabilize the slope surface during and after slope construction.

(4) Determine the required reinforcement lengths (L_e) at different elevations to provide adequate pullout resistance.

$$L_e = \frac{T_{allowable} \cdot FS}{2F^* \cdot \alpha \cdot \sigma_v' \cdot C \cdot R_c}$$

where:

Design tension of each reinforcement layer, $T_{allowable} = 9.5 \, \text{kN/m}$

Factor of safety for pullout resistance, $FS = 1.5$

The reinforced backfill peak friction angle $\varphi_r = \phi = 34°$

So the pullout friction factor of geosynthetics is (Equation 8.28):

$$F^* = 0.67 \tan \varphi_r = 0.67 \times \tan 34° = 0.45$$

Scale effect correction factor for geogrid: $\alpha = 0.8$

Effective vertical stress at the reinforcement level in the reinforcement zone: $\sigma'_v = \gamma z = 19z$

Overall reinforcement surface area geometry factor for geogrid: $C = 2$

Reinforcement coverage ratio: $R_c = 0.2$

So

$$L_e = \frac{T_{allowable} \cdot FS}{2F^* \cdot \alpha \cdot \sigma_v' \cdot C \cdot R_c} = \frac{9.5 \times 1.5}{2 \times 0.45 \times 0.8 \times 19z \times 2 \times 0.2} = \frac{2.60}{z}$$

$$z_{min} = S_v = 0.4 \ \text{m}, \ L_{e(max)} = 6.50 \ \text{m}$$

$$z = H/2 = 5.0 \ \text{m}, \ L_e = 0.5 \ \text{m}$$

$$z_{max} = H = 10.66 \ \text{m}, \ L_{e(min)} = 0.24 \ \text{m}, \ \text{use} \ L_{e(min)} = 3.0 \ \text{m}$$

For ease of construction and inspection, lengthen the reinforcement layers in the bottom zone to create two sections of equal reinforcement length. The configuration of the reinforcement in the slope is shown in Figure 8.18.

Step 7: Check external stability
      (1) Sliding resistance

$$FS_{slide} = \frac{\text{Resisting force}}{\text{Sliding force}} = \frac{(W + P_a \sin \phi_b) \tan \phi_{min}}{P_a \cos \phi_b}$$

where:
      Internal friction angle of the reinforced backfill, $\phi_b = 34°$
      Unit weight of the reinforced backfill, $\gamma_b = 19$ kN/m^3
      Friction angle between the reinforced soil and the reinforce-
      ment or between the slope and the foundation soil, $\phi_{min} =$
      20°
      Height of the potential sliding portion of the slope, $H =$
      10.66 m (considering surcharge)
      Rankine active earth pressure coefficient:

$$K_a = \tan^2 \left(45° - \frac{\phi_b}{2} \right) = \tan^2 \left(45° - \frac{34°}{2} \right) = 0.283$$

Active soil force per unit length of the slope:

$$P_a = \frac{1}{2} \gamma_b H^2 K_a = \frac{1}{2} \times 19 \times 10.66^2 \times 0.283 = 305.5 \text{ kN/m}$$

Reinforced portion of the slope per unit length of the slope:

$$W = A\gamma_b = (3.0 \times 5.2 + 6.5 \times 4.8) \times 19$$
$$= 889.2 \text{ kN/m}$$

$$FS_{slide} = \frac{\text{Resisting force}}{\text{Sliding force}} = \frac{(W + P_a \sin \phi_b) \tan \phi_{min}}{P_a \cos \phi_b}$$
$$= \frac{(889.2 + 305.5 \times \sin 34°) \tan 20°}{305.5 \times \cos 34°}$$
$$= 1.52 > 1.3 \text{ OK}$$

Deep-seated slope failure was not found by the computer pro-
gram. It is concluded that deep-seated failure will not occur.
      (2) Deep-seated global stability
      Weak soil layer beneath the embankment was not found. It is
      concluded that lateral squeeze will not occur.
      (3) Local bearing capacity at the toe (lateral squeeze)
      (4) Foundation settlement analysis is omitted in this sample prob-
      lem.
Step 8: Seismic stability analysis is omitted in this sample problem.
Step 9: Surface water runoff control and subsurface drainage design are
      omitted in this sample problem.

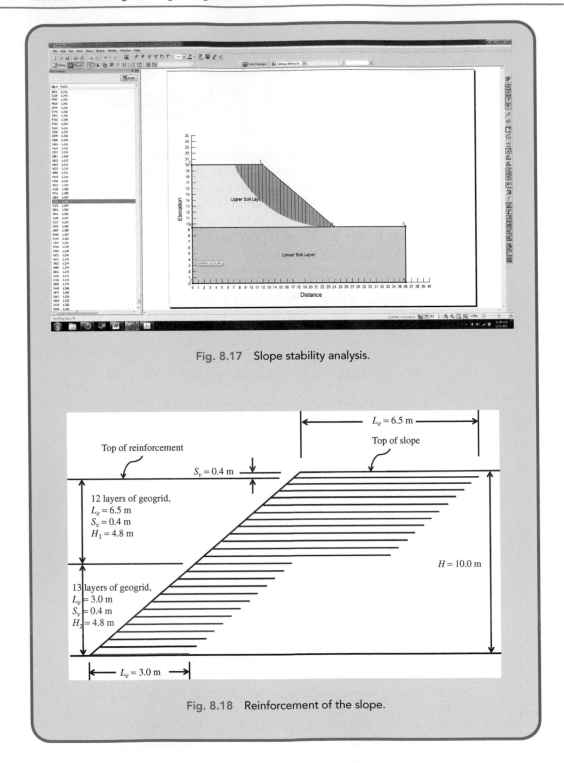

Fig. 8.17   Slope stability analysis.

Fig. 8.18   Reinforcement of the slope.

8.4   Filtration and drainage design using geotextiles

Geotextiles have been successfully used to replace granular filters and drainage layers in almost all drainage applications, because of their comparable performance, improved economy, consistent properties, and ease of placement. The designs of geotextiles as filters and drains use the fundamental concepts of water flow in saturated porous media – Darcy's law and the flow net, which were reviewed in Chapter 6. In addition, specific hydraulic properties of geotextiles are used.

8.4.1   Hydraulic properties of geotextiles

- Porosity (n)

$$n = \frac{V_v}{V_t} \tag{8.33}$$

where V_v is the volume of void and V_t is the total volume of the specimen. As direct measurement of porosity is difficult, the porosity of geotextile is usually obtained by:

$$n = 1 - \frac{m}{\rho \cdot t} \tag{8.34}$$

where:

m = mass of geotextile per unit area, (g/m²),
ρ = density of geotextile,
t = thickness of geotextile under a specified normal stress. As geotextiles are usually embedded in the subsoil and subjected to some normal stress, the actual thickness depends on the applied overburden stress.

- Percent open area (POA) (only for woven monofilament geotextile)
POA represents the openness of the filaments of a geotextile. It is measured by projecting light through the geotextile onto a cardboard and measuring the area of the light penetrating through the open spaces.

$$\text{POA} = \frac{\text{Total open area}}{\text{Total specimen area}} \times 100\% \tag{8.35}$$

The POA ranges from 0% for closed geotextile to 36% for extremely open geotextile (Koerner 2005).
- Apparent opening size (AOS) or equivalent open size (EOS)
AOS is used to approximately represent the pore sizes of geotextiles. It is measured by sieving successively smaller uniform glass beads through the geotextile; when 5% or less (by mass) of the glass beads pass the geotextile, the glass beads' diameter is denoted as O_{95}, which represents the AOS or EOS.
- Cross-plane permeability and permittivity
When geotextiles are installed as a filter, the seepage should flow perpendicular to the geotextile sheet, that is, the flow is *cross-plane*. The permeability of the geotextile as a filter is the cross-plane permeability, k_n.

Use Darcy's law:

$$q = k \cdot i \cdot A = k_n \frac{\Delta h}{t} A \tag{8.36}$$

Then:

$$k_n = \frac{q \cdot t}{\Delta h \cdot A} \tag{8.37}$$

where

t = thickness of geotextile under a specified normal stress.

As geotextiles have various thicknesses, the cross-plane permeability per unit *thickness* of the geotextile is often used to describe geotextile filters and is called permittivity ($\sec^{-1}$):

$$\Psi = \frac{k_n}{t} = \frac{q}{\Delta h \cdot A} \tag{8.38}$$

- In-plane permeability and transmissivity

  When geotextiles are used as a drain, the seepage should flow parallel to the geotextile sheet, that is, the flow is *in-plane*. The permeability of the geotextile as a drain is the in-plane permeability, k_p.

  Use Darcy's law:

$$q = k \cdot i \cdot A = k_p \frac{\Delta h}{L} (w \cdot t) \tag{8.39}$$

Then

$$k_p = \frac{q \cdot L}{\Delta h \cdot w \cdot t} \tag{8.40}$$

where:

L = the length of the geotextile in the flow direction,
w = the width of the geotextile,
v = the thickness of the geotextile under a specified normal stress.

As the seepage capacity of a geotextile also depends on the thickness, transmissivity ($cm^2/\sec$) is used to describe geotextile drains:

$$\theta = k_p \cdot t = \frac{q \cdot L}{\Delta h \cdot w} \tag{8.41}$$

8.4.2 Filtration and drainage criteria

Two typical examples of geotextile filters are shown in Figures 8.19 and 8.20.
Geotextile filter design and geotextile drainage design follow the same criteria.

- *Soil retention*: the pore sizes of the geotextile should be sufficiently small to retain the particles carried by the seepage. AOS (or O_{95}) is used to evaluate the soil retention.
- *Drainage*: the permeability should be sufficiently large, that is, the pore sizes of the filter should be sufficiently large to permit the seepage without impedance. Permittivity is used to evaluate the drainage capacity.
- *Long-term flow compatibility (clogging resistance)*: the filter should continue to function with adequate soil retention and drainage capabilities in the long term without excessive clogging.

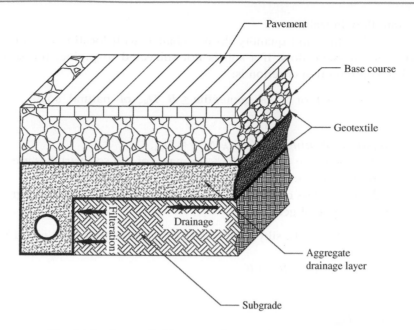

Fig. 8.19   Geotextile filter around a drain beneath pavement.

Fig. 8.20   Geotextile filter behind a Gabion basket wall. (Photo courtesy of Midwest Construction Products, Fort Myers, FL.)

• *Survivability and durability during construction*: to ensure the geotextile will not be damaged during installation, certain geotextile strength and durability properties are required for the selected geotextile filters and drainage layers.

(a) Soil retention criteria
    The soil retention criteria are based on the following three flow and soil conditions.

Chapter 8

i. Steady-state flow in stable soils:

If the flow's direction and quantity do not change with location and time, the flow is defined as steady-state flow. If a soil is not subjected to piping, the soil is internally stable. The soil retention criterion is:

$$\text{AOS or } O_{95} \text{ of geotextile} \leq B \cdot D_{85} \text{ of natural soil} \qquad (8.42)$$

where:

AOS = *apparent opening size (mm)*; $\text{AOS} \approx O_{95}$,

D_{85} = soil particle size of which 85% (in mass) of the soil particles are smaller than the particle size,

B = a dimensionless coefficient that depends on the soil type and gradation, and the type of geotextile.

- For sands, gravelly sands, silty sands, and clayey sands (per the Unified Soil Classification system):
  - $C_u \leq 2$ or $C_u \geq 8$: $B = 1.0$
  - $2 \leq C_u \leq 4$: $B = 0.5C_u$
  - $4 < C_u < 8$: $B = 8/C_u$
    where $C_u = D_{60}/D_{10}$
- For silts and clays:
  - For woven geotextile: $B = 1$
  - For nonwoven geotextile: $B = 1.8$
  - If using both types of geotextile: $B = 0.3$.

ii. Dynamic (transient) flow in stable soils:

If a flow's direction or quantity changes with location or time, the flow is defined as dynamic or transient flow. The soil retention criterion is:

$$O_{95} \leq 0.5 \ D_{85} \qquad (8.43)$$

iii. Unstable soils:

Unstable soil is susceptible to piping. If unstable soil is encountered, filtration or drainage tests should be conducted to select a suitable geotextile.

(b) Drainage criteria

Drainage considerations include the following three criteria.

i. Permeability criteria:

The geotextile's permeability requirements depend on the type of applications and the soil and hydraulic conditions.

- For less critical applications and less severe conditions:

$$k_{\text{geotextile}} \geq k_{\text{soil}} \qquad (8.44)$$

- For critical applications and severe conditions:

$$k_{\text{geotextile}} \geq 10 \, k_{\text{soil}} \qquad (8.45)$$

where:

$k_{\text{geotextile}}$ = cross-plane permeability of geotextile,

k_{soil} = permeability of soil.

ii. Permittivity criteria:

The permittivity criteria depend on the percentage of fines in the soils to be filtered. The more fines are in the soil, the greater permittivity is required.

- For less than 15% passing No. 200 sieve (0.075 mm): $\Psi \geq 0.5$ sec^{-1}
- For 15%–50% passing No. 200 sieve (0.075 mm): $\Psi \geq 0.2$ sec^{-1}
- For more than 50% passing No. 200 sieve (0.075 mm): $\Psi \geq 0.1$ sec^{-1}

iii. Flow criteria:

In the applications where portions of the geotextile may not be available for flow, the flow capacity of the geotextile should follow:

$$q_{required} = q_{geotextile} \left(\frac{A_g}{A_t} \right) \tag{8.46}$$

where:

$q_{required}$ = required seepage rate,
$q_{geotextile}$ = needed geotextile seepage rate,
A_g = geotextile area available for flow,
A_t = total geotextile area.

(c) Long-Term Flow Compatibility (Clogging Resistance)

The geotextile's clogging resistance criteria also depend on the type of application and the soil and hydraulic conditions, as in the permeability criteria.

i. For less critical applications and less severe conditions:

- For $C_u > 3$: O_{95} of geotextile $\geq 3D_{85}$ of soil
- For $C_u \leq 3$: Use the maximum O_{95} from the retention criteria.

In situations where clogging is a possibility such as in gap-graded or silty soils, the following optional qualifiers may be applied:

- For nonwoven geotextile: porosity of the geotextile $n \geq 50\%$
- For woven monofilament and slit film woven geotextile: POA $\geq 4\%$

ii. For critical applications and severe conditions:

Select geotextiles that meet the soil retention and drainage criteria above; then perform filtration tests using samples of the in situ soil and hydraulic conditions, following ASTM D5101: "Measuring the Soil-Geotextile System Clogging Potential by the Gradient Ratio" and the FHWA publication "Geosynthetic Design and Construction Guidelines Reference Manual" (FHWA 2008).

(d) Survivability and durability criteria

Finally, to ensure the geotextile will not be damaged during the construction process, certain geotextile strength and durability properties are required for the selected geotextile filters and drainage layers. The FHWA publication "Geosynthetic Design & Construction Guidelines Reference Manual" (FHWA 2008) and the AASHTO specification "Standard Specifications for Transportation Materials and Methods of Sampling and Testing" (AASHTO 2006) can be used. Table 8.5 shows the geotextile strength property requirements for geotextile filters and drains.

Chapter 8

Table 8.5   Geotextile minimum strength property requirements for geotextile filters and drains (after AASHTO 2006).

Strength properties	Test methods	Woven geotextile, elongation <50% N (lb)	Nonwoven geogextile, elongation >50% N (lb)
Grab strength	ASTM D 4632	1100 (250)	700 (157)
Sewn seam strength	ASTM D 4632	990 (220)	630 (140)
Tear strength	ASTM D 4533	400 (90)*	250 (56)
Puncture strength	ASTM D 6241	2200 (495)	1375 (309)

*If using monofilament geotextile, 250 N (or 56 lb) is used.

Sample Problem 8.3: Geotextile filter design

A geotextile filter is to be selected to protect the stone aggregate drain behind a cantilever retaining wall. The wall stem is 7.5 m high, retaining a sandy soil with $k = 2.5 \times 10^{-2}$ cm/s. The grain size distribution of the backfill is shown in Figure 8.21.

Design the geotextile filter and specify (1) the apparent opening size, (2) permeability, and permittivity, and (3) survivability.

Solution:

1. Soil retention criteria

   Assume the flow is steady state in stable soils, on the basis of Equation (8.42):

   $$AOS \leq B \cdot D_{85}$$

   From Figure 8.21, the backfill soil retained on the No. 200 sieve (0.075 mm) > 50%, so the soil is coarse material. Find $D_{10} = 0.20$ mm, $D_{30} = 0.42$ mm, $D_{60} = 0.70$ mm, $D_{85} = 1.4$ mm.
   $C_u = D_{60}/D_{10} = 3.5$,
   So: $B = 0.5 C_u = 0.5 \times 3.5 = 1.75$

   $$AOS \leq B \cdot D_{85} = 1.75 \times 1.4 = 2.45 \text{ mm}$$

2. Drainage Criteria

   (a) Permeability criteria:
       It is judged that the filtration behind the retaining wall is a critical application under severe conditions. On the basis of Equation (8.45):

   $$k_{geotextile} \geq 10 k_{soil} = 10 \times (2.5 \times 10^{-2}) \text{ cm/sec} = 0.25 \text{ cm/sec}$$

(b) Permittivity criteria:
    The backfill soil passing the No. 200 sieve (0.075 mm): about 2%, less than 15%, so,

$$\text{Permittivity } \Psi \geq 0.5 \text{ sec}^{-1}$$

(c) Flow criteria
    It is assumed the entire geotextile area is available for flow, so the optional criteria are not checked.

3. Long-term flow compatibility (clogging resistance):
    It is judged that filtration behind the retaining wall is a critical application under severe condition, so geotextile is first selected on the basis of the above specifications (AOS $\leq$ 2.45 mm, $k_{\text{geotextile}} \geq$ 0.25 cm/s, and $\Psi \geq$ 0.5 sec^{-1}); then laboratory filtration tests should be conducted using the selected geotextile and the in situ soil samples and hydraulic conditions, per ASTM D5101, "Measuring the Soil-Geotextile System Clogging Potential by the Gradient Ratio" and the FHWA publication "Geosynthetic Design & Construction Guidelines Reference Manual" (FHWA 2008), so the geotextile's clogging resistance can be evaluated.

4. Survivability and durability criteria:
    Nonwoven geotextile should be used as a filter. On the basis of Table 8.6, the selected geotextile should have the following minimum strength values for construction survivability and durability.

    Grab strength: 700 N
    Sewn seam strength: 630 N
    Tear strength: 250 N
    Puncture strength: 1375 N

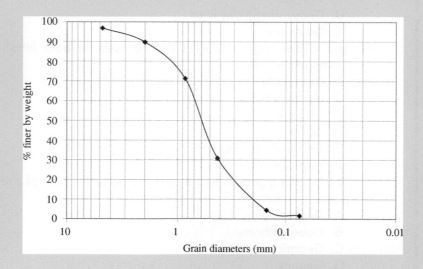

Fig. 8.21   Grain size distribution of the backfill.

Homework Problems

1. What are the major families of geosynthetics and the applications of each type of geosynthetics in civil engineering?
2. What are the polymers used to manufacture geosynthetics? List the types of geosynthetics manufactured from each of the polymers.
3. For each of the following multiple-choice questions, select the correct answers.

    (1) The function of a geonet is primarily:
       A. Reinforcement
       B. Drainage
       C. Filtration
       D. Containment
       E. Separation

    (2) The following geosynthetics can be used for reinforcement purposes:
       A. Geotextile
       B. Geomembrane
       C. Geogrid
       D. Geonet
       E. Geosynthetic clay liner
       F. Geofoam

    (3) The following geosynthetics can be used for containment purposes:
       A. Geotextile
       B. Geomembrane
       C. Geonet
       D. Geosynthetic clay liner
       E. Geofoam

    (4) The following geosynthetics can be used for **both** filtration and drainage purposes:
       A. Geotextile
       B. Geomembrane
       C. Geogrid
       D. Geonet
       E. Geosynthetic clay liner
       F. Geofoam

    (5) The following geosynthetics can be used as the liner of a retention pond:
       A. Geotextile
       B. Geomembrane
       C. Geogrid
       D. Geonet

E. Geosynthetic clay liner

F. Compacted clay liner

(6) The advantages of MSE walls are:

A. Increased internal integrity because of the geosynthetics' tensile strength and the friction between the soil and the reinforcement.

B. Increased shear resistance to resist slope failure.

C. Rapid construction.

D. Flexible wall system can accommodate large differential settlement.

E. Suited for seismic region.

4. Briefly describe the geosynthetics market nowadays.

5. Geotextile filtration is distinguished from drainage function. Describe these two functions and how they are different. Two different parameters (permittivity and transmissivity) are defined. What are the definitions of the two parameters? Which parameter is used for which function?

6. A geotextile is 50 mm in diameter and 0.50 mm in thickness. It is tested as a filter in the lab. The following data are obtained using a constant-head cross-plane flow of water through the geotextile. Calculate the geotextile's average permittivity (s^{-1}) and cross-plane coefficient of permeability, k_n(cm/s).

Δh(mm)	q(cm^3/min)
45	450
91	978
180	1476
275	2530

7. A geotextile was tested as a drain in the lab. The geotextile is 2.0 mm thick, 300 mm wide, and 600 mm long. The following constant-head data for planar flow of water in the longitudinal direction in the geotextile were obtained. Calculate the geotextile's average transmissivity $(m^3/\text{min} - m)$ and the planar coefficient of permeability, k_p(cm/s).

Δh(mm)	q(cm^3/min)
45	450
91	978
180	1476
275	2530

Chapter 8

8. Constant-head data for transmissivity measurements are given under different normal pressures on the geotextile. The geotextile is 4-mm thick, 50-mm wide, and 60-mm long. The tests were conducted at a constant head of 360 mm. Plot the transmissivity ($m^3/min-m$) versus applied normal pressure (kN/m^2) response curve.

Pressure (kPa)	Flow rate, $q(cm^3/min)$
25	1725
50	1240
75	1100

9. A nonwoven, needle-punched geotextile is installed to accommodate cross-plane flow, that is, the flow is normal to the geotextile sheet. The geotextile is 3.0 mm thick, 1000 mm long, and 1000 mm wide. The total hydraulic head loss across the geotextile is 5.0 mm, and the design flow rate is 2000 ml/min. Is the geotextile used as a filter or a drain? Calculate the appropriate hydraulic properties of the geotextile, such as the transmissivity or permittivity, in-plane permeability or cross-plane permeability, whichever is applicable.

10. Design a 5-m high wraparound woven geotextile reinforced soil wall carrying a road consisting of 25-cm stone base ($\gamma = 21.2\,kN/m^3$) and 12.5-cm asphalt ($\gamma = 24.0\,kN/m^3$). The wall is to be backfilled with sand of $\gamma = 18.0\,kN/m^3$, $\phi' = 35°$, and $c' = 0$. The external friction angle between the soil and geotextile is $\delta = 25°$. The geotextile has an ultimate tensile strength of 50 kN/m. Use reduction factors for installation damage, creep, and chemical/biological degradation. The factor of safety for tensile strength is 1.5.
Determine:
(1) How many layers of geotextile are needed and the vertical spacing of each layer.
(2) Geotextile embedment length of each layer.
(3) Geotextile overlap length of each layer.
(4) The external stability in terms of overturning, sliding, and bearing capacity of the MSE wall that you designed in (1)–(3). The foundation soil is ML-CL with $\gamma = 18.8\,kN/m^3$, $\phi = 15°$, $d = 0.9\phi$, $c = 500$ psf, and $c_a = 0.90c$.

11. A mechanically stabilized earth (MSE) wall is needed to retain a 10-m-high earth fill. Geogrid is used as reinforcement. Lab testing shows the ultimate tensile strength of the geogrid is 80 kN/m, and

the external friction angle between the backfill and the geogrid is 30°. The earth fill material will be compacted at 95% of its maximum dry unit weight ($\gamma_{dmax} = 21$ kN/m^3) at the optimum water content. Under typical field conditions, the moisture content of the backfill is 5%. Direct shear tests indicate the effective friction angle of the backfill is 30 degrees, and the effective cohesion is zero. No surcharge is on the backfill. Use reduction factors of 1.5, 2.5, 1.2 for installation damage, creep, and chemical/biological degradation. The foundation soil beneath the MSE wall is silty clay; its moist unit weight in its natural condition is 19 kN/m^3, the internal friction angle is 10 degrees, and the cohesion is 25 kN/m^2. Between the geogrid and the foundation soil, the external friction angle is 9 degrees, and the adhesion is 22 kN/m^2.

Determine:

(1) How many layers of geogrid are needed and the vertical spacing of each layer.

(2) Geotextile embedment length of each layer.

(3) Geotextile overlap length of each layer.

(4) The external stability in terms of overturning, sliding, and bearing capacity of the MSE wall you designed in (1)–(3).

12. An existing roadside embankment needs to be reinforced using geogrid sheets. The embankment is 15-m high and the slope angle is 40°. The in situ soil will be used as the reinforced backfill. The soil strength parameters are $f = 22^0$ and $c = 15$ kN/m^2 in both the embankment and foundation sections. The unit weight is 17.3 kN/m^3. The critical failure surface is assumed to be a toe circle, and the center is located at (+3 m, +18 m) with respect to the toe at (0,0). Assume the groundwater table is far below the foundation. The ultimate tensile strength of the geogrid is 75 kN/m. The vertical spacing is chosen to be 40 cm. For the critical failure circle, how many layers of geogrid are required to raise the factor of safety to 1.4?

13. Geosynthetic reinforced slope design. A 1-km long, 5-m high, 2.5H:1V side-slope road embankment is to be widened to add one more lane. At least a 6-m-wide extension is required to allow the additional lane plus shoulder improvements. A 1H:1V RSS from the toe of the existing slope will provide a 7.5-m width to the alignment. The following design parameters have been provided for the first three design steps.

Step 1: Establish the geometric, loading, and performance requirements for design.

  1. Geometric and loading requirements:
     ◦ Slope height, $H = 5$m
     ◦ Slope angle, $\theta = 45°$

- Surcharge load, $q = 10$ kPa (for pavement)
- Temporary live load, $\Delta q = 0$
- Design seismic acceleration, $A_m = 0$

2. Performance requirements:
   - External stability
     - Sliding: FS ≥ 1.3
     - Deep-seated (overall stability): FS ≥ 1.3
     - Local bearing failure: FS ≥ 1.3
     - Dynamic loading: no requirement
   - Compound failure: FS ≥ 1.3
   - Internal slope stability: FS ≥ 1.3

Step 2: Determine the engineering properties of the in situ soils.

- Subsurface exploration revealed that the slope and the foundation contain stiff and low-plasticity silty clay with traces of sand.
- Strength parameters of the slope:
  -
  $$c_u = 100 \text{ kPa}, \phi' = 28°, c' = 0$$
  -
  $$\gamma_{dry} = 19 \text{ kN/m}^3, w_{opt} = 15\%$$
- The groundwater table is 45 m below the toe level

Step 3 Determine the properties of reinforced fill and the retained natural soil.

- Gradation and plasticity index.
- Compaction results: $\gamma_{dry} = 21 \text{ kN/m}^3$, $w_{opt} = 15\%$
- $\phi' = 33°$, $cI = 0Z$
- Soil pH $= 7.5$

Follow the remaining steps (4 ~ 7) in this book to design the reinforcements and check the external stability.

14. A geotextile is preferred as the filter that wraps around a perforated pipe for a highway trench drain. The depth of the trench drain is 1.0 m. Along the highway, three types of soil are encountered. The sieve analyses of the soils are listed in the following table. The permeability of the soils are measured as follows: $k_{\text{soil A}} = 3.0 \times 10^{-2}$ cm/sec, $k_{\text{soil B}} = 3.6 \times 10^{-3}$ cm/sec, $k_{\text{soil C}} = 2.0 \times 10^{-2}$ cm/sec. Select a geotextile filter that can be used as the filter along the entire highway section.

Sieve	Percent passing by mass (%)		
number	Soil A	Soil B	Soil C
4	95	100	100
12	90	96	100
20	78	86	93
40	55	74	70
100	10	40	11
200	1	15	0

15. Geotextile filter design (Figure 8.22)

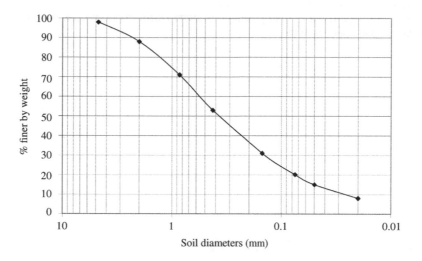

Fig. 8.22   Grain size distribution in problem 8.15.

A geotextile is used as a filter behind a geonet drain in a reinforced slope. The grain size distribution of the retained natural soil is shown in Figure 8.22. The soil's permeability is measured to be 4.8×10^{-4} cm/sec. Select a geotextile that meets the soil retention, drainage, clogging resistance, and survivability criteria and its type (woven or nonwoven).

Chapter 8

References

AASHTO (2006). Standard Specifications for Geotextiles-M 288, *Standard Specifications for Transportation Materials and Methods of Sampling and Testing*, 26th Edition, American Association of State Transportation and Highway Officials, Washington, DC.

Federal Highway Administration (FHWA), U.S. Department of Transportation (2001). *Mechanically Stabilized Earth Walls and Reinforced Soil Slopes Design & Construction Guidelines*. Publication No. FHWA-NHI-00-043. National Highway Institute, Office of Bridge Technology, Washington, DC. March 2001.

Federal Highway Administration (FHWA), U.S. Department of Transportation (2008). *Geosynthetic Design and Construction Guidelines Reference Manual*. Publication No. FHWA-NHI-07-092, by Holtz, R.D., Christopher, B.R., and Berg, R.R. National Highway Institute, Washington, DC.

Holtz, R.D. (2001). Geosynthetics for Soil Reinforcement. The Ninth Spencer J. Buchanan Lecture, College Station, TX. Nov 9 2001.

Koerner, R.M. (2005). *Designing with Geosynthetics*. 5th Edition. Pearson Prentice Hall, Upper Saddle River, NJ.

Leshchinsky, D. (2002). *Stability of Geosynthetic Reinforced Soil Structures*. ADAMA Engineering, Inc., Newark, DE, USA. www.GeoPrograms.com.

Silvestri, V. (1983). "The Bearing Capacity of Dikes and Fills Founded on Soft Soils of Limited Thickness." *Canadian Geotechnical Journal*, Vol. 20 No. 3, 428–436.

South Carolina Department of Transportation (SCDOT) (2010). "SCDOT Geotechnical Design Manual." Appendix D, *Reinforced Soil Slopes*. SCDOT.

Chapter 9

Introduction to Geotechnical Earthquake Design

9.1 Basic seismology and earthquake characteristics

9.1.1 Seismic faults and earthquake terminology

An earthquake is caused by a sudden release of energy from a geologic fault rupture in the earth's crust, and the energy propagates through the earth's body to the ground surface, where damages to civil infrastructure can occur.

A geologic fault is a fracture in the earth's crust. Along the fracture, two tectonic plates move relatively to each other. There are different types of faults, depending on their relative movements, or slip. They are listed in Table 9.1 and illustrated in Figure 9.1.

A well-known fault is the San Andreas Fault (Figure 9.2), which is between two tectonic plates: the North American Plate and the Pacific Plate. The San Andreas Fault is a right lateral fault, as the North American Plate moves south and the Pacific Plate moves north, at approximately 33–37 mm per year. The fault is about 1,300 km long, 15–20 km deep, and the width varies from a few meters to over a thousand meters. Large earthquakes caused by the San Andreas Fault are the 1906 San Francisco earthquake (estimated magnitude 7.8) and the 1989 Loma Prieta earthquake (magnitude 6.9).

As two tectonic plates slowly move in opposite directions along a fault, the earth's crust along the fault is subjected to shear stress and incurs elastic deformation, and a significant amount of strain energy is accumulated. As illustrated in Figure 9.3(b), the strain energy per unit volume of material is defined as the area beneath the stress-strain curve. When the shear stress exceeds the shear resistance, the earth's crust ruptures and the accumulated strain energy is released (Figures 9.3(a) and 9.4).

The location where a fault rupture first occurs is called the focus. The epicenter is the point vertically above the focus and on the ground surface. The focal depth is the vertical distance from the focus to the ground surface. The epicentric distance is the horizontal distance between the epicenter and a given site on the ground surface. The hypocentric distance is the distance between the focus and a given site on the ground surface.

9.1.2 Seismic waves

When energy is released from the focus, it propagates in the body of the earth in different forms of seismic waves. There are two major types of seismic waves: body waves and surface waves.

Geotechnical Engineering Design, First Edition. Ming Xiao.
© 2015 John Wiley & Sons, Ltd. Published 2015 by John Wiley & Sons, Ltd.
Companion Website: www.wiley.com/go/Xiao

Table 9.1   Geologic fault types.

Major types	Subtypes	Description
Dip-slip fault	Normal fault	The movement is vertical, the hanging wall moves down relative to the foot wall.
	Reverse fault (thrust fault)	The movement is vertical, the hanging wall moves up relative to the foot wall.
Strike-slip fault (transform fault)	Right lateral fault	The movement is horizontal (or lateral), each block moves to the right of the opposite block.
	Left lateral fault	The movement is horizontal (or lateral), each block moves to the left of the opposite block.
Oblique fault	Left-lateral normal fault	The movement is a combination of normal fault movement and left-lateral fault movement.
	Left-lateral reverse fault	The movement is a combination of reverse fault movement and left-lateral fault movement.
Oblique fault with rotational motion		The movement is a combination of dip-slip movement, strike-slip movement, and a rotational movement of one block.

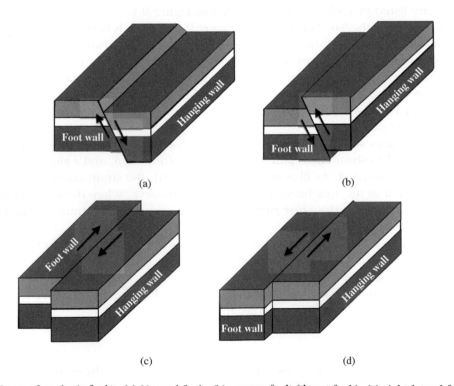

Fig. 9.1   Types of geologic faults. (a) Normal fault, (b) reverse fault (thrust fault), (c) right lateral fault, (d) left lateral fault.

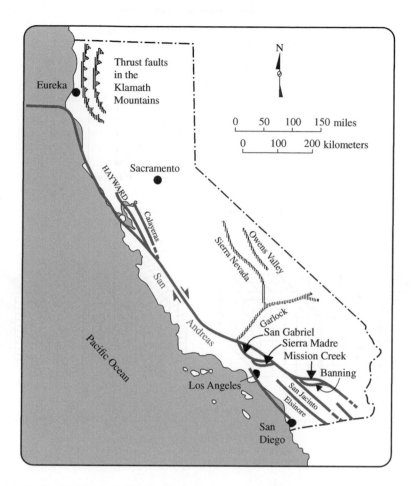

Fig. 9.2 San Andreas Fault. (Photo courtesy of California Geological Survey.)

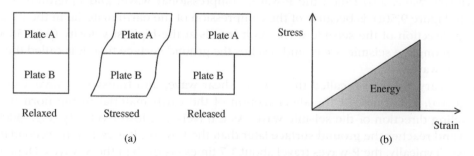

Fig. 9.3 Earthquake energy accumulation and release. (a) Energy accumulation and release, (b) strain energy stored following elastic deformation.

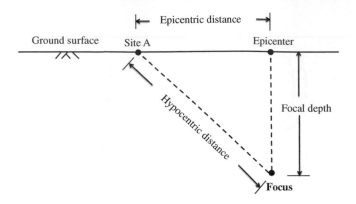

Fig. 9.4   Earthquake terminology.

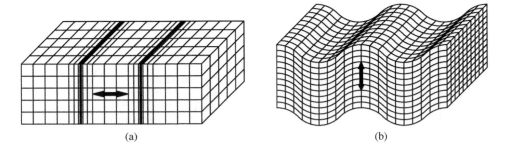

Fig. 9.5   Seismic body waves. (a) Compression wave (P-wave), (b) shear wave (S-wave).

Body waves

Body waves are the seismic waves that travel inside the earth's interior. The body waves have the primary wave and the secondary wave.

- The primary wave, also called the P-wave, compressional wave, and longitudinal wave, as shown in Figure 9.5(a), is because of the compression of the earth material in the longitudinal traveling direction of the seismic wave. As it travels at the highest velocity (e.g., 5800 m/s in granite) among the seismic waves and reaches the ground surface first, it is called the primary wave, or P-wave.
- The secondary wave, also called the S-wave, shear wave, and transverse wave, as shown in Figure 9.5(b), is because of the shear motion of the earth material in the normal plane to the traveling direction of the seismic wave. As it travels at a lower velocity (e.g., 3000 m/s in granite) and reaches the ground surface later than the P-wave, it is called the secondary wave, or S-wave. Typically, the P-waves travel about 1.7 times faster than the S-waves. Depending on the direction of movement of the S-waves in the normal plane to the longitudinal direction, the S-waves are divided into SV wave (vertical plane movement) and SH-wave (horizontal plane movement).

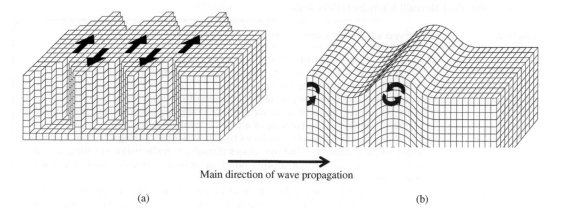

Main direction of wave propagation

(a)   (b)

Fig. 9.6   Seismic surface waves. (a) Love wave, (b) Rayleigh wave.

Surface waves

When body waves reach the ground surface and interact with the surficial layers, they become surface waves. Surface waves include the Love wave and the Rayleigh wave, as shown in Figure 9.6.

- Love waves (L waves) were named after their discoverer, British mathematician Augustus Love, in 1911. They are a type of seismic surface wave in which particles move with a side-to-side motion perpendicular to the main direction of wave propagation. Love waves are caused by the P-wave and the SH-wave. The amplitude of this motion decreases with depth. Love waves cause the rocks they pass through to change in shape.
- Rayleigh waves were named after Lord Rayleigh, an English physicist, who predicted this type of wave in 1885. They are a type of seismic surface wave that moves with a rolling motion that consists of a combination of particle motion perpendicular and parallel to the main direction of wave propagation. Rayleigh waves are caused by the P-wave and the SV-wave. The amplitude of this motion decreases with depth. Like primary waves, Rayleigh waves are alternatingly compressional and extensional and they cause changes in the volume of the rocks they pass through. Rayleigh waves travel slower than Love waves.

9.1.3   Earthquake characteristics

Modified Mercalli intensity (MMI) scale

Classifications of earthquake intensity can be based on the damages and other observed effects on humans, objects of natures, and man-made structures. An intensity scale that is most often used in the United States is the Modified Mercalli intensity (MMI) scale, which was developed by Harry O. Wood and Frank Neumann in 1931, on the basis of the previous versions of the Mercalli scale. The MMI scale contains 12 degrees, as listed in Table 9.2. The lower degrees on the MMI scale generally deal with the manner in which an earthquake is felt by people. The higher numbers on the scale are based on observed structural damage.

Table 9.2 Modified Mercalli intensity (MMI) scale.

Intensity	Observed effects of earthquake
I. Instrumental	Generally not felt by people except in special conditions.
II. Weak	Felt only by a few people at rest, especially on the upper floors of buildings. Delicately suspended objects may swing.
III. Slight	Felt quite noticeably by people indoors, especially on the upper floors of buildings. Many do not recognize it as an earthquake. Standing vehicles may rock slightly. Vibration similar to the passing of a truck. Duration estimated.
IV. Moderate	Felt indoors by many people, outdoors by a few people during the day. At night, some awaken. Dishes, windows, doors disturbed; walls make cracking sound. Sensation like heavy truck striking building. Standing vehicles rock noticeably. Dishes and windows rattle alarmingly.
V. Rather Strong	Felt inside by most, may not be felt by some outside in certain conditions. Dishes and windows may break, and large bells will ring. Vibrations like a large train passing nearby.
VI. Strong	Felt by all; many are frightened and run outdoors, walk unsteadily. Windows, dishes, glassware broken; books fall off shelves; some heavy furniture moved or overturned; a few instances of fallen plaster. Damage slight.
VII. Very Strong	Difficult to stand; furniture broken; damage negligible in building of good design and construction; slight to moderate in well-built ordinary structures; considerable damage in poorly built or badly designed structures; some chimneys broken. Noticed by people driving vehicles.
VIII. Destructive	Damage slight in specially designed structures; considerable in ordinary substantial buildings with partial collapse. Damage great in poorly built structures. Fall of chimneys, factory stacks, columns, monuments, walls. Heavy furniture moved.
IX. Violent	General panic; damage considerable in specially designed structures, well-designed frame structures thrown out of plumb. Damage great in substantial buildings, with partial collapse. Buildings shifted off foundations.
X. Intense	Some well-built wooden structures destroyed; most masonry and frame structures destroyed with foundation. Rails bent slightly. Large landslides.
XI. Extreme	Few, if any, masonry structures remain standing. Bridges destroyed. Rails bent greatly. Numerous landslides, cracks and deformation of the ground.
XII. Catastrophic	Total destruction – Everything is destroyed. Lines of sight and level distorted. Objects are thrown into the air. The ground moves in waves or ripples. Large amounts of rock move position. Landscape altered or leveled by several meters. In some cases, even the routes of rivers changed.

Richter magnitude scale

In 1935, Charles Richter and Beno Gutenberg from California Institute of Technology (Caltech) developed the local magnitude (M_L) scale (popularly known as the Richter scale) with the goal of quantifying medium-sized earthquakes (between magnitude 3.0 and 7.0) in Southern California. Their original formula is:

$$M_L = \log_{10}\left[\frac{A}{A_0(\delta)}\right] \tag{9.1}$$

where: A is the maximum amplitude measured by the Wood-Anderson seismometer, and $A_0(\delta)$ is an empirical function of the epicentral distance (δ) of the station. In practice, readings from all observing stations are averaged after adjustment with station-specific corrections to obtain the M_L value.

Because of the logarithmic basis of the scale, each whole number increase in magnitude represents a 10-fold increase in measured amplitude; in terms of energy, each whole number increase corresponds to an increase of about 31.6 times the amount of energy released. This scale was based on the ground motion measured by a particular type of seismometer at a distance of 100 km (62 mi) from the earthquake's epicenter. Therefore, there is an upper limit on the highest measurable magnitude, and all large earthquakes will tend to have a local magnitude of around 7. The magnitude becomes unreliable for measurements taken at an epicentral distance of more than about 600 km (370 mi) from the epicenter (USGS 2010).

Moment magnitude scale (MMS)

The moment magnitude (M_w) scale was introduced in 1979 by Caltech seismologists Thomas C. Hanks and Hiroo Kanamori to address the shortcomings in the Richter scale while maintaining consistency. Thus, for medium-sized earthquakes, the moment magnitude values should be similar to the Richter values. For example, a magnitude 5.0 earthquake is about 5.0 on both scales. This scale was based on the physical properties of earthquakes, specifically the seismic moment. Unlike other scales, the moment magnitude scale does not saturate at the upper end, and there is no upper limit to the possible measurable magnitudes. However, this has the side effect that the scales diverge for smaller earthquakes. The MMS is now the scale used to estimate magnitudes for all modern large earthquakes by the United States Geological Survey (USGS 2010).

The formula for the dimensionless moment magnitude scale is:

$$M_w = \frac{2}{3}(\log_{10} M_0 - 9.1) \tag{9.2}$$

where the subscript w means mechanical work accomplished, and M_0 is the magnitude of the seismic moment in [N·m]. The constant values in the equation are chosen to achieve consistency with the magnitude values produced by the earlier Richter scales, including the local magnitude scale and the surface wave magnitude scale.

Table 9.3 Significant recent earthquakes.

Year	Earthquake	Magnitude	PGA (g)	Fatalities
1960	Alaska earthquake, USA	9.2	0.18	143
1971	San Fernando earthquake, USA	6.6	1.25	65
1976	Tangshan earthquake, China	7.8	N/A	242,769
1989	Loma Prieta earthquake, USA	6.9	0.65	63
1992	Cape Mendocino earthquake, USA	7.2	1.85	0
1994	Northridge earthquake, USA	6.7	1.7	57
1995	Kobe earthquake, Japan	6.8	0.8	6,434
1999	Jiji earthquake, Taiwan	7.3	1.01	2,415
2008	Wenchuan earthquake, China	7.9	0.98	69,195 (18,392 missing)
2010	Chile earthquake, Chile	8.8	0.78	521
2010	Haiti earthquake	7.0	0.5	92,000–316,000
2011	Tōhoku earthquake and tsunami, Japan	9.0	2.7	15,861

Chapter 9

Ground motion characteristics

- *Peak ground acceleration*

  The peak (maximum) ground acceleration (PGA) is a measure of earthquake acceleration of the ground by a seismometer. PGA is the most commonly used type of ground acceleration in engineering applications and is used to set building codes and assess hazard risks. In an earthquake, damage to buildings and infrastructure is related more closely to ground motion than the magnitude of the earthquake. PGA is commonly expressed in terms of "g's," gravitational acceleration. Some significant earthquakes in recent history are listed in Table 9.3.

- *Peak vertical and horizontal acceleration*

  The peak horizontal acceleration (PHA) is the largest absolute value of horizontal acceleration. It is derived by adding the two horizontal, orthogonal acceleration components measured by an accelerogram. PHA is directly related to the inertial forces, the largest dynamic forces induced in a certain earthquake. Therefore, many seismic design codes, for example, the International Building Code, are generally based on the PHA alone.

  The peak vertical acceleration (PVA), despite occurring in almost all earthquakes, has received less consideration in earthquake engineering design, primarily because of the assumption that structures with horizontal seismic resistance automatically have adequate vertical seismic resistance. PVA is generally taken 1/3–2/3 of the PHA.

- *Amplitude, period, and frequency*

  In a seismic wave as shown in Figure 9.7, the amplitude is the peak displacement; the period is the duration of one complete oscillation; and the frequency is the number of oscillations per second and is the reciprocal of the period. The earth's surficial materials (rock and soil) also have their own period and frequency, known as the natural period and natural frequency, respectively. Earthquakes generally cover a wide range of frequency from 0.1 to 25 Hz. Site (soil) period is a significant factor contributing to structural damage. When a site has a natural frequency of vibration that is equal to the predominant earthquake frequency, site movement can be greatly magnified. This is known as resonance. Thus, buildings can experience much greater ground motions than would be predicted on the basis of only the seismic energy released.

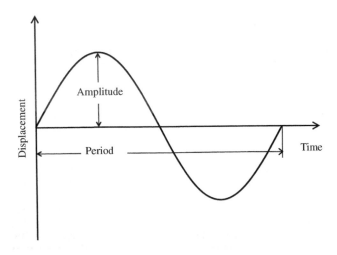

Fig. 9.7   Sinusoidal seismic wave.

9.2   Dynamic Earth pressures

9.2.1   Dynamic active earth pressure

Okabe (1926) and Mononobe and Hatsuo (1929) developed a pseudostatic method to determine the dynamic active earth force behind an earth retaining structure. The method is popularly known as Mononobe–Okabe method. It is widely accepted in existing design codes. The method considers the horizontal and vertical seismic forces that act at the centroid of the failure wedge. Then the seismic forces are used in the static force equilibrium, that is, pseudostatic method. Figure 9.8(a) shows the forces acting on a failure wedge behind a retaining wall, including the active earth force under earthquake loading, P_{AE}, weight W, horizontal seismic force $k_h W$, vertical seismic force $k_h W$, and the reaction force from the soil beneath the failure wedge F. Figure 9.8(b) shows the force equilibrium. The Mononobe–Okabe method is based on the static Coulomb theory. The total dynamic active earth force per unit length of a wall is:

$$P_{AE} = \frac{1}{2}\gamma H^2 (1 - k_v) K_{AE} \tag{9.3}$$

where the dynamic active earth pressure coefficient, K_{AE}, is:

$$K_{AE} = \frac{\cos^2(\phi - \theta - \beta)}{\cos\beta \cos^2\theta \cos(\delta + \theta + \beta)\left[1 + \sqrt{\dfrac{\sin(\delta + \phi)\sin(\phi - \alpha - \beta)}{\cos(\delta + \theta + \beta)\cos(\alpha - \theta)}}\right]^2} \tag{9.4}$$

where:

α = inclination angle of the backfill surface,
θ = inclination angle of the back of the wall with respect to the vertical direction,
δ = external friction angle between the soil and the wall; if partial factors of safety are used according to Eurocode EN 1998-5:2004, a design value should be used, that is,

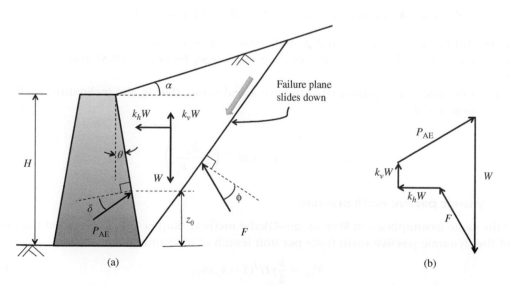

Fig. 9.8 Dynamic active earth force in the Mononobe–Okabe method. (a) Forces acting on the failure wedge, (b) force polygon illustrating equilibrium.

$$\delta'_d = \tan^{-1}\left(\frac{\tan \delta'_k}{\gamma_{\phi'}}\right) \tag{9.5}$$

ϕ = internal friction angle of the soil; if partial factors of safety are used according to EN 1998-5:2004, a design value should be used, that is,

$$\phi'_d = \frac{\phi'_k}{\gamma_{\phi'}} \tag{9.6}$$

$$\beta = \tan^{-1}\left[\frac{k_h}{1 - k_v}\right] \tag{9.7}$$

k_h = horizontal acceleration coefficient,
k_v = vertical acceleration coefficient.

It is noted that if $\beta = 0$, the above equation reduces to Equation (7.28), the static Coulomb's active earth pressure coefficient.

To calculate the point of application of P_{AE}, the total dynamic earth force can be separated into the static force, P_A, that can be determined using Coulomb's theory and the earth force caused by earthquake, ΔP_{AE}:

$$P_{AE} = P_A + \Delta P_{AE} \tag{9.8}$$

P_A acts at $H/3$ from the bottom of the wall. Seed and Whitman (1970) recommended that ΔP_{AE} acts at approximately $0.6H$ from the bottom of the wall. Using moment equilibrium, the point of application of P_{AE} from the bottom of the wall is:

$$z_0 = \frac{P_A\dfrac{H}{3} + \Delta P_{AE}(0.6H)}{P_{AE}} \tag{9.9}$$

The Mononobe–Okabe method is based on the following assumptions:

- The backfill is a homogeneous and granular soil, that is, $c = 0$.
- No surcharge is on the backfill. If there is a surcharge, Equations (9.3) and (9.4) no longer apply.
- $\phi - \alpha \geq \beta$ in order for Equation (9.4) to have a real value, that is, the inclination angle of the backfill must satisfy:

$$\alpha \leq \phi - \beta = \phi - \tan^{-1}\left(\frac{k_h}{1 - k_v}\right) \tag{9.10}$$

9.2.2   Dynamic passive earth pressure

Using the same assumptions in Mononobe–Okabe method, Kapila (1962) provided the calculation of the dynamic passive earth force per unit length of a retaining wall:

$$P_{PE} = \frac{1}{2}\gamma H^2(1 - k_v)K_{PE} \tag{9.11}$$

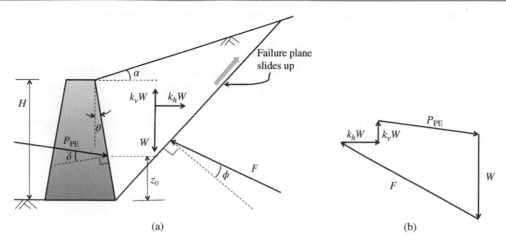

Fig. 9.9   Dynamic passive earth pressure. (a) Forces acting on the failure wedge, (b) force polygon illustrating equilibrium.

where the dynamic passive earth pressure coefficient, K_{PE}, is:

$$K_{PE} = \frac{\cos^2(\phi + \theta - \beta)}{\cos \beta \cos^2 \theta \cos(\delta - \theta + \beta) \left[1 - \sqrt{\dfrac{\sin(\delta + \phi)\sin(\phi + \alpha - \beta)}{\cos(\delta - \theta + \beta)\cos(\alpha - \theta)}}\right]^2} \qquad (9.12)$$

where the symbols retain the same definitions as in the dynamic active earth pressure. The point of application of the total dynamic force can be determined using the same approach as in the dynamic active earth pressure. It is noted that if $\beta = 0$, the above equation reduces to Equation (7.30), the static Coulomb's passive earth pressure coefficient. Figure 9.9 illustrates the forces acting on a failure soil wedge in dynamic passive soil pressure conditions.

To calculate the point of application of P_{PE}, the total dynamic earth force can be separated into a static force, P_P, that can be determined using Coulomb's theory and the earth force caused by an earthquake, ΔP_{PE}:

$$P_{PE} = P_A + \Delta P_{PE} \qquad (9.13)$$

P_P acts at $H/3$ from the bottom of the wall, and it is assumed that ΔP_{PE} acts at approximately $0.6H$ from the bottom of the wall. It is important to note that the dynamic component, ΔP_{PE}, acts in the opposite direction of the static component, P_P, as P_P points toward the wall, and ΔP_{PE} points away from the wall. Using moment equilibrium, the point of application of P_{PE} from the bottom of the wall is:

$$z_0 = \frac{P_P \dfrac{H}{3} + \Delta P_{PE}(0.6H)}{P_{PE}} \qquad (9.14)$$

Sample Problem 9.1: Dynamic active earth pressure

A soil-retaining wall is shown in Figure 9.10. Under the design earthquake, the horizontal acceleration coefficient is 0.2, and the vertical acceleration coefficient is 0.1. The external friction angle between the granular backfill and the wall is 20°. Calculate the total dynamic earth thrust (force) under the design earthquake and determine its point of application.

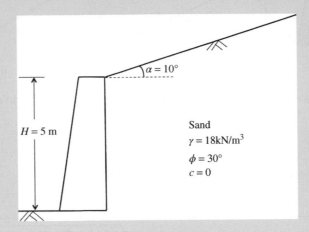

$\alpha = 10°$

$H = 5\ m$

Sand
$\gamma = 18 kN/m^3$
$\phi = 30°$
$c = 0$

Fig. 9.10   Sample problem for dynamic active earth pressure.

Solution:

The given parameters are:

$$\alpha = 10°, \theta = 0, \delta = 20°, \phi = 30°, k_h = 0.2, k_v = 0.1,$$

so: $\beta = \tan^{-1}\left(\frac{k_h}{1-k_v}\right) = \tan^{-1}\left(\frac{0.2}{1-0.1}\right) = 12.5°$

The dynamic active earth pressure coefficient, K_{AE}, is:

$$K_{AE} = \frac{\cos^2(\phi - \theta - \beta)}{\cos\beta\cos^2\theta\cos(\delta + \theta + \beta)\left[1 + \sqrt{\dfrac{\sin(\delta + \phi)\sin(\phi - \alpha - \beta)}{\cos(\delta + \theta + \beta)\cos(\alpha - \theta)}}\right]^2}$$

$$= \frac{\cos^2(30 - 0 - 12.5)}{\cos 12.5 \times \cos^2 0 \times \cos(20 + 0 + 12.5)}$$
$$\left[1 + \sqrt{\dfrac{\sin(20 + 30)\sin(30 - 10 - 12.5)}{\cos(20 + 0 + 12.5)\cos(10 - 0)}}\right]^2$$

$$= 0.609$$

The total dynamic active earth force per unit length of a wall is:

$$P_{AE} = \frac{1}{2}\gamma H^2(1 - k_v)K_{AE} = 0.5 \times 18 \times 5^2 \times (1 - 0.1) \times 0.609 = 123.3\,kN/m$$

To calculate the point of application, z_0, from the bottom of the wall:

$$P_{AE} = P_A + \Delta P_{PE}$$

The active earth force under static condition, P_A, can be determined using the Coulomb theory (Equation 7.27):

$$P_A = \frac{1}{2}\gamma H^2 K_a$$

where Coulomb's active earth pressure coefficient is:

$$K_a = \frac{\cos^2(\phi - \theta)}{\cos^2\theta \cos(\delta + \theta)\left[1 + \sqrt{\dfrac{\sin(\phi + \delta)\sin(\phi - \alpha)}{\cos(\theta + \delta)\cos(\theta - \alpha)}}\right]^2}$$

$$= \frac{\cos^2(30 - 0)}{\cos^2 0 \times \cos(20 + 0)\left[1 + \sqrt{\dfrac{\sin(30 + 20)\sin(30 - 10)}{\cos(0 + 20)\cos(0 - 10)}}\right]^2}$$

$$= 0.340$$

$$P_A = \frac{1}{2}\gamma H^2 K_a = 0.5 \times 18 \times 5^2 \times 0.340 = 76.5\,kN/m$$

So : $\Delta P_{AE} = P_{AE} - P_A = 123.3 - 76.5 = 46.8\,kN/m$

So : $z_0 = \dfrac{P_P\dfrac{H}{3} + \Delta P_{AE}(0.6H)}{P_{AE}} = \dfrac{76.5 \times \dfrac{5}{3} + 46.8 \times 0.6 \times 5}{123.3} = 2.17\,m$

In the limit state design using partial factors of safety, an alternative solution to this problem is available. The same approach is followed, except for the substitution of ϕ and δ with corresponding design values (i.e., ϕ_d and δ_d), while the partial factors of safety are selected according to eurocode EN 1998-5:2004.

Chapter 9

Sample Problem 9.2: Dynamic passive earth pressure

A soil-retaining structure is shown in Figure 9.11. Under the design earthquake, the horizontal acceleration coefficient is 0.2, and the vertical acceleration coefficient is 0.1. The external friction angle between the granular backfill and the wall is 20°. The retaining structure is pushed into the backfill by a horizontal strut at the upper portion of the wall. Calculate the total dynamic earth thrust (force) under the design earthquake and its point of application.

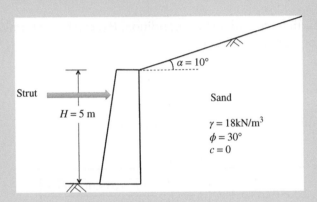

Fig. 9.11   Sample problem for dynamic passive earth pressure.

Solution:

The given parameters are:

$$\alpha = 10°, \theta = 0, \delta = 20°, \phi = 30°, k_h = 0.2, k_v = 0.1,$$

so: $\beta = \tan^{-1}\left(\dfrac{k_h}{1-k_v}\right) = \tan^{-1}\left(\dfrac{0.2}{1-0.1}\right) = 12.5°$

The dynamic passive earth pressure coefficient, K_{PE}, is:

$$K_{PE} = \frac{\cos^2(\phi + \theta - \beta)}{\cos\beta\cos^2\theta\,\cos(\delta - \theta + \beta)\left[1 - \sqrt{\dfrac{\sin(\delta + \phi)\sin(\phi + \alpha - \beta)}{\cos(\delta - \theta + \beta)\cos(\alpha - \theta)}}\right]^2}$$

$$= \frac{\cos^2(30 + 0 - 12.5)}{\cos 12.5 \times \cos^2(20 - 0 + 12.5)\left[1 - \sqrt{\dfrac{\sin(20 + 30)\sin(30 + 10 - 12.5)}{\cos(20 - 0 + 12.5)\cos(10 - 0)}}\right]^2}$$

$$= 9.152$$

The total dynamic passive earth force per unit length of the wall is:

$$P_{PE} = \frac{1}{2}\gamma H^2(1 - K_v)K_{PE} = 0.5 \times 18 \times 5^2 \times (1 - 0.1) \times 9.152 = 1853.3\,\text{kN/m}$$

To calculate the point of application, z_0, from the bottom of the wall:

$$P_{PE} = P_P + \Delta P_{PE}$$

The passive earth force under static conditions, P_P, can be determined using Coulomb's theory (Equation 7.29):

$$P_P = \frac{1}{2}\gamma H^2 K_p$$

where Coulomb's passive earth pressure coefficient is:

$$K_p = \frac{\cos^2(\phi + \theta)}{\cos^2\theta \cos(\delta - \theta)\left[1 - \sqrt{\dfrac{\sin(\phi + \delta)\sin(\phi + \alpha)}{\cos(\theta - \delta)\cos(\theta - \alpha)}}\right]^2}$$

$$= \frac{\cos^2(30 + 0)}{\cos^20 \times \cos(20 - 0)\left[1 - \sqrt{\dfrac{\sin(30 + 20)\sin(30 + 10)}{\cos(0 - 20)\cos(0 - 10)}}\right]^2}$$

$$= 10.903$$

$$P_P = \frac{1}{2}\gamma H^2 K_p = 0.5 \times 18 \times 5^2 \times 10.903 = 2453.2\,\text{kN/m}$$

$$\text{So}: \Delta P_{PE} = P_{PE} - P_P = 1853.3 - 2453.2 = -599.9\,\text{kN/m}$$

$$\text{So}: z_0 = \frac{P_P\dfrac{H}{3} + \Delta P_{PE}(0.6H)}{P_{PE}} = \frac{2453.2 \times \dfrac{5}{3} + (-599.9) \times 0.6 \times 5}{1853.3} = 1.23\,\text{m}$$

In the limit state design using partial factors of safety, an alternative solution to this problem is available. The same approach is followed, except for the substitution of ϕ and δ with corresponding design values (i.e., ϕ_d and δ_d), while the partial factors of safety are selected according to Eurocode EN 1998-5:2004.

9.3 Seismic slope stability

Earthquake-induced slope failure is one of the most damaging seismic hazards. This section introduces three widely used analyses: pseudostatic analysis, Newmark sliding block analysis, and Makdisi–Seed analysis. Evaluation of seismic slope stability should be based on thorough

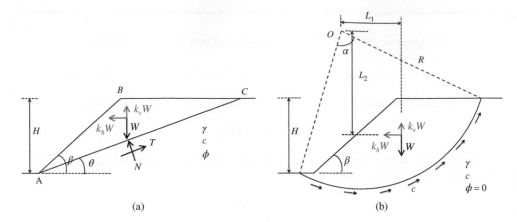

Fig. 9.12 Pseudostatic analyses of seismic slope stability. (a) Plane (translational) failure surface, (b) circular (rotational) failure surface.

subsurface investigation to obtain detailed geologic, hydrologic, topographic, geometric, and engineering information as input for any analysis method.

9.3.1　Pseudostatic analysis

The pseudostatic analysis is based on the limit equilibrium, that is, the forces or moments on the sliding soil mass are in equilibrium. In the pseudostatic analysis, the dynamic earthquake stresses on the soil mass of a slope are simplified as horizontal and vertical inertial forces, which are assumed to act at the centroid of the sliding mass (Figure 9.12). Then the inertial forces are used in static slope stability analysis methods to determine the seismic factor of safety. In Figure 9.12, k_h is the coefficient of horizontal acceleration; k_v is the coefficient of vertical acceleration.

For the plane or translational failure surface in Figure 9.12(a), the factor of safety for an assumed failure plane is:

$$\text{FS} = \frac{\Sigma(\text{Resisting forces})}{\Sigma(\text{Driving forces})} = \frac{c.L_{\text{AC}} + \left[(W - K_v W)\cos\theta - k_h W \sin\theta\right]\tan\phi}{(W - K_v W)\sin\theta + k_h W \cos\theta} \tag{9.15}$$

where:

k_h = coefficient of horizontal acceleration,
k_v = coefficient of vertical acceleration,
W = weight of the failure wedge ABC, per unit length of the slope, W = area of triangle ABC × γ.
L_{AC} = length of AC,
θ = angle of the assumed failure plane,
c = cohesion of the soil,
ϕ = internal friction angle.

Multiple failure planes may be assumed to obtain the minimum factor of safety for the slope.

For a circular (rotational) failure surface in a saturated clay slope with $\phi = 0$, as shown in Figure 12(b), the factor of safety for an assumed failure plane is:

$$FS = \frac{\Sigma(\text{Resisting moment})}{\Sigma(\text{Driving moment})} = \frac{c(R \cdot \alpha)R}{(W - K_v W)L_1 + k_h W L_2} \tag{9.16}$$

where:

W = weight of the rotational failure mass, per unit length of the slope, W = area of rotational section $\times \gamma$,

a = angle of the arc, in radian,

R = radius of the rotational failure surface,

L_1 and L_2 = arms of the vertical and horizontal forces, respectively, with regard to the rotational center O.

The vertical inertial force because of the vertical acceleration has less influence on the factor of safety than the horizontal inertial force. Therefore, the vertical acceleration is often neglected in the pseudostatic analysis.

A similar approach can be followed according to BS EN 1998-5:2004. However, in this standard, the horizontal (F_H) and vertical (F_V) inertial forces should be calculated as follows:

$$F_H = 0.5 \times \alpha \times S \times W \tag{9.17}$$

$$F_V = \pm 0.5 F_H \ \text{(if ratio } a_{vg}/a_g > 0.6) \tag{9.18}$$

or

$$F_V = \pm 0.33 F_H \ \text{(if ratio } a_{vg}/a_g \leq 0.6) \tag{9.19}$$

where:

α is the ratio of the design ground acceleration on type A ground (i.e. rock), a_g, to the acceleration of gravity g;

a_{vg} is the design ground acceleration in the vertical direction;

a_g is the design ground acceleration for type A ground;

S is the soil parameter of EN 1998-1:2004, 3.2.2.2, a value depending on the response spectra for a given soil type.

W is the weight of the sliding mass.

In addition, a topographic amplification factor might be needed in the calculation of these forces depending on the structure importance (i.e. for hospitals and other critical infrastructure). Consistently, instead of using a global factor of safety (FS), in the limit state design approach, the design resistance must be compared against the design effect of the actions (which include the inertial forces), that is:

$$E_d \leq R_d \tag{9.20}$$

In the case of a rotational failure and following the notations above, it shall be verified that:

$$\left(W - F_V\right) L_1 + F_H L_2 \leq c \left(R \cdot \alpha\right) R \tag{9.21}$$

The *methods of slices* in Chapter 5 can also be used to evaluate soil slopes. In these methods, the horizontal seismic force is assumed to act at the centroid of each slice and contributes to the driving moment. The vertical seismic force can be neglected. Figure 9.13 illustrates a soil slope involving two strata (layer 1 and layer 2). The slope can be divided into n slices. To easily calculate the resisting moment of each slice, the bottom of each slice should rest in only one layer of soil.

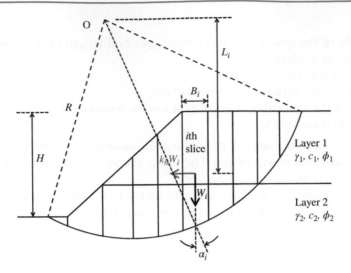

Fig. 9.13 Method of slices considering the seismic forces.

The factor of safety using the "ordinary method of slices" *without* seismic forces is:

$$\text{FS} = \frac{\Sigma\,(\text{Resisting moment})}{\Sigma\,(\text{Driving moment})} = \frac{\sum\limits_{i=1}^{n}\left(c.B_i\sec\alpha_i + W_i\cos\alpha_i\tan\phi\right)}{\sum\limits_{i=1}^{n}\left(W_i\sin\alpha_i\right)} \tag{9.22}$$

where:

c = cohesion of the soil,
ϕ = internal friction angle,
W_i = weight of the ith slice,
i = $1, \ldots, n$, where n is the number of slices,
α_i = angle of each slice as shown in Figure 9.13,
B_i = width of the ith slice.

The factor of safety using the "ordinary method of slices" *with* seismic forces is:

$$\text{FS} = \frac{\Sigma\,(\text{Resisting moment})}{\Sigma\,(\text{Driving moment})} = \frac{\sum\limits_{i=1}^{n}\left(c.B_i\sec\alpha_i + W_i\cos\alpha_i\tan\phi\right)}{\sum\limits_{i=1}^{n}\left[W_i\sin\alpha_i + k_h W_i\left(\dfrac{L_i}{R}\right)\right]} \tag{9.23}$$

where:

L_i = length arm from the horizontal seismic force to the rotation center O for the ith slice.

Alternative solutions using limit state design similar to those explained in Chapter 5 can also be obtained by comparing the numerator and denominator of equation (9.23) and replacing $k_h W_i$ with $F_{\text{H}i}$.

Selection of coefficients of acceleration

The results of pseudostatic analyses are critically dependent on the values of the horizontal acceleration coefficient, also referred to as the pseudostatic coefficient, k_h. If the slope material is rigid, the inertial force because of the horizontal acceleration should be equal to the product of the actual horizontal acceleration and the mass of the unstable material. If the actual horizontal acceleration reaches its maximum, a_{max}, the horizontal inertial force should reach its maximum too. As the actual soil slope is not rigid and the peak horizontal acceleration exists only for a short period, the horizontal accelerations used in the pseudostatic methods are well below a_{max} (Kramer, 1996). Marcuson (1981) and Hynes-Griffin and Franklin (1984) suggested the following relationship for appropriate pseudostatic coefficient, k_h:

$$k_h = 0.5 \frac{a_{max}}{g} \qquad (9.24)$$

Many researchers have indicated that the pseudostatic approach has significant shortcomings: the approach can be unreliable for soils that build up large pore pressures or show more than about 15% weakening of strength because of earthquakes (Seed et al. 1975; Marcuson et al. 1979; Kramer 1996). However, the pseudostatic approach can be used to preliminarily evaluate seismic slope stability.

Chapter 9

Sample Problem 9.3: Pseudostatic analysis of seismic slope stability with assumed plane failure surface.

A natural slope has a slope angle of 45 degrees and a height of 12 meters. The soil has unit weight of $19\,kN/m^3$, cohesion of $34\,kN/m^2$, and internal friction angle of 30 degrees. The slope is subjected to a horizontal acceleration of $0.25\,g$ and a vertical acceleration of $0.1\,g$. What is the factor of safety of the slope on a potential plane failure surface inclined at an angle of 25 degrees with respect to the horizontal?

Solution:

The above problem statement can be illustrated in Figure 9.14.
First calculate the area of the triangle of ABC. Using trigonometry:

$$AB = \frac{BD}{\cos 45°} = 16.97\,m$$

As: $\dfrac{AB}{\sin 25°} = \dfrac{BC}{\sin 20°} = \dfrac{AC}{\sin 135°}$

Find: $BC = 13.76\,m$, $AC = 28.39\,m$

Area $ABC = \dfrac{1}{2} \times 13.76 \times 12 = 82.56\,m^2$

$$W = 82.56 \times 19 = 1568.64\,kN/m$$

Use Equation (9.15):

$$FS = \frac{\Sigma\,(\text{Resisting forces})}{\Sigma\,(\text{Driving forces})} = \frac{c.L_{AC} + \left[(W - K_v W)\cos\theta - k_h W \sin\theta\right]\tan\phi}{(W - K_v W)\sin\theta + k_h W \cos\theta}$$

$$= \frac{34 \times 28.39 + \left[(1568.64 - 0.1 \times 1568.64) \times \cos 25 - 0.25 \times 1568.64 \times \sin 25\right] \times \tan 30}{(1568.64 - 0.1 \times 1568.64) \times \sin 25 + 0.25 \times 1568.64 \times \cos 25}$$

$$= 1.68 > 1.5$$

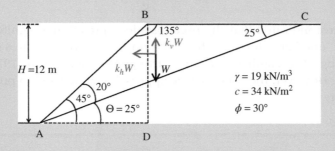

Fig. 9.14   Sample problem of pseudostatic analysis on plane failure surface.

Sample Problem 9.4: Pseudostatic analysis of seismic slope stability with assumed circular failure surface.

A natural slope has a slope angle of 45 degrees and a height of 12 meters. The soil is saturated clay, its unit weight is 19 kN/m³, and cohesion is 34 kN/m². The slope is subjected to a horizontal acceleration of 0.25 g and a vertical acceleration of 0.1 g. A potential sliding surface is shown in Figure 9.15. Determine the factor of safety for the seismic slope stability along the circular failure surface.

Solution:

The area of the sliding section can be found to be 148.60 m², and the location of the centroid is shown in Figure 9.15. The weight of the sliding section per unit length of the slope is:

$$W = \gamma A = 19 \times 148.60 = 2823.40 \text{ kN/m}$$

Assume: $\phi = 0$. Therefore, the factor of safety is:

$$FS = \frac{\Sigma \, (\text{Resisting moment})}{\Sigma \, (\text{Driving moment})} = \frac{c \left(R \cdot \dfrac{\alpha \cdot \pi}{180} \right) R}{(W - k_v W) \, L_1 + k_h W L_2}$$

$$= \frac{34 \times \left(30 \cdot \dfrac{65 \times \pi}{180} \right) \times 30}{(2823.4 - 0.1 \times 2823.4) \times 11.83 + 0.25 \times 2823.4 \times 23.26}$$

$$= 0.75 < 1.0$$

The factor of safety is less than 1.0, and slope failure will occur under the design earthquake.

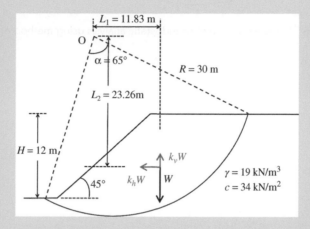

Fig. 9.15 Sample problem of pseudostatic analysis on circular failure surface.

Sample Problem 9.5: Pseudostatic analysis of seismic slope stability using ordinary method of slices.

A natural slope has a slope angle of 45 degrees and a height of 12 meters. The soil has unit weight of 19 kN/m³, cohesion of 34 kN/m², and internal friction angle of 30 degrees. The slope is subjected to a horizontal acceleration of 0.25 g and a vertical acceleration of 0.1 g. A potential sliding surface is shown in Figure 9.16. Using the ordinary method of slices, determine the factor of safety for the seismic slope stability along the circular failure surface.

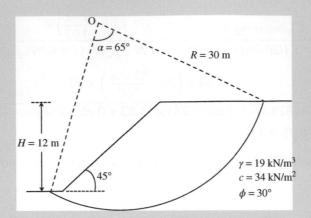

Fig. 9.16   Sample problem of pseudostatic analysis using method of slices.

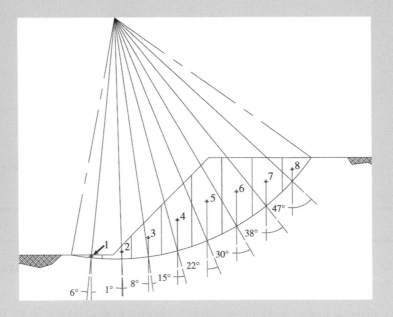

Fig. 9.17   Solution for pseudostatic analysis using method of slices.

Solution:

As shown in Figure 9.17, the assumed failure portion is divided into 8 slices.

The factor of safety using the ordinary method of slices is given by Equation (9.16):

$$FS = \frac{\Sigma \left(\text{Resisting moment} \right)}{\Sigma \left(\text{Driving moment} \right)} = \frac{\sum\limits_{i=1}^{n} \left(c.B_i \sec \alpha_i + W_i \cos \alpha_i \tan \phi \right)}{\sum\limits_{i=1}^{n} \left[W_i \sin \alpha_i + k_h W_i \left(\frac{L_i}{R} \right) \right]}$$

The weight, W_i, angle, α_i, width, B_i, and length of arm from horizontal seismic force to the rotation center O, L_i, for each slice can be obtained from computer aided design (CAD) software and are listed in Table 9.4, along with the calculations.

Table 9.4   Ordinary method of slices calculation approach.

Slice number	Area of each slice (m²)	Weight of each slice (kN/m)	α_i (deg)	B_i (m)	L_i (m)	$cB_i\sec\alpha_i + W_i\cos\alpha_i\tan\phi$ (kN/m)	$W_i\sin\alpha_i + k_hW_i\left(\frac{L_i}{R}\right)$ (kN/m)
1	1.017	19.32	−6	3.781	18.598	138.94	2.26
2	4.644	88.24	−1	3.752	19.033	178.48	18.45
3	16.828	319.73	8	3.779	19.285	310.04	117.90
4	21.167	402.17	15	3.884	18.997	351.84	195.04
5	36.234	688.44	22	4.054	18.074	496.33	406.03
6	31.584	600.09	30	4.344	17.004	427.95	421.52
7	21.830	414.77	38	4.830	15.145	318.11	330.14
8	8.226	156.29	47	5.610	14.683	191.62	141.62
					Σ	2413.32	1632.97

$$FS = \frac{\Sigma \left(\text{Resisting moment} \right)}{\Sigma \left(\text{Driving moment} \right)} = \frac{2413.32}{1632.97} = 1.48 < 1.5$$

9.3.2   Newmark sliding block analysis

The Newmark sliding block analysis provides calculation of the *displacement* of a slope under a certain earthquake. As a slope failure is indicated by the magnitude of displacement, this method gives more direct and useful evaluation of the seismic slope stability. When a slope slides, the sliding mass is analogous to a block sliding along an inclined plane, as shown in Figure 9.18. Newmark (1965) considered the force equilibrium of the sliding block and proposed a method to predict the permanent displacement of a slope subjected to ground acceleration.

Consider the force equilibrium of the block in Figure 9.18. The block is subjected to a horizontal inertial force, $k_h W$. Assuming the resistance to the sliding is only because of friction, and there is no cohesion, the factor of safety for the sliding is:

$$FS = \frac{\text{Resisting force}}{\text{Sliding force}} = \frac{\left(W \cos \beta - k_h W \sin \beta \right) \tan \phi}{W \sin \beta + k_h W \cos \beta} = \frac{\left(\cos \beta - k_h \sin \beta \right) \tan \phi}{\sin \beta + k_h \cos \beta} \qquad (9.25)$$

where ϕ = friction angle, β = inclination angle of the plane, and k_h = horizontal acceleration coefficient.

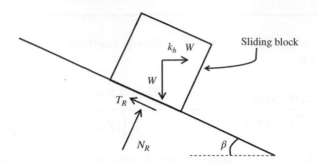

Fig. 9.18   Sliding block on an inclined plane.

The critical condition for sliding is FS = 1.0, which corresponds to a critical horizontal acceleration coefficient called the yield acceleration coefficient. The yield acceleration coefficient, k_y, can be obtained by setting FS = 1.0 in Equation (9.25) and solving for k_y:

$$k_y = \frac{\tan \phi - \tan \beta}{1 + \tan \phi \cdot \tan \beta} = \tan (\phi - \beta) \tag{9.26}$$

Accordingly, the yield acceleration is defined by:

$$a_y = k_y g \tag{9.27}$$

Displacement occurs when the force equilibrium is not satisfied, that is, when the actual acceleration exceeds the yield acceleration. By analyzing the acceleration, velocity, displacement, and duration of four earthquakes, Newmark (1965) proposed a conservative upper-bound permanent displacement, u_{max}:

$$u_{max} = \frac{v^2_{max'}}{2a_y} \cdot \frac{a_{max}}{a_y} \tag{9.28}$$

where:

v_{max} = maximum velocity of ground motion,
a_{max} = peak horizontal ground acceleration.

Sample Problem 9.6: Newmark sliding block analysis

During the 1994 Northridge earthquake, the ground motion was recorded at the City Hall in Santa Monica. The peak acceleration was 866 cm/sec^2 that occurred at 9.8 sec of the earthquake, and the peak velocity of 41.75 cm/sec was recorded at 14.0 seconds of the earthquake. An adjacent earth embankment is 12 meters high, the slope inclination is 1:2 (vertical:horizontal). The soil has unit weight of 19 kN/m^3, cohesion of 34 kN/m^2, and internal friction angle of 38 degrees. Evaluate the permanent displacement of the slope using the Newmark sliding block method.

Solution:

The problem statement gives:

Maximum velocity of ground motion: $v_{max} = 41.75 \, \text{cm/sec}$
Peak horizontal ground acceleration: $a_{max} = 866 \, \text{cm/sec}^2 = 0.883 \, g$.
($g = 9.81 \, \text{m/sec}^2$)

The yield acceleration coefficient can be approximated using Equation (9.26):

$$k_y = \tan(\phi - \beta)$$

where: internal friction angle $\phi = 38°$, and the slope angle $\beta = \tan^{-1}(0.5) = 26.6°$.
So: $k_y = \tan(\phi - \beta) = \tan(38 - 26.6) = 0.202$
So $a_y = 0.202 \, g$
The upper-bound permanent displacement of the slope is:

$$u_{max} = \frac{v^2{}_{max}}{2a_y} \cdot \frac{a_{max}}{a_y} = \frac{(41.75 \, \text{cm/sec})^2}{2 \times 0.202 \times 981 \, \text{cm/sec}^2} \times \frac{0.883 \, g}{0.202 \, g} = 19.2 \, \text{cm}$$

9.3.3 Makdisi–Seed analysis

The Makdisi–Seed analysis is based on the sliding block method for earthen dams and embankments. Makdisi and Seed (1978) provided simple charts to estimate earthquake-induced permanent soil displacement. Figure 9.19 shows the variation of normalized permanent displacement with the normalized yield acceleration.

In the figure:

M = earthquake magnitude in the Richter scale,
U = permanent displacement of an earthen dam or embankment,
a_{max} = peak horizontal ground acceleration,
a_y = yield acceleration of the earthen dam or embankment,
T_0 = natural period of the earthen dam or embankment. It can be estimated by the first natural period, on the basis of the circular frequency, ω_n, given by Gazetas (1982).

$$\omega_n = \frac{\bar{v}_s}{H} \frac{\beta_n}{8} (4 + m)(2 - m) \tag{9.29}$$

where:
$\bar{v}_s$ = average shear wave velocity,
H = height of earthen dam or embankment.

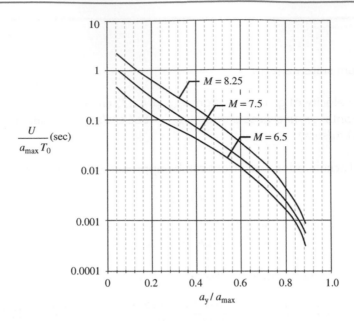

Fig. 9.19   Makdisi–Seed chart to estimate permanent displacement of earthen dams and embankments (After Makdisi and Seed 1978. Reprinted with permission of ASCE.)

For the first natural frequency: $m = 0$, $\beta_n = 2.404$. So:

$$\omega_1 = 2.404\frac{\overline{v}_s}{H} \tag{9.30}$$

$$T_0 = \frac{2\pi}{\omega_1} \tag{9.31}$$

Sample Problem 9.7: Makdisi–Seed analysis of earthquake-induced permanent displacement of an embankment.

Use the same problem statement in Sample Problem 9.6 and an average shear wave velocity of 7800 cm/sec. Evaluate the earthquake-induced permanent displacement of the embankment using the Makdisi–Seed analysis.

Solution:

The earthquake magnitude in the Richter scale is 6.7,
The maximum velocity of ground motion $v_{max} = 41.75$ cm/sec
The peak horizontal ground acceleration $a_{max} = 866$ cm/sec$^2 = 0.883$ g.
The yield acceleration $a_y = 0.202$ g
   The first natural frequency:

$$\omega_1 = 2.404\frac{\overline{v}_s}{H} = 2.404 \times \frac{7800\,\text{cm/sec}}{1200\,\text{cm}} = 15.6\,\text{rad/sec}$$

The natural period of the embankment:

$$T_0 = \frac{2\pi}{15.6} = 0.402 \text{ sec}$$

$$\frac{a_y}{a_{max}} = \frac{0.202\,g}{0.883\,g} = 0.229$$

From Figure 9.19, using $M = 6.5$, find: $U/a_{max}T_0 = 0.09$ sec
So: $U \approx 0.09$ sec $\times$ 866 cm/sec^2 $\times$ 0.402 sec = 31.4 cm

9.4 Liquefaction analysis

9.4.1 Liquefaction hazard

Liquefaction is defined as the transformation of a granular material from a solid to a liquefied state as a consequence of increased pore-water pressure and reduced effective stress (Marcuson 1978). During ground shaking, shrinkage of pore spaces of loose to medium-compact granular soils squeezes the pore water; when the pore water cannot easily drain, the pore-water pressure, u, significantly increases, thus reducing the effective stress, σ':

$$\sigma' = \sigma - u \qquad (9.32)$$

where $\sigma =$ total stress. When effective stress decreases to a certain value, the soil losses grain-to-grain contact and tends to behave like a liquid. Figure 9.20 illustrates the liquefaction phenomenon. Liquefaction may cause the reduction or loss of bearing capacity, large settlement, and horizontal displacement because of lateral spreads of liquefied soils. Liquefaction can

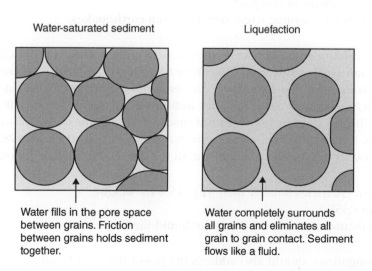

Water-saturated sediment — Water fills in the pore space between grains. Friction between grains holds sediment together.

Liquefaction — Water completely surrounds all grains and eliminates all grain to grain contact. Sediment flows like a fluid.

Fig. 9.20 Saturated sand condition during liquefaction. (Photo courtesy of Dr. Stephen A. Nelson, Tulane University.)

(a)   (b)

Fig. 9.21   Liquefaction in the 2010 Baja California earthquake ($M_w = 7.2$). (a) Sand boil (Photo courtesy of Prof. Scott J. Brandenberg, University of California, Los Angeles.), (b) Land lateral spread because of liquefaction. (Photo courtesy of Timothy P. McCrink, California Geological Survey.)

be exhibited in the forms of sand boils (Figure 9.21(a)) or lateral spread of surficial soils (Figure 9.21(b)). Examples of this type of damage were observed in many earthquakes, such as the 1964 Niigata, the 1964 Alaska, the 1971 San Fernando, the 1985 Mexico City, the 1994 Northridge, the 1994 Kobe, the 1999 Taiwan, the 1999 Turkey, the 2010 Baja California, and the 2011 Tōhoku earthquakes.

The conditions required for liquefaction to occur are:

1. the soil deposit is sandy or silty soil;
2. the soil is saturated or nearly saturated (usually below groundwater table);
3. the soil is loose or medium compact;
4. the soil is subjected to seismic stress (such as from earthquake).

Relatively free-draining soils such as well or poorly graded gravels (GW, GP) are much less likely to liquefy than sand or silty sand (SW, SP, or SM). Dense granular soils are less likely to liquefy than looser soils. Granular soils under higher initial confining effective stress (e.g., deeper soils) are less likely to liquefy. Case histories indicate that liquefaction usually occurs within a depth of 15 m. Cohesive soils are generally not susceptible to liquefaction. To qualitatively evaluate cohesive soils, the so-called Chinese criteria defined by Seed and Idriss (1982) can be used – liquefaction can occur in cohesive soils only if all three of the following conditions are met:

1. the clay content (particles smaller than 5 μ) <15% by weight;
2. the liquid limit <35%;
3. the natural moisture content >0.9 times the liquid limit.

Screening investigations should also address the possibility of a locally perched groundwater table, which may be caused by changes in local or regional water management patterns that can significantly raise the groundwater table.

9.4.2   Evaluations of liquefaction hazard

Liquefaction is commonly evaluated using a factor of safety (Equation 9.33), which is defined as the ratio between the available liquefaction resistance, expressed in terms of the cyclic stresses required to cause liquefaction, and the cyclic stresses generated by the design earthquake. Both of these stress parameters are commonly normalized with respect to the effective overburden stress at the depth in question. They are referred to as cyclic resistance ratio (CRR) and cyclic stress ratio (CSR).

$$FS = \frac{CRR}{CSR} \qquad (9.33)$$

The National Earthquake Hazards Reduction Program (NEHRP) Recommended Provisions for Seismic Regulations for New Buildings and Other Structures (2004) stated that a factor of safety of 1.2–1.5 is usually appropriate for building sites, and the actual FS selected is based on the importance of the structure and the potential for ground displacement. The California Geological Survey (1997) suggested FS greater than 1.3 can be considered an acceptable level of risk; and the DOD handbook on Soil Dynamics and Special Design Aspects (1997) stipulated that FS of 1.2 is appropriate in engineering design.

In calculating the factor of safety, the empirical methods are most widely used in practice. Seed and Idriss (1971) first developed and published the "simplified procedure" for evaluating liquefaction resistance. This empirical procedure has evolved and become a standard of practice in North America and much of the world. In 1996, a workshop sponsored by the National Center for Earthquake Engineering Research (NCEER) was convened by Professors T.L. Youd and I.M. Idriss with 20 experts to update the simplified procedure and incorporate the research findings from the previous decade. The NCEER workshop resulted in another milestone report (NCEER, 1997; Youd et al., 2001), which is the most updated liquefaction evaluation reference to date. The following evaluations of CSR and CRR are based on this report.

9.4.3   Evaluation of CSR

Seed and Idriss (1971) formulated the following equation for the calculation of the cyclic stress ratio (CSR), and this equation is still the most widely used empirical method:

$$CSR = \frac{\tau_{av}}{\sigma'_{vo}} = 0.65 \frac{a_{max}}{g} \cdot \frac{\sigma_{vo}}{\sigma'_{vo}} \cdot r_d \qquad (9.34)$$

where:

τ_{av} = average cyclic shear stress induced by design ground motion,

σ'_{vo} = initial vertical effective stress at the depth under consideration in static condition,

σ_{vo} = initial vertical total stress at the depth under consideration in static condition,

a_{max} = peak horizontal acceleration at the ground surface generated by the earthquake,

r_d = stress reduction coefficient. The NCEER workshop (1997) recommended the following equations by Liao and Whitman (1986a) for routine practice and noncritical projects:

$$r_d = 1.0 - 0.00765z \quad \text{for } z \leq 9.15 \, \text{m} \qquad (9.35a)$$

$$r_d = 1.174 - 0.0267z \quad \text{for } 9.15 \, \text{m} < z \leq 23 \, \text{m} \qquad (9.35b)$$

The NCEER workshop (1997) cautioned users that there is considerable variability in r_d, and the r_d calculated from Equation (9.35) is the mean of a wide range of possible r_d, and the range of r_d increases with depth.

Chapter 9

9.4.4   Evaluation of CRR

Empirical methods for the evaluation of the CRR commonly employ the following four field tests: the standard penetration test (SPT), the cone penetration test (CPT), shear wave velocity measurements, and the Becker penetration test (BPT). The SPT and CPT methods are generally preferred because of the more extensive database and past experience, but the other tests may be applied at sites underlain by gravelly sediment or where access by large equipment is limited. Comparisons of the SPT and the CPT are summarized in Chapter 2. In this section, only the SPT and CPT methods for evaluating the CRR are explained.

SPT method

The CRR is graphically determined from the SPT blow count as shown in Figure 9.22. This CRR curve – the SPT clean-sand base curve – is for fines content $\leq 5\%$ under magnitude 7.5 earthquakes. The SPT blow count is first corrected to consider overburden stress, equipment used to conduct the SPT, and the fines content ($>5\%$); then Figure 9.22 is used to derive the CRR, which in turn is corrected for other earthquake magnitudes.

Step 1:   Corrections to overburden stress and various SPT equipment

To account for the effect of overburden stress and various equipment used for SPT, the following equation is used:

$$(N_1)_{60} = N_m C_N C_E C_B C_R C_S \tag{9.36}$$

where:

N_m = measured standard penetration blow count,
C_N = correction factor based on the effective overburden stress, σ'_{vo}, it can be evaluated using either of the following two equations:

$$C_N = \left(\frac{P_a}{\sigma'_{vo}} \right)^{0.5} \quad \text{(Liao and Whitman, 1986b)} \tag{9.37}$$

$$C_N = \frac{2.2}{1.2 + \dfrac{\sigma'_{vo}}{P_a}} \quad \text{(Seed and Idriss, 1982)} \tag{9.38}$$

where:

σ'_{vo} = initial vertical effective stress at the depth under   static condition, consideration in
P_a = 1 atm, or approximately 101 kPa, or 2116 lb/ft^2.
C_E = correction for SPT hammer energy ratio (ER). An energy efficiency of 60% is usually accepted as the approximate average for the US SPT equipment. C_E is calculated using:

$$C_E = \frac{\text{ER}}{60\%} \tag{9.39}$$

where: ER is the actual energy ratio (in percentage) based on different hammers; it can be measured during the SPT.

C_B = correction factor for borehole diameter,
C_R = correction factor for SPT rod length,
C_S = correction factor for samplers with or without liners.

C_N should not exceed 1.7. The variations of C_N with σ'_{vo} for Equations (9.37) and (9.38) are plotted in Figure 9.23. The NCEER workshop (1997) concluded that Equation (9.38) provides a better fit for overburden pressure up to 300 kPa. For pressures >300 kPa, C_N should be estimated by other methods.

The ranges of the correction factors are summarized in Table 9.5, which was originally suggested by Skempton (1986) and updated by Robertson and Wride (1998).

Step 2: *Corrections to fines content*

The corrected $(N_1)_{60}$ in Equation (9.36) is further corrected for the fines content (FC) in the soil. The following equations were developed by I.M. Idriss with the assistance of R.B. Seed (Youd et al. 2001):

$$(N_1)_{60CS} = \alpha + \beta(N_1)_{60} \tag{9.40}$$

where:

$(N_1)_{60CS}$ = the $(N_1)_{60}$ for equivalent clean sand; $(N1)_{60CS}$ is used in Figure 9.22 to find the CRR under magnitude 7.5 earthquakes,
$(N_1)_{60}$ = corrected SPT blow count calculated in Equation (9.36),
α and β = coefficients determined from the following relationships:

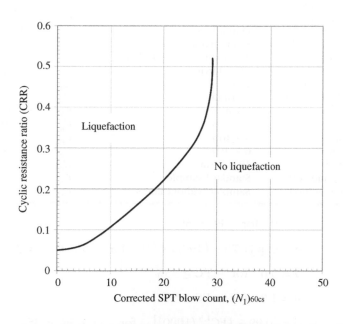

Fig. 9.22   SPT clean-sand (percent fines ≤ 5%) base curve for magnitude 7.5 earthquakes. (Reproduced from Youd et al. 2001, Figure 2 in the paper. Used with permission of ASCE.)

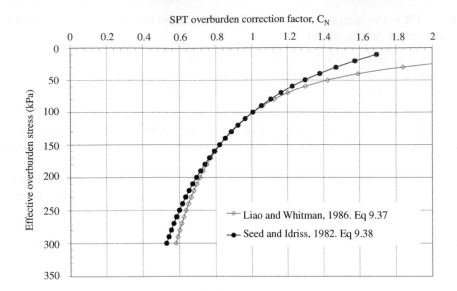

Fig. 9.23   Comparison of Equations (9.37) and (9.38).

Table 9.5   Correction factors for SPT $(N_1)_{60}$ (reproduced from Youd et al. 2001, Figure 2 in the paper. Used with permission of ASCE).

Factor	Equipment variable	Term	Correction
Overburden pressure	–	C_N	$(P_a/\sigma'_{vo})^{0.5}$
Overburden pressure	–	C_N	$C_N \leq 1.7$
Energy ratio	Donut hammer	C_E	0.5 – 1.0
Energy ratio	Safety hammer	C_E	0.7 – 1.2
Energy ratio	Automatic trip donut hammer	C_E	0.8 – 1.3
Borehole diameter	65 – 115 mm	C_B	1.0
Borehole diameter	150 mm	C_B	1.05
Borehole diameter	200 mm	C_B	1.15
Rod length	<3 m	C_R	0.75
Rod length	3 – 4 m	C_R	0.8
Rod length	4 – 6 m	C_R	0.85
Rod length	6 – 10 m	C_R	0.95
Rod length	10 – 30 m	C_R	1.0
Sampling method	Standard sampler	C_S	1.0
Sampling method	Sampler without liners	C_S	1.1 – 1.3

$$\alpha = 0 \quad \text{for} \quad FC \leq 5\% \tag{9.41a}$$

$$\alpha = \exp\left[1.76 - \left(190/FC^2\right)\right] \quad \text{for} \quad 5\% < FC < 35\% \tag{9.41b}$$

$$\alpha = 5.0 \quad \text{for} \quad FC \geq 35\% \tag{9.41c}$$

$$\beta = 1.0 \quad \text{for} \quad FC \leq 5\% \tag{9.42a}$$

$$\beta = \left[0.99 + \left(FC^{1.5}/1000\right)\right] \quad \text{for} \quad 5\% < FC < 35\% \tag{9.42b}$$

$$\beta = 1.2 \quad \text{for} \quad FC \geq 35\% \tag{9.42c}$$

Step 3: *Magnitude scaling factors (MSFs)*

A magnitude scaling factor (MSF) is used to correct the factor of safety (FS) when the earthquake magnitude is not 7.5:

$$FS = \left(\frac{CRR_{7.5}}{CSR} \right) \cdot MSF \qquad (9.43)$$

where:

$CRR_{7.5}$ = the cyclic resistance ratio for a magnitude 7.5 earthquake.

The NCEER workshop (1997) summarized the MSFs proposed by various investigators (Figure 9.24). The NCEER workshop (1997) had the following recommendations:

1. For magnitudes <7.5, the lower bound for the recommended range is the new MSF proposed by Idriss; the suggested upper bound is the MSF proposed by Andrus and Stokoe (1997). The upper-bound values are consistent with the MSFs suggested by Ambraseys (1988), Arango (1996), and Youd and Noble (1997) for $P_L < 20\%$.
2. For magnitudes >7.5, the new factors recommended by Idriss should be used for engineering practice.

CPT method

The CRR can be graphically obtained from Figure 9.25 using the corrected CPT tip resistance, $(q_{c1N})_{cs}$, for clean sands (fines content ≤5%) under magnitude 7.5 earthquakes. The curve was

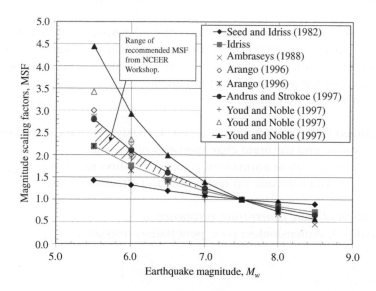

Fig. 9.24 Magnitude scaling factors derived by various investigators. (Reproduced from Youd and Noble, 1997.)

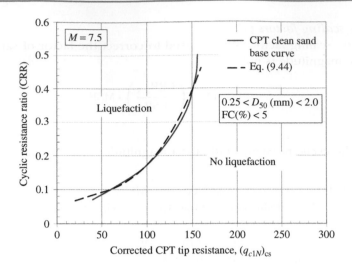

Fig. 9.25 Curve recommended for the calculation of CRR from CPT data. (From Robertson and Wride 1998.)

derived from empirical liquefaction data from compiled case histories. The clean-sand base curve can be approximated using the following equations (Robertson and Wride 1998):

$$\text{IF } (q_{c1N})_{cs} < 50 \quad \text{CRR}_{7.5} = 0.883 \left[\frac{(q_{c1N})_{cs}}{1000}\right] + 0.05 \tag{9.44a}$$

$$\text{IF } 50 \leq (q_{c1N})_{cs} < 160 \quad \text{CRR}_{7.5} = 93 \left[\frac{(q_{c1N})_{cs}}{1000}\right]^3 + 0.08 \tag{9.44b}$$

The approximation using Equation (9.44) is also plotted with the CPT clean-sand base curve. Although slight discrepancies can be noticed from the two curves, the NCEER workshop (1997) recommended either the "CPT clean-sand base curve" in Figure 9.25 or Equation (9.44) be used in finding the CRR under magnitude 7.5 earthquake.

To find $(q_{c1N})_{cs}$, two steps are taken:

1. Find the normalized CPT tip resistance, q_{c1N}, on the basis of the measured tip resistance, the in-situ effective stress, and the grain characteristics of the soil.
2. Find the clean-sand-equivalent normalized cone penetration resistance, $(q_{c1N})_{cs}$, on the basis of the CPT sleeve resistance, cone tip resistance, and effective stress.

Step 1: *Normalization of cone penetration resistance, q_{c1N}*

The normalized, dimensionless cone penetration resistance, q_{c1N}, is calculated by:

$$q_{c1N} = C_Q \left(\frac{q_c}{P_a}\right) \tag{9.45}$$

$$C_Q = \left(\frac{P_a}{\sigma'_{vo}}\right)^n \tag{9.46}$$

where:

C_Q = normalized factor for cone penetration resistance,
q_c = measured cone penetration resistance at the depth in question,
P_a = 1 atm or approximately 101 kPa,
σ'_{vo} = in-situ effective vertical stress at the depth in question,
n = a coefficient that varies from 0.5 to 1.0, depending on the grain characteristics of the soil; for clean sand, $n = 0.5$ is appropriate; for clayey soils, $n = 1.0$ is appropriate; and for silts and sandy silts, n is between 0.5 and 1.0.

To appropriately evaluate n and $(q_{c1N})_{cs}$, Robertson (1990) and Robertson and Wride (1998) proposed a soil behavior type index, I_c:

$$I_c = \left[(3.47 - \log Q)^2 + (1.22 + \log F)^2\right]^{0.5} \tag{9.47}$$

where:

$$Q = \left[\frac{(q_c - \sigma_{v0})}{P_a}\right] \cdot \left[\left(\frac{P_a}{\sigma'_{vo}}\right)^n\right] \tag{9.48}$$

and:

$$F = \left[\frac{f_s}{(q_c - \sigma_{vo})}\right] \times 100\% \tag{9.49}$$

where:

F = normalized friction ratio,
f_s = measured sleeve resistance.

As the soil type is initially unknown (therefore n value is unknown), Robertson and Wride (1998) recommended the following trial-and-error procedure for calculating I_c:

(1) Assuming the soil is clay, $n = 1.0$, use Equations (9.47)–(9.49) to find I_c.
(2) If $I_c > 2.6$, the soil should be clayey and it is considered nonliquefiable. A soil sample should be retrieved and tested to confirm the soil type. The aforementioned Chinese criteria can be used to confirm the liquefaction resistance.
(3) If $I_c < 2.6$, the soil is most likely granular, and $n = 0.5$ should be used in Equation (9.48). Recalculate I_c.
(4) If the recalculated $I_c > 2.6$, the soil is likely to be very silty and possibly plastic. In this case, $n = 0.7$ should be used. A soil sample should be retrieved and tested to confirm the soil type. The aforementioned Chinese criteria can be used to confirm the liquefaction resistance.

The NCEER workshop (1997) also recommended that all soils with $I_c \geq 2.4$ are sampled and tested to confirm the soil type and to test the liquefiability using other criteria such as the SPT method, V_s method, or BPT method. The workshop also pointed out that soil layers with $I_c > 2.6$ but $F < 1\%$ may be very sensitive and should be sampled and tested; such sensitive soils may suffer from softening and strength loss during earthquake shaking.

Step 2: *Calculation of clean-sand equivalent normalized cone penetration resistance, $(q_{c1N})_{cs}$*
Fines content is reflected in the soil behavior type index, I_c. With the calculated I_c, the clean-sand-equivalent normalized cone penetration resistance, $(q_{c1N})_{cs}$, can be calculated:

$$(q_{c1N})_{cs} = K_c q_{c1N} \tag{9.50}$$

where q_{c1N} is calculated using Equation (9.45), and K_c is the correction factor for grain characteristics, defined by Robertson and Wride (1998):

$$\text{for} \quad I_c \leq 1.64 \quad K_c = 1.0 \tag{9.51}$$

$$\text{for} \quad I_c > 1.64 \quad K_c = -0.403I_c^4 + 5.581I_c^3 - 21.63I_c^2 + 33.75I_c - 17.88 \tag{9.52}$$

$(q_{c1N})_{cs}$ is then used in Figure 9.25 to find $CRR_{7.5}$ at magnitude 7.5. To adjust the CRR to magnitudes other than 7.5, the same magnitude scaling factors are used as in the SPT method.

Sample Problem 9.8: Liquefaction analysis using SPT method

It is proposed to build a new bridge across a river. The construction site contains poorly graded sandy soil with fines content (passing #200 sieve) of 18%. The soil deposit of the riverbed is fully saturated with $\gamma_{sat} = 19.5 \, kN/m^3$. The nearby Foot Hill fault system could generate a peak (horizontal) ground acceleration, a_{max}, of 0.25 g at this construction site. Caissons are used as the bridge foundation. The bottom of the caissons is at a depth of 5 m below the riverbed. SPT were performed in a 10.2-cm (4-inch) diameter borehole using a safety trip hammer with a blow count of 6 for the first 15 cm (6 inches), 7 blows for the second 15 cm (6 inches), and 9 blows for the third 15 cm (6 inches) of driving penetration. During the design earthquake of magnitude 6.0, will the saturated sand located at the bottom of the caisson liquefy?

Solution:

1. Calculate CSR:

$$CSR = \frac{\tau_{av}}{\sigma'_{vo}} = 0.65 \frac{a_{max}}{g} \cdot \frac{\sigma_{vo}}{\sigma'_{vo}} \cdot r_d$$

where:

Total overburden stress: $\sigma_{vo} = 19.5 \times 5 = 97.5 \, kN/m^2$
Effective overburden stress: $\sigma'_{vo} = 97.5 - 9.81 \times 5 = 48.5 \, kN/m^2$
Peak horizontal acceleration: $a_{max} = 0.25 \, g$
As $z = 5m \leq 9.15m$,
Stress reduction resistance: $r_d = 1.0 - 0.00765 \times 5 = 0.965$

So: $CSR = \dfrac{\tau_{av}}{\sigma'_{vo}} = 0.65 \dfrac{a_{max}}{g} \cdot \dfrac{\sigma_{vo}}{\sigma'_{vo}} \cdot r_d = 0.65 \times \dfrac{0.25g}{g} \times \dfrac{97.5}{48.5} \times 0.965 = 0.304$

2. Calculate CRR using the SPT method:
First calculate the corrected SPT blow count:

$$(N_1)_{60} = N_m C_N C_E C_B C_R C_S$$

N_m is the measured SPT blow count, which is the total blow count of the second and third six-inch penetrations. So $N_m = 7 + 9 = 16$.

The correction factor on the basis of effective stress (note: $P_a = 1 \text{atm} = 101.3 \text{ kN/m}^2$):

$$C_N = \frac{2.2}{1.2 + \dfrac{\sigma'_{vo}}{P_a}} = \frac{2.2}{1.2 + \dfrac{48.5}{101.3}} = 1.31$$

From Table 9.5, given the automatic trip hammer, the bore hole diameter of 102 mm (4 inch), and the rod length of 5m (15 ft), choose the correction factor for SPT hammer energy ratio $C_E = 1.0$, the correction factor for borehole diameter $C_B = 1.0$, the correction factor for SPT rod length $C_R = 0.85$, and the correction factor for samplers with liner $C_S = 1.0$.

$$\text{So} : (N_1)_{60} = 16 \times 1.31 \times 1.0 \times 1.0 \times 0.85 \times 1.0 \approx 18$$

Then calculate the equivalent clean-sand SPT blow count $(N_1)_{60}$:

$$(N_1)_{60CS} = \alpha + \beta(N_1)_{60}$$

As fines content (FC) = 18%,

$$\alpha = \exp\left[1.76 - (190/FC^2)\right] = \exp\left[1.76 - \left(190/18^2\right)\right] = 3.234$$

$$\beta = \left[0.99 + (FC^{1.5}/1000)\right] = \left[0.99 + \left(18^{1.5}/1000\right)\right] = 1.066$$

$$(N_1)_{60cs} = 3.234 + 1.066 \times 18 = 23$$

From Figure 9.22, find $CRR_{7.5} = 0.26$.

3. Calculate MSF:
   From Figure 9.24, at M=6.0, MSF is between 1.76 and 2.1. On the basis of the critical nature of the project (foundation of a bridge), the MSF is chosen as the lower bound, 1.76.
4. Factor of safety against liquefaction:

$$FS = \left(\frac{CRR_{7.5}}{CSR}\right) \cdot MSF = \left(\frac{0.26}{0.304}\right) \cdot 1.76 = 1.5 > 1.3$$

Conclusion: The SPT analysis concludes that the site will not liquefy under the design earthquake.

Chapter 9

Sample Problem 9.9: Liquefaction analysis using CPT method

The liquefaction potential is evaluated for the same project at the same site as in Sample Problem 9.8, using the CPT method. The site condition is the same, but the subsoil characteristics are initially unknown. At the depth of 5 m (15 ft), which is the depth of the caisson bottom, the penetration resistances at the cone tip and the sleeve were measured to be 5 MPa and 100 kPa, respectively. During the design earthquake of magnitude 6.0, will the saturated sand located at the bottom of the caisson liquefy?

Solution:

1. Calculate CSR, using the same approach and calculation as in sample problem 9.8:

$$CSR = \frac{\tau_{av}}{\sigma'_{vo}} = 0.65\frac{a_{max}}{g} \cdot \frac{\sigma_{vo}}{\sigma'_{vo}} \cdot r_d = 0.304$$

2. Calculate CRR using the CPT method:
   First calculate normalized cone penetration resistance, q_{c1N}:
   As the soil characteristics are unknown, first assume the soil is clay, so:

$$n = 1.$$
$$q_c = 5\,MPa = 5000\,kPa$$
$$f_s = 100\,kPa$$

Normalized friction ratio:

$$F = [f_s/(q_c - \sigma_{vo})] \times 100\% = [100/(5000 - 97.5)] \times 100\% = 2\%$$

$$Q = [(q_c - \sigma_{vo})/p_a][(p_a/\sigma'_{vo})^n] = \frac{5000 - 97.5}{p_a} \times \frac{p_a}{48.5} = 101$$

So:

$$I_c = \left[(3.47 - \log Q)^2 + (1.22 + \log F)^2\right]^{0.5}$$

$$= \left[(3.47 - \log 101)^2 + (1.22 + \log 2)^2\right]^{0.5}$$

$$= 2.11$$

As $I_c < 2.6$, the soil is most likely granular, use $n = 0.5$, recalculate I_c:

$$Q = [(q_c - \sigma_{vo})/p_a][(p_a/\sigma'_{vo})^n] = \frac{5000 - 97.5}{101.3} \times \left(\frac{101.3}{48.5}\right)^{0.5} = 70$$

$$I_c = \left[(3.47 - \log Q)^2 + (1.22 + \log F)^2\right]^{0.5}$$

$$= \left[(3.47 - \log 70)^2 + (1.22 + \log 2)^2\right]^{0.5}$$

$$= 2.22 < 2.6$$

So:

$$q_{c1N} = C_Q(q_c/P_a)$$

where $C_Q = (p_a/\sigma'_{vo})^n = \left(\dfrac{101.3}{48.5}\right)^{0.5} = 1.444$

$$q_{c1N} = C_Q(q_c/P_a) = 1.444 \times \dfrac{5000}{101.3} = 71$$

Then calculate the clean-sand equivalent normalized cone penetration resistance:

$$(q_{c1N})_{cs} = K_c q_{c1N}$$

As $I_c = 2.22 > 1.64$,

$$K_c = -0.403 I_c^4 + 5.581 I_c^3 - 21.63 I_c^2 + 33.75 I_c - 17.88 = 1.717$$

So: $(q_{c1N})_{cs} = K_c q_{c1N} = 1.717 \times 71 \approx 122$

From Figure 9.25, find $CRR_{7.5} = 0.24$

3. Calculate the MSF using the same approach and calculation as in Sample Problem 9.8:

   MSF is chosen to be 1.76.

4. Factor of safety against liquefaction:

$$FS = \left(\dfrac{CRR_{7.5}}{CSR}\right) \cdot MSF = \left(\dfrac{0.26}{0.304}\right) \times 1.76 = 1.39 > 1.3$$

Conclusion: The CPT analysis concludes that the site will not liquefy under the design earthquake.

Homework Problems

1. For each of the following multiple-choice questions, select the correct answers.

   (1) The movement of a fault is horizontal (or lateral) and each block moves to the left of the opposite block. Which of the following faults fit(s) this type of movement?

Chapter 9

A. Normal fault.
B. Reverse fault (thrust fault).
C. Right lateral fault.
D. Left lateral fault.
E. Left-lateral normal fault.

(2) Which of the following movements describes the "left-lateral normal fault"?

A. The movement is vertical, the hanging wall moves up relative to the foot wall.
B. The movement is horizontal (or lateral), each block moves to the right of the opposite block.
C. The movement is horizontal (or lateral), each block moves to the left of the opposite block.
D. The movement is combined by the normal fault movement and left-lateral fault movement.
E. The movement is combined by the reverse fault movement and left-lateral fault movement.

(3) An epicenter is:

A. The location where a fault rupture first occurs.
B. The point vertically above the focus and on the ground surface.
C. The location where the seismometer is installed.
D. The position after the fault rupture.
E. The location where the body wave originates.

(4) A primary wave is a:

A. Shear wave.
B. Body wave.
C. Surface wave.
D. Compressional wave.
E. Longitudinal wave.

(5) A surface wave is a(n):

A. Interaction between P-waves and S-waves.
B. Interaction between SV-waves and SH-waves.
C. Body wave.
D. Love wave and Rayleigh wave.
E. Primary wave.

(6) The following conditions contribute to the occurrence of liquefaction:

A. Sandy or silty soils.
B. Saturation.
C. Loose or medium compaction.
D. Seismic stress.
E. High plasticity index of soil.

2. A soil-retaining wall is shown in Figure 9.26. Under the design earth-quake, the horizontal acceleration coefficient is 0.25, and the vertical acceleration coefficient is 0.1. The external friction angle between the granular backfill and the wall is two-thirds of the internal friction angle.

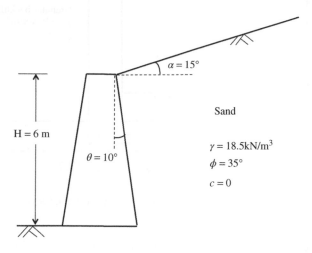

Fig. 9.26   Schematic for problems 2 and 4.

(1) Calculate the total static earth force behind the retaining wall and determine the location and direction of the resultant force.
(2) Calculate the total dynamic earth force under the design earth-quake and determine its point of application.

3. A cantilever retaining wall is shown in Figure 9.27.
The backfill is granular soil. The wall geometry and the characteristics of the backfill and foundation soils are shown in the figure. Under the design earthquake, the horizontal acceleration coefficient is 0.2, and the vertical acceleration coefficient is 0.05. The external friction angle between the granular backfill and the wall is two-thirds of the internal friction angle.

(1) Calculate the total static earth force behind the retaining wall and determine the location and direction of the resultant force.
(2) Calculate the total dynamic earth force under the design earth-quake and determine its point of application.

4. The problem statement is the same as in Problem 2. Assume the lateral soil pressure is passive.

(1) Calculate the total static earth force behind the retaining wall and determine the location and direction of the resultant force.
(2) Calculate the total dynamic earth force under the design earth-quake and determine its point of application.

Chapter 9

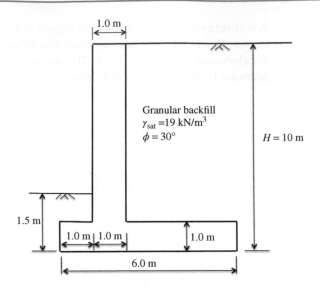

Fig. 9.27   Schematic for problems 3 and 5.

5. The problem statement is the same as in Problem 3. Assume the lateral soil pressure is passive.
   (1) Calculate the total static earth force behind the retaining wall and determine the location and direction of the resultant force.
   (2) Calculate the total dynamic earth force under the design earth-quake and determine its point of application.
6. An earth embankment has a slope angle of 38 degrees and a height of 10 m. The soil has unit weight of 18.8 kN/m³, cohesion of 43 kN/m², and internal friction angle of 30 degrees. The slope is subjected to a horizontal acceleration of 0.25 g and vertical acceleration of 0.1 g. What is the factor of safety of the slope on a potential plane failure surface inclined at 25 degrees with respect to the horizontal?
7. As shown in Figure 9.28, a natural slope has a slope angle of 45 degrees and a height of 12 meters. The soil is saturated clay, its unit weight is 18 kN/m³, and cohesion is 50 kN/m². The slope is subjected to a horizontal acceleration of 0.3 g and vertical acceleration of 0.1 g. A potential sliding surface is shown in the figure. Determine the factor of safety for the seismic slope stability along the circular failure surface.
8. The problem statement is the same as Problem 7. Using the ordinary method of slices, determine the factor of safety for the seismic slope stability along the specified circular failure surface.
9. As shown in Figure 9.29, a 15 m high embankment with a slope angle of $\beta = 40°$ is to be constructed in an earthquake-active region. The embankment soil is sandy silt with strength parameters of $\phi' = 25°$ and

$c' = 15\,\text{kPa}$, and the unit weight is $17\ \text{kN/m}^3$. The foundation soil is sandy clay with strength parameters of $\phi' = 20°$ and $c' = 20\,\text{kPa}$, and the unit weight is $18\ \text{kN/m}^3$. Assume a toe failure with circular failure surface as shown in the figure. Given $\alpha = 30°$ and $R = 20\,\text{m}$, the center O can be determined. Assume the horizontal coefficient of acceleration $k_h = 0.2$. Determine the factor of safety of the slope on the circular failure plane.

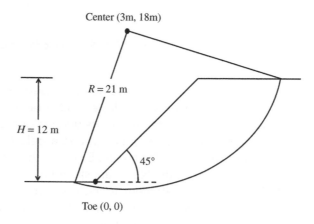

Fig. 9.28 Schematic for problem 7.

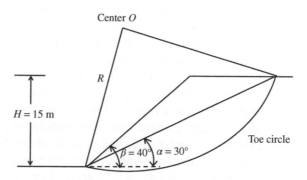

Fig. 9.29 Schematic for problem 9.

10. A natural slope is shown in Figure 9.30. The slope angle is 40 degrees. The slope is subjected to a horizontal acceleration of 0.2 g. Using the ordinary method of slices, determine the factor of safety for the seismic slope stability. Several trial failure surfaces may be needed.

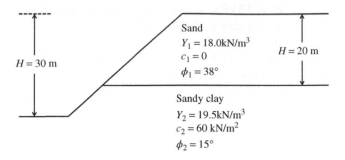

Fig. 9.30   Schematic for problem 10.

11. A natural slope is to be evaluated for permanent displacement under a design earthquake. The slope is 33 m tall and comprises mainly sandy soil. The slope angle is 38°. Its unit weight is 18.2 kN/m³, and its friction angle is 40°. The peak ground acceleration is 0.5 g, and the peak ground velocity is 0.5 m/sec. Evaluate the permanent displacement of the slope using the Newmark sliding block method.

12. The problem statement is the same as in Problem 11. The average shear wave velocity in the soil slope was measured to be 90 m/sec. The earthquake magnitude in Richter scale is 7.5. Evaluate the earthquake-induced permanent displacement of the slope using the Makdisi–Seed method.

13. What are the conditions necessary for liquefaction to occur?

14. Is it true that cohesive soils do not liquefy? How to qualitatively evaluate the liquefaction potential of cohesive soils?

15. A multistory building is to be built on a level ground at a seismically active site. The SPT is performed during the subsoil exploration phase. A safety hammer is used, and the outside diameter of the modified California sampler used is 3 inches. The peak ground horizontal acceleration that was recorded in the seismic history in this area is 0.2 g, and the design earthquake magnitude is 6.5. The mat foundation is designed to be 8 m below ground where the fine content is 15%. At 8 m depth, the soil sample is retrieved by the sampler. The blow counts at the first, second, and third six-inch penetrations are 8, 9, and 8, respectively. The boring log indicates the following subsurface profile. Assess the liquefaction hazard at the footing of the building at this site (Figure 9.31).

16. A hospital is built in an earthquake-prone region. To mitigate the liquefaction hazard, dynamic compaction is performed that also improves the soil. Then the CPT is conducted to evaluate the mitigation effort. The peak ground horizontal acceleration is 0.35 g and the design earthquake magnitude is 8.0. The design depth of the hospital's foundation

is at 6 m; at this depth, the cone penetration resistance at the tip is 11 MPa and the sleeve resistance is measured to be 200 kPa. Assume the groundwater table is at the ground surface, and saturated unit weight is 19.6 kN/m^3. Calculate the factor of safety against liquefaction.

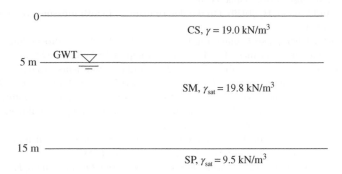

Fig. 9.31   Subsoil Profile of the site for liquefaction evaluation, problem 15.

References

Ambraseys, N.N. (1988). "Engineering seismology." *Earthquake Engineering and Structural Dynamics*, Vol. 17, pp. 1–105.

Andrus, R.D., and Stokoe, K.H., II, (1997). Liquefaction resistance based on shear wave velocity. *Proceedings of NCEER Workshop on Evaluation of Liquefaction Resistance of Soils*, Nat. Ctr. For Earthquake Engrg. Res., State University of New York at Buffalo, 89–128.

Arango, I. (1996). "Magnitude scaling factors for soil liquefaction evaluations." *ASCE Journal of Geotechnical Engineering*, Vol 122, No. 11, pp. 929–936.

Building Seismic Safety Council of National Institute of Building Sciences (2004). *NEHRP (National Earthquake Hazards Reduction Program) Recommended Provisions for Seismic Regulations for New Buildings and Other Structures*. FEMA 450-2/2003 Edition. Part 2: Commentary. Building Seismic Safety Council, National Institute of Building Sciences. Washington, D.C. 2004

California Geological Survey (1997). *Guidelines for Evaluation and Mitigating Seismic Hazards in California*. Special publication 117, Offices of the California Geological Surveys. Sacramento, CA.

Gazetas, G. (1982). "Shear vibrations of vertically inhomogeneous earth dams." *International Journal for Numerical and Analytical Methods in Geomechanics*, Vol. 6, No. 1, pp 219–241.

Hynes-Griffin, M.E., and Franklin, A.G. (1984). Rationalizing the seismic coefficient method. Miscellaneous Paper GL-84-13, U.S. Army Corps of Engineers Waterways Experiment Station, Vicksburg, Mississippi, 21 pp.

Kapila, J.P. (1962). Earthquake resistant design of retaining walls. *Proceedings, 2nd Earthquake Symposium*, University of Roorkee, Roorkee, India.

Kramer, S.L. (1996). *Geotechnical Earthquake Engineering*, Prentice-Hall, Inc., Upper Saddle River, NJ.

Liao, S.S.C., and Whitman, R.V. (1986a). Catalogue of liquefaction and non-liquefaction occurrence during earthquake. Res. Rep., Dept. of Civil Engrg., Massachusetts Institute of Technology, Cambridge, MA.

Liao, S.S.C., and Whitman, R.V. (1986b). "Overburden correction factors for SPT in sand." *ASCE Journal of Geotechnical Engineering*, Vol. 112, No. 3, pp 373–377.

Makdisi, F.I., and Seed, H.B. (1978). "Simplified procedure for estimating dam and embankment earthquake-induced deformations." *ASCE Journal of Geotechnical Engineering*, Vol. 104, No. GT7, pp 849–867.

Chapter 9

Marcuson, W.F., III (1978). "Definition of terms related to liquefaction." *ASCE Journal of Geotechnical Engineering Division*, Vol. 104, No. 9, pp 1197–1200.

Marcuson, W.F., III (1981). Moderator's report for session on 'Earth dams and stability of slopes under dynamic loads'. *Proceedings, International Conference on Recent Advances in Geotechnical Earthquake Engineering and Soil Dynamics*, St. Louis, Missouri, Vol. 3, pp. 1175.

Marcuson, W.F., III, Ballard, R.F., Jr., and Ledbetter, R.H. (1979). Liquefaction failure of tailings dams resulting from the Near Izu Oshima earthquake, 14th and 15th January, 1978. *Proceedings, 6th Pan American Congress on Soil Mechanics and Foundation Engineering*, Lima, Peru.

Mononobe, N., and Hatsuo, H. (1929). On the determination of earth pressures during earthquakes. Proceedings, World Engineering Congress.

National Center for Earthquake Engineering Research (NCEER). (1997). Proceedings of the NCEER Workshop on Evaluation of Liquefaction Resistance of Soils, Technical Report NCEER-97-0022, 276p.

Newmark, N. (1965). "Effects of earthquakes on dams and embankments." *Geotechnique*, Vol. 15, No. 2, pp. 139–160.

Okabe, S. (1926). "General theory of earth pressures." *Journal of Japan Society of Civil Engineers*, Vol. 12, No. 1, 123–134.

Robertson, P.K. (1990). "Soil classification using CPT." *Canadian Geotechnical Journal*, Vol. 27, No. 1, pp 151–158.

Robertson, P.K., and Wride, C.E. (1998). "Evaluating cyclic liquefaction potential using the cone penetration test." *Canadian Geotechnical Journal*, Vol. 35, No. 3, pp 442–459.

Seed, H.B., and Idriss, I.M. (1971). "Simplified procedure for evaluating soil liquefaction potential." *ASCE Journal of Geotechnical Engineering Division*, Vol. 97, No. 9, pp 1249–1273.

Seed, H.B., and Idriss, I.M. (1982). *Ground Motions and Soil Liquefaction During Earthquakes.*" Earthquake Engineering Research Institute Monograph, Oakland, CA.

Seed, H.B., Lee, K.L., Idriss, I.M., and Makdisi, F.I. (1975). "The slides in the San Fernando Dams during the earthquake of February 9, 1971." *ASCE Journal of Geotechnical Engineering Division*, Vol. 101, No. GT7, pp 651–688.

Seed, H.B., and Whitman, R.V. (1970). Design of earth retaining structures for dynamic loads. *Proceedings, ASCE Specialty Conference on Lateral Stresses in the Ground and Design of Earth Retaining Structures*, pp. 103–147.

Skempton, A.K. (1986). "Standard penetration test procedures and the effects in sands of overburden pressure, relative density, particle size, aging, and overconsolidation." *Geotechnique*, Vol. 36, No. 3, pp 425–447.

U.S. Geological Survey (USGS). (2010). USGS Earthquake Magnitude Policy. USGS. March 29, 2010.

Youd, T.L. et al. (2001). "Liquefaction resistance of soils: Summary report from the 1996 NCEER and 1998 NCEER/NSF workshops on evaluation of liquefaction resistance of soils." *ASCE Journal of Geotechnical and Geoenvironmental Engineering*, Vol. 127, No. 10, pp 817–833.

Youd, T.L., and Noble, S.K. (1997). Magnitude scaling factors. *Proceedings of NCEER Workshop on Evaluation of Liquefaction Resistance of Soils*, National Center for Earthquake Engineering Research (NCEER), State Univ. of New York at Buffalo, 149-165.

Chapter 9

Index

Geotechnical Engineering Design, First Edition. Ming Xiao.
© 2015 John Wiley & Sons, Ltd. Published 2015 by John Wiley & Sons, Ltd.
Companion Website: www.wiley.com/go/Xiao

FSC C018575

Printed and bound by CPI Group (UK) Ltd, Croydon, CR0 4YY

27/10/2024

14580143-0002